Ubiquitin Proteasome System - Current Insights into Mechanism Cellular Regulation and Disease

Edited by Matthew Summers

Published in London, United Kingdom

IntechOpen

Supporting open minds since 2005

Ubiquitin Proteasome System - Current Insights into Mechanism Cellular Regulation and Disease
http://dx.doi.org/10.5772/intechopen.78430
Edited by Matthew Summers

Contributors
Gali Prag, Tal Keren-Kaplan, Ilan Attali, Olga Levin-Kravets, Shay Ben-Aroya, Oded Kleifeld, Pai-Sheng Chen, Michael Massiah, Jorge Torres, Joseph Ong, Michael Emanuele, Taylor Enrico, Thao Dang, Bryce Paschal, Teddy Kamata, Jeffrey Parvin, Alexander Hare, Paola Pontrelli, Francesca Conserva, Loreto Gesualdo, Elena Santonico

First published in London, United Kingdom, 2019 by IntechOpen
IntechOpen is the global imprint of INTECHOPEN LIMITED, registered in England and Wales, registration number: 11086078, The Shard, 25th floor, 32 London Bridge Street
London, SE19SG - United Kingdom
Printed in Croatia

British Library Cataloguing-in-Publication Data
A catalogue record for this book is available from the British Library

Additional hard and PDF copies can be obtained from orders@intechopen.com

Ubiquitin Proteasome System - Current Insights into Mechanism Cellular Regulation and Disease
Edited by Matthew Summers
p. cm.
Print ISBN 978-1-83880-490-9
Online ISBN 978-1-83880-491-6
eBook (PDF) ISBN 978-1-83880-717-7

For EU product safety concerns:
IN TECH d.o.o., Prolaz Marije Krucifikse Kozulić 3, 51000 Rijeka, Croatia, info@intechopen.com or visit our website at intechopen.com.

Meet the editor

Matthew Summers received his BS degree in Biochemistry from the University of Delaware and his PhD in Cell and Molecular Biology from the University of Pennsylvania for his work on the E3 ubiquitin ligase and tumor suppressor CHFR. As a postdoctroral fellow at Stanford University and Genentech, Inc., his research focused on the mechanisms and regulation of the anaphase promoting complex/cyclosome ubiquitin ligase. He is currently an associate professor at Ohio State University and a member of the OSU-Comprehensive Cancer Center. Research in the Summers lab focuses on the interplay between cellular checkpoints and the ubiquitin proteasome system to regulate cellular growth and genome stability.

Contents

Preface

It has been over three decades since the modification of proteins by covalent attachment (primarily to lysine residues) of the small, 76 amino acid protein ubiquitin, and the ability of this modification to target proteins for destruction by a protease, now known as the 26S proteasome, was first discovered. Subsequent discoveries of an E1 ubiquitin-activating enzyme, E2 ubiquitin-conjugating enzymes, and E3 ubiquitin ligases that function as an enzymatic cascade culminating in the ubiquitination of substrates revealed that the vast majority of protein degradation depends on this ubiquitin proteasome system (UPS). Today, the human UPS is comprised of nearly 1000 proteins and has expanded beyond ubiquitin to include ubiquitin-like modifiers (UBLs, including SUMOs, NEDD8, ATG8, ATG12, URM1, UFM1, FAT10, HUB1, and ISG15), a handful of E1s, tens of E2s, hundreds of E3s, nearly 100 deubiquitinating enzymes, and UBL-specific interacting proteins that regulate the UBL-modified proteasome within our cells. Not surprisingly, the pathways that comprise the UPS rely on the coordinated activities of multiple pathways.

Indeed, the canonical protein turnover function of the UPS is a highly dynamic process that is regulated on multiple levels. For example, E3 ligases, which provide substrate specificity within the system, may be regulated at the level of expression both transcriptionally and at the protein level by ubiquitination-mediated degradation that may be self-catalyzed or mediated by an antagonistic E3. For some multi-subunit E3s the formation of an active holoenzyme is regulated by posttranslational modification, including phosphorylation and conjugation of the UBL NEDD8. Once the E3 is present in the cell, its ability to recognize substrates may be regulated by posttranslational modification of the substrate, while the ability to ubiquitinate the substrate may be further regulated by the expression or modification of the cognate E2 enzyme. Then, once ubiquitinated, the fate of a substrate destined for the proteasome is still not sealed because the modification may be removed by deubiquitinating enzymes or its delivery to the proteasome may be regulated by the modification or availability of ubiquitin receptors.

Conjugation with ubiquitin itself is also a multifaceted modification and we now know that degradation is but one outcome resulting from the covalent addition of ubiquitin to a protein. It is now accepted that polyubiquitin linkages occur via all of seven lysines of ubiquitin, branched heterotypic chains, as well as linear ubiquitin modifications and monoubiquitination. Although the roles of some of these modifications are not yet well established, monoubiquitination of histones has emerged as an important feature of the "histone code," and regulating many chromatin-related processes and K63-linked polyubiquitin chains is a fundamental part of many signaling pathways. In addition to the diversity of ubiquitin signals created by specific linkages, posttranslational modification of ubiquitin, including phosphorylation, acetylation, and ADP-ribosylation, have recently been shown to impact chain stability and chain elongation, among other effects.

In keeping with the diversity of the components within the UPS, we now know that the UPS mediates central signaling events in myriad processes involved in both cellular and organismal health and homeostasis. For example, numerous pathways within the UPS are implicated in disease, ranging from cancer to neurodegenerative diseases such as Parkinson's. An in-depth understanding of these signaling cascades will significantly enhance our knowledge of their pathological roles while identifying potential therapeutic targets.

It would require a veritable encyclopedia of review articles to fully encompass the current view of UPS research. The goal of this book is to deliver a collection of synopses of current areas of UPS research that highlight the importance of understanding the biology of the UPS to identify disease-relevant pathways, and the need to elucidate the molecular machinations within the UPS to develop methods for therapeutic modulation of these pathways. Specifically, the chapters of this book provide up-to-date views on cellular regulation in the context of control of the cell cycle by ubiquitin-mediated proteolysis, the role of ubiquitin-modified chromatin in DNA repair, and transcriptional elongation and mitotic bookmarking of epigenetic information during mitosis. The UPS is also examined from a disease perspective with regard to altered E3 activities in cancer (including developing therapeutic strategies), the role of K63-linked polyubiquitin in multiple disease settings, and the role of UCHL1 in Parkinson's disease. Current snapshots of several molecular aspects of the UPS are also provided in discussions of how cells discriminate between ubiquitin and UBLs, regulation of proteasome function by ubiquitin, the function of B-box domains in ubiquitin ligases, and regulation of ubiquitin ADP-ribosylation. The poignant overviews encompassed within this book are provided by researchers at the forefronts of their respective areas and are intended to serve as an informative primer for researchers who are new to the field and as a concise state of the research for those who are already entrenched in the field of ubiquitin.

Matthew K. Summers
The Ohio State University,
Department of Radiology and Comprehensive Cancer Center,
Columbus, OH, USA

Section 1

Roles of the UPS in Cellular Regulation and Disease

Chapter 1

Ubiquitin Signaling in Regulation of the Start of the Cell Cycle

Michael James Emanuele and Taylor Paige Enrico

Abstract

The small protein ubiquitin plays a vital role in virtually all aspects of cellular life. Among the diverse signaling outcomes associated with ubiquitination, the most well-established is the targeted degradation of substrates via the proteasome. During cell growth and proliferation, ubiquitin plays an outsized role in promoting progression through the cell cycle. In particular, ubiquitin-mediated degradation is critically important at transition points where it provides directionality and irreversibility to the cell cycle, which is essential for maintaining genome integrity. Specifically, the boundary between G1 and S-phase is tightly regulated by the ubiquitin proteasome system. Notably, the G1/S boundary represents a major barrier to cell proliferation and is universally dysfunctional in cancer cells, allowing for the unbridled proliferation observed in malignancy. Numerous E3 ubiquitin ligases, which facilitate the ubiquitination of specific substrates, have been shown to control G1/S. In this chapter, we will discuss components in the ubiquitin proteasome system that are implicated in G1/S control, how these enzymes are interconnected, gaps in our current knowledge, and the potential role of these pathways in the cancer cycle and disease proliferation.

Keywords: cell cycle, ubiquitin, cullin RING ligase, anaphase promoting complex/cyclosome (APC/C), G1, S-phase, SCF

1. Introduction

Progression through the cell cycle is driven by the oscillating activity of Cyclin Dependent Kinases (CDKs). The activity of CDKs is controlled by their binding to coactivator subunits termed Cyclins, as well as by CDK inhibitory proteins termed CKIs. The accumulation of both Cyclin and CKI proteins is tightly regulated at the level of transcription. In addition, Cyclin and CKI proteins are controlled at the level of their destruction. Remarkably, during each and every passage through the cell cycle, Cyclins, CKIs, and hundreds of other proteins, accumulate and are subsequently destroyed via a highly regulated process of programmed degradation. This degradation is controlled by ubiquitin.

Ubiquitin is conjugated to substrate lysines, and because ubiquitin itself contains seven lysine residues to which ubiquitin can be added, the repetitive addition of ubiquitin can result in the formation of polyubiquitin chains on substrates. These chains can be formed through each of the different lysines in ubiquitin, as well as through the amino-terminal methionine, leading to chain formations that adopt distinct topological features [1, 2]. The most well-characterized of these are chains linked through lysine 48 in ubiquitin, so-called K48-linked ubiquitin chains, which target substrates to the proteasome for destruction. More recently, K11-linked chains were also shown to target substrates to the proteasome [3, 4]. Alternatively, ubiquitin chains linked through other lysines (or through methionine 1) lead to diverse signaling outputs by altering protein-protein interactions, protein localization, enzyme activity, etc. This already complex picture is further complicated by the recent discovery of branched ubiquitin chains, which contain non-homogeneous lysine linkages. For example, branched K11/K48 chains likely represent remarkably strong degradative signals [5, 6].

Protein degradation through the ubiquitin proteasome system (UPS) is the major regulator of programmed protein destruction in human cells and plays an outsized role in controlling cell cycle progression [7]. Importantly, the targeted degradation and/or stabilization of specific proteins at transition points (e.g. mitosis/G1 and G1/S boundaries) promotes cell cycle progression, provides directionality and irreversibility to the cell cycle and maintains genome integrity [8]. Accordingly, numerous enzymes in the ubiquitin system have been implicated in these transition points.

The start of DNA replication represents a tightly controlled barrier to proliferation in normal cells. As such, nearly all of the non-dividing cells in the human body are arrested prior to the start of S-phase, in either G1, or in quiescence (G0), where they maintain the equivalent of G1-phase (2C) DNA content. In diseases of uncontrolled proliferation, and most notably in cancer, the S-phase boundary is perturbed. Thus, cancer cells are able to aberrantly enter S-phase due to a weakening of the G1/S border [9]. The retinoblastoma tumor suppressor pathway plays a key role in controlling G1/S. However, the ubiquitin system is also tightly linked to G1/S regulation in normal and cancer cells. Below, we will discuss the particular enzymes and pathways associated with ubiquitin signaling that have been implicated in regulating the start of S-phase.

2. Introduction to cell cycle ubiquitin ligases

2.1 Cullin RING E3 ubiquitin ligases

The RING domain family of E3 ubiquitin ligases is the largest family of E3s in higher eukaryotes, and in humans it is represented by several hundred unique enzymes and/or enzyme complexes. The cullin RING ligases (CRLs) are the largest subfamily of RING E3s, encoding nearly 300 unique enzymes. The CRL E3s all share a common molecular architecture [10]. CRLs utilize a cullin protein backbone, which simultaneously binds to both an E2 ubiquitin conjugating enzyme and substrate, positioning E2 and substrate in close proximity, and enabling the rapid transfer of ubiquitin onto substrates (**Figure 1A**).

The human genome encodes several cullin proteins, including Cul1, Cul2, Cul3, Cul4A, Cul4B, Cul5, Cul7, Cul9 and the related cullin-like protein APC2. With the exception of APC2, each cullin is thought to assemble into a ligase with a similar architecture, where the amino terminus of the cullin engages targets and functions as a substrate targeting module, and the carboxy terminus engages the E2, functioning as a ubiquitin transfer module (**Figure 1B**). Cullin binding to substrates and E2-ubiquitin conjugating enzymes is indirect. Most cullins first bind to an adaptor protein which in turn binds to a family of substrate receptors that then recruit substrates for ubiquitination (**Figure 1**). Similarly, cullin proteins indirectly interact with one of two RING domain containing proteins (Roc1/Rbx1 or Roc2/Rbx2)

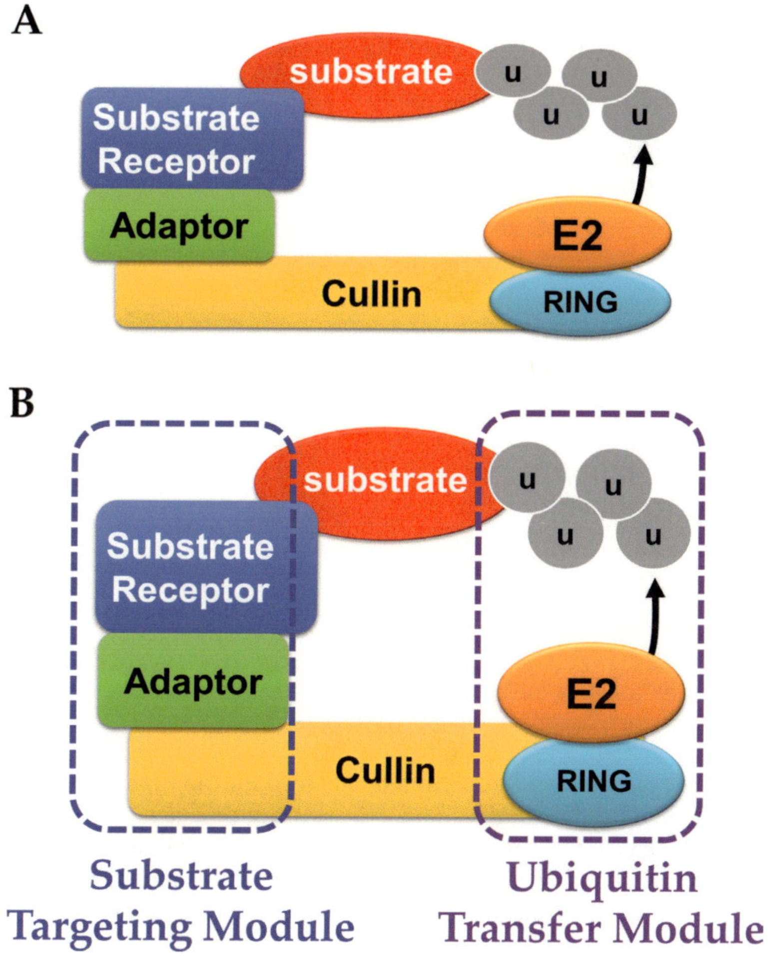

Figure 1.
Architecture of the cullin RING E3 ubiquitin ligases. (A) Architecture of a canonical CRL E3 ligase. (B) Boxes highlighting the substrate targeting (dark blue) and ubiquitin transfer (purple) modules.

which in turn bind to E2 ubiquitin conjugating enzymes. This architecture is shared among all known CRL complexes.

The archetypical CRL sub-family, and one which will be discussed in greater detail, is the Skp1-Cul1-F-box family of CRLs. These ligases, commonly referred to as SCF or CRL1 ligases, utilize a family of 69 interchangeable substrate receptor proteins, termed F-box proteins, which designate substrates for ubiquitination and degradation. F-box proteins rely on an F-box domain to interact with an adaptor protein termed Skp1, which bridges F-box proteins to Cul1 (**Figure 2**). The CRL nomenclature dictates that specific ligase complexes are depicted with the F-Box protein as a superscript, following the name of the cullin complex. Thus, Cul1-based CRLs, in complex with the F-box substrate receptor Skp2, are designated as SCF^{Skp2} or $CRL1^{Skp2}$ (hereafter, Cul1-based CRL complexes will be referred to as SCF).

Importantly, substrate receptors recognize proteins for degradation based on short, linear sequence motifs, called degrons. Degron sequences are shared among the substrates of a specific E3. In addition, degrons are transferrable, and the addition of degron sequences to non-substrates is often sufficient to trigger their recognition by the E3 and subsequent ubiquitination and degradation. Also, many substrate receptors, although not all, require post-translation modification (e.g.

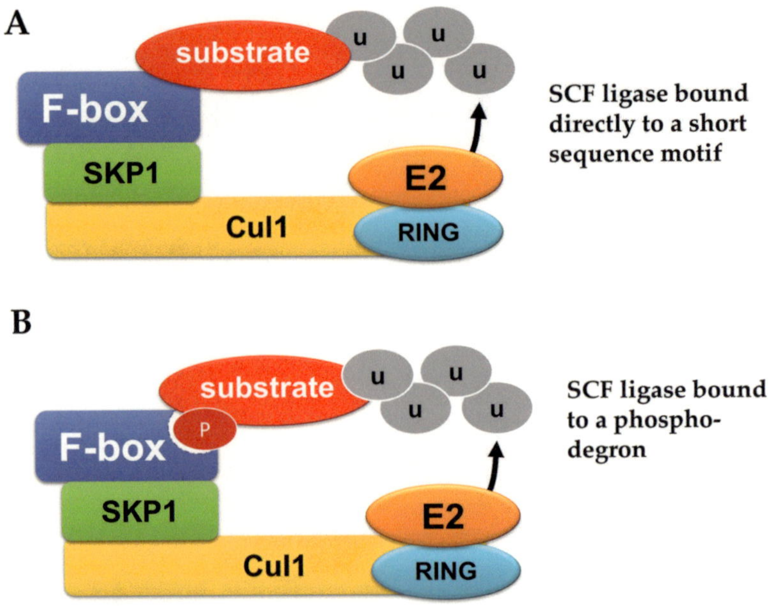

Figure 2.
Architecture of the SCF ligase. (A) Example of an SCF-type ligase bound to a short linear degron sequence motif in a substrate. (B) Example of an SCF-type ligase bound to phosphorylation-dependent degron in a substrate.

phosphorylation) of the substrate within the degron for the substrate to be recognized, ubiquitinated, and degraded. Thus, the degradation of many SCF substrates is regulated at the level of the substrate and is a two-step process. First, the substrate must be present and modified, and second, the ligase must also be available, thereby enabling substrate recognition and degradation. It is important to note that each substrate receptor can have many substrates. Furthermore, individual substrates can be controlled by multiple ligases. Finally, distinct substrate adaptors can themselves be targeted for degradation by other E3 ligases.

The Cul1-based SCF ligases are the founding members of the CRL family. They were first discovered in yeast based on their role in controlling cell cycle progression. Their discovery grew out of gain-of-function screens performed by Elledge and colleagues, which identified suppressors of the yeast cell cycle mutant Cdc4. This screen uncovered a new protein, whose mRNA and protein levels oscillated during the cell cycle. Moreover, the amino acid sequence of this new protein included a Cyclin homology domain, similar to that found in the previously identified Cyclins A, B, D, and E. Thus, this new protein was named Cyclin F [11]. Significantly, Cyclin F contained a domain with sequence similarity to Cdc4, which they named the F-box domain. They found that the F-box domains in Cyclin F and Cdc4 were essential for tethering both proteins to the ubiquitin machinery via binding to Skp1 [12]. Shortly thereafter, the Harper lab, in collaboration with Elledge, as well as the Deshaies lab, showed that SCF complexes could trigger the ubiquitination and degradation of the yeast CDK inhibitor Sic1. Moreover, these studies demonstrated that the F-box protein Cdc4 preferentially bound to the phosphorylated version of Sic1, thereby triggering its ubiquitination and degradation [13, 14].

2.2 The Anaphase Promoting Complex/Cyclosome

Like other E3 ubiquitin ligases, the Anaphase Promoting Complex/Cyclosome (APC/C) plays an important role in protein degradation. APC/C regulates the ubiquitination and degradation of the CDK activator proteins Cyclin A and Cyclin B, in addition to many other cell cycle regulated proteins. As such, it is a core component of the cell cycle oscillator. As its name suggests, the APC/C is activated in metaphase of mitosis, during which time it triggers the ubiquitination and degradation of numerous proteins including two critical substrates, Cyclin B and securin, thereby "promoting anaphase" and mitotic exit. In addition to its essential function in mitosis, APC/C also plays an evolutionarily conserved role in G1-phase. The APC/C remains active throughout G1, where in contrast to its role in promoting progression through mitosis, the APC/C restrains progression through G1-phase into S-phase [17], and is not turned off until immediately prior to the start of DNA replication [15, 16]. Significantly, APC/C inactivation at the G1/S boundary is required for the start of S-phase.

Similar to the CRLs discussed above, the APC/C has both a cullin-like subunit (APC2) and a RING subunit (APC11). However, the APC/C is significantly different than the CRL ligases discussed above. Notably, the APC/C is composed of 18 polypeptide subunits and is a remarkable 1.2 mDa in size (**Figure 3**). The cullin subunit, APC2, is the most divergent of the cullins, and lacks features that are common among other cullin proteins. For example, while other cullin proteins are post-translationally modified and activated by the small, ubiquitin-like protein Nedd8, this process is not thought to be involved in APC/C activity.

The APC/C utilizes either of two substrate receptors during somatic cell cycles. First, during mitosis, the APC/C binds to the substrate receptor/coactivator Cdc20, which brings Cyclin B and Securin to the APC/C for ubiquitination. Immediately following mitotic exit, APC/C shifts to using a second substrate adaptor, the Cdc20-related protein Cdh1/Fzr1 (hereafter referred to as Cdh1). The Cdh1-bound form of APC/C remains active throughout G1-phase and targets a myriad of cell cycle regulators for degradation, including proteins involved in transcription, nucleotide metabolism, and CDK activation. Thus, it is APC/C^{Cdh1} that must be inactivated prior to the beginning of S-phase. Both Cdc20 and Cdh1 recognize substrates via short, linear degron motifs in substrates. The most well-characterized and widespread of these degron motifs among APC/C substrates are the D-box (amino acid sequence R-X-X-L, where X is any amino acid) and the KEN box (amino acid sequence K-E-N). Thus, the ability of Cdc20 or Cdh1 to recruit substrate proteins harboring D- or KEN-box motifs to the APC/C is required for the subsequent ubiquitination and destruction of APC/C targets.

Like the SCF, the APC/C was identified by virtue of its key role in cell cycle. It had been known that the key CDK activator Cyclin B is controlled by degradation,

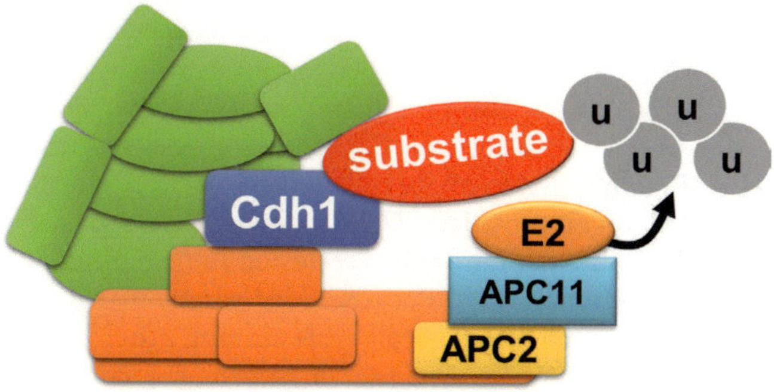

Figure 3.
Architecture of the APC/C ubiquitin ligase. The color scheme is the same as above for SCF ligases. Several proteins are specifically shown, including the cullin subunit APC2, the RING subunit APC11, and the substrate receptor Cdh1. Note that there are many more components.

and that both the accumulation and degradation of Cyclin B play a vital role in cell cycles, particularly in early frog embryos [18]. In 1995, the regulator of Cyclin B was discovered by the Kirschner and Hershko labs, who named it the Anaphase Promoting Complex and Cyclosome, respectively [19, 20].

3. Role and regulation of SCF ligases in G1/S control

The SCF complexes can assemble from any one of 69 well-established substrate receptor F-box proteins in humans. A subset of SCF ubiquitin ligase complexes have been directly implicated in G1/S control. Here we will discuss the role of each of these distinct complexes and/or substrate receptors, aspects of their regulation and function, and their contribution to G1 progression and S-phase initiation.

3.1 CDC4

The yeast specific Cell Division Control gene/protein 4, called Cdc4, was one of the original cell cycle mutants identified by Hartwell and colleagues, who later received the Nobel Prize for the analysis of cell cycle in budding yeast. They showed that Cdc4 mutant yeast arrest at the G1/S boundary, prior to the start of DNA replication [21]. However, it took another 20 years for the essential molecular function of Cdc4 in promoting cell cycle progression to become clear, and in doing so, laid the foundation for the discovery of CRL ligases.

As the analysis of cell cycle control became increasingly popular in the late 1980s and early 1990s, researchers revisited the role of Cdc4. Nasmyth and colleagues showed that the budding yeast Cdc4 mutants, which arrest before the start of DNA replication when grown at their restrictive temperature, lack appreciable CDK activity [22]. Interestingly, cell cycle arrest is caused by an inability of Cdc4 mutant cells to downregulate the yeast CKI Sic1, which normally decreases at the end of G1. The decrease in Sic1 allows the increase in CDK activity needed to enter S-phase. Thus, yeast cells cannot enter S-phase when Cdc4 is inactivated [11].

As discussed above, Cdc4 is an F-box protein that binds to Sic1, promoting its ubiquitination by the SCF^{Cdc4} complex. The mechanism by which Cdc4 recognizes Sic1 to promote its degradation provides a clear example of the interplay between phosphorylation and ubiquitination cascades. Interestingly, Sic1 must first be phosphorylated by Cyclin-CDK complexes, and this phosphorylation enables the binding of Cdc4 to Sic1 [13, 14]. Once phosphorylated and bound to Cdc4, Sic1 is recruited to the SCF complex for ubiquitination (**Figure 4**). Thus, CDKs promote their own activity at the G1/S boundary by triggering the degradation of their inhibitor, Sic1 (**Figure 4B**). This implies a positive feedback loop in control of S-phase entry. While the mechanism by which Cdc4 controls G1/S is largely attributed to its role in destroying Sic1, Cdc4 has also been linked to other cell cycle regulators and proteins involved in proliferative control. Cdc4 substrates include numerous proteins involved in MAPK signaling that mediate cell cycle arrest in response to pheromone [23–26], the replication regulator Cdc6 [27], the sirtuin deacetylase Hst3 [28], as well as proteins involved in sister chromatid cohesion [29], regulation of calcineurin [30], and mating-type switching [31]. Because Cdc4 has many substrates, it plays a complex and multi-faceted role in yeast cell cycle, among other processes.

3.2 Skp2

The F-box protein Skp2 has been well-characterized in human cells and plays an important role in the G1/S transition. Similar to Cdc4, Skp2 plays a key role

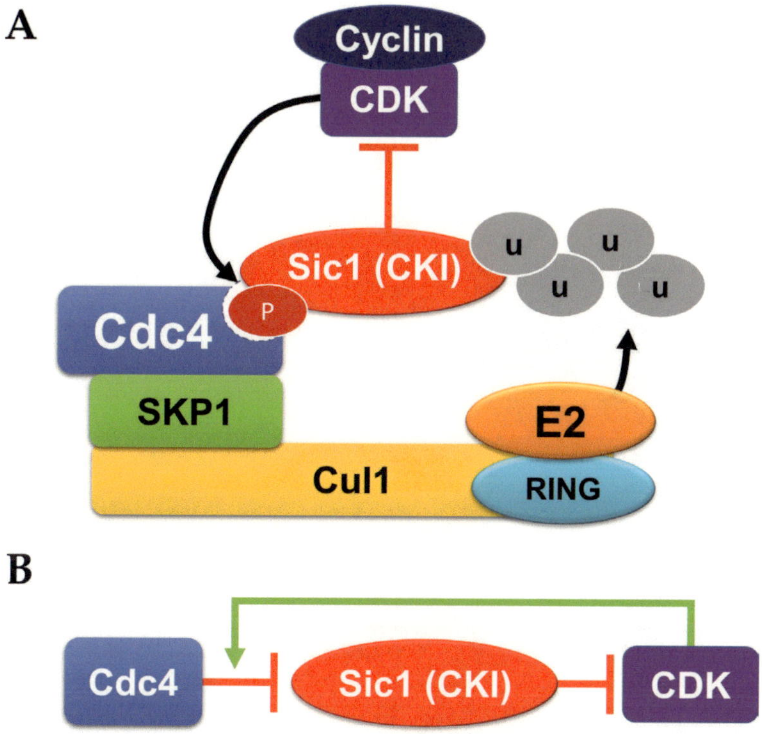

Figure 4.
SCF^{Cdc4} promotes S-phase entry in yeast by triggering the degradation of the CKI Sic1. (A) Binding between Sic1 and Cdc4 is triggered by phosphorylation of Sic1, which then promotes Sic1 ubiquitination and degradation. (B) A positive feedback loop between Cdc4, Sic1 and CDK promotes S-phase entry.

in regulating CDKs by promoting the destruction of CKI proteins. In particular, Skp2 plays an important role in promoting the destruction of the human CKI p27 [32, 33]. Moreover, the ubiquitination of p27 by SCF^{Skp2} requires that it first be phosphorylated by CDK, and this subsequently targets p27 for destruction, suggesting a similar positive feedback loop in G1/S regulation (**Figure 5**) [34]. Similarly, SCF^{Skp2} can target two other CKI proteins for degradation. These are p21 and p57, both of which are degraded in proliferating cells going through the cell cycle [35, 36], although p21 is also degraded by a second Cul4-based CRL ligase once DNA replication has begun [37]. Finally, Skp2 has been linked to the degradation of the retinoblastoma related protein RBL2/p130 [38, 39]. Like RB, RBL2/p130 restrains the activity of a cell cycle E2F transcription factor that promotes proliferation and cell cycle progression.

As might be expected, Due to its role in promoting S-phase via the degradation of CKIs, Skp2 is often overexpressed in cancers, which likely contributes to cancer cell proliferation [40]. Chemical approaches aimed at identifying Skp2 inhibitors have been undertaken, with some success [41, 42].

In addition to its role in regulating several target proteins, including the CKIs discussed above, Skp2 plays a complex and more paradoxical role in regulating proliferation. The Myc transcription factor is a potent oncogene, that is activated in many cancers and which drives proliferation through myriad mechanisms [43]. Myc is ubiquitinated by Skp2 [44, 45]. However, remarkably, the ubiquitination and degradation of Myc catalyzed by SCF^{Skp2} triggers an *increase* in Myc activity. This is consistent with prior work implicating proteolysis in the activation of several transcription factors in both yeast and humans [46]. Accordingly, a stable allele of Myc that cannot be ubiquitinated is more abundant, localized to target promoters,

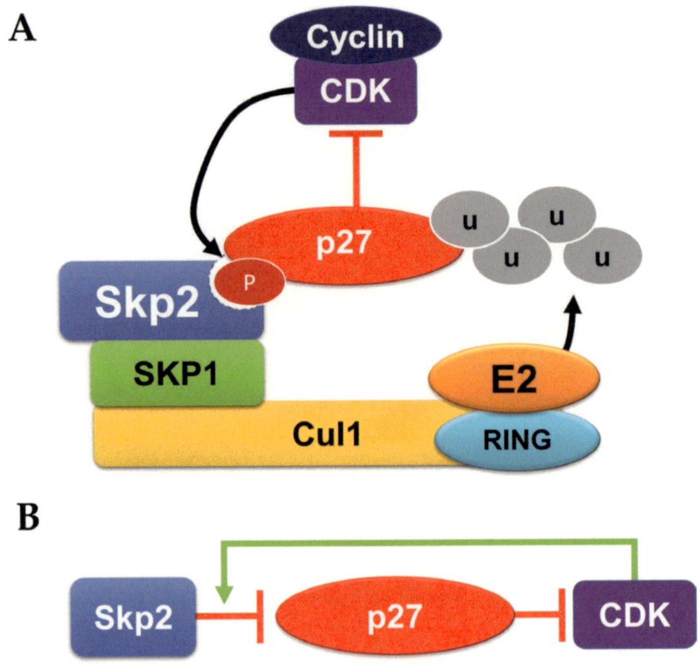

Figure 5.
SCF^{Skp2} promotes S-phase entry in humans by triggering the degradation of the CKI p27. (A) Binding between p27 and Skp2 is triggered by phosphorylation of p27, which then promotes p27 ubiquitination and degradation. (B) A positive feedback loop between Skp2, p27 and CDK promotes S-phase entry.

but it is less active [47]. Taken together, these studies paint a complex picture of the role of Skp2 in cell cycle progression but suggest an important role in proliferation and likely in the pathogenesis of cancer.

Interestingly, Skp2 is itself regulated by ubiquitin mediated proteolysis. Skp2 is targeted for degradation by the Anaphase Promoting Complex/Cyclosome during G1-phase of the cell cycle [48, 49]. The degradation of p27 requires the upregulation of Skp2. This degradation would presumably occur after Skp2 levels accumulate, following the inactivation of APC/C, which occurs in late G1. That is, APC/C inactivation should lead to an increase in Skp2 levels, since Skp2 would no longer be degraded. Only then could Skp2 promote the degradation of p27. However, this complex order of events remains unclear and has not yet been tested directly. Since the abundance of CKIs, like p27, should prevent the activation of G1/S CDKs, this also implies that APC/C inactivation precedes CDK activation. As discussed below, this too remains unknown, and recent evidence suggests, in fact, that APC/C inactivation occurs after CDK activation in G1 [16].

In addition to its regulation by ubiquitination, Skp2 is also regulated by phosphorylation. This phosphorylation is mediated, in part, by the oncogenic kinase AKT [50]. Notably, AKT kinase activity is cell cycle regulated, and begins to increase in late G1-phase [51]. Skp2 phosphorylation by AKT increases Skp2 stability and alters its localization. Surprisingly, SCF^{Skp2} also ubiquitinates AKT, and enhances AKT activation [52]. The degradation of p27 and activation of AKT and Myc, by Skp2, are likely to play an important role in tumor biology and treatment. The degradation of p27, a negative cell cycle regulator, creates an environment more permissive to proliferation because cells lacking p27 can progress through the cell cycle more rapidly. In addition, the activation of AKT and Myc could contribute significantly to cancer cell cycles.

3.3 Cyclin F/FBX01

The eponymous Cyclin F is the founding member of the F-box family of E3 ubiquitin ligases [11, 12, 53]. Cyclin F is unique among F-box proteins in that it contains a Cyclin homology domain, similar to canonical Cyclins that bind and activate CDKs. However, unlike those other Cyclins, Cyclin F neither binds nor activates a CDK [53]. In addition, Cyclin F levels oscillate strongly throughout the cell cycle, and this is the result of both changes in its transcription and degradation. Notably, Cyclin F is the only F-box protein that was identified as cell cycle regulated in all global studies of human cell cycle transcriptional dynamics [54]. Accordingly, Cyclin F knockout mouse embryonic fibroblasts showed a strong defect in cell cycle entry following synchronization in quiescence [55]. Nevertheless, despite this strong cell cycle phenotype and being the first described F-box protein in higher eukaryotes, Cyclin F went a long time without having a bona fide substrate.

The first two substrates described for Cyclin F were the centrosome protein CP110 and the spindle associated, mitotic phospho-protein NUSAP1 [56, 57], further supporting a role in cell cycle, and pointing to a function in organizing the microtubule cytoskeleton. In addition, Cyclin F regulates the RRM2 subunit of ribonucleotide reductase [58], histone mRNA stem loop binding protein SLBP [59], and the DNA replication protein Cdc6 [60], highlighting a role in S-phase progression and genome stability.

Importantly, Cyclin F regulates the degradation of Cdh1, the substrate receptor for the APC/C ubiquitin ligase (**Figure 6**). APC/C^{Cdh1} is activated throughout G1-phase and its inactivation is critical for S-phase entry. Thus, Cyclin F-mediated degradation of Cdh1 was shown to play a critical role in entry into S-phase [61]. Interestingly, in addition to targeting the APC/C substrate receptor Cdh1 for degradation, Cyclin F is also a substrate of APC/C in mitosis and early G1-phase [61]. Thus, Cyclin F exists in a double-negative feedback loop with APC/C, where it is a substrate in mitosis and early G1, and then the regulator of Cdh1 degradation in late G1 and S-phase (**Figure 6**).

Like Skp2, Cyclin F is also phosphorylated by the oncogenic kinase AKT [62]. Similar to Skp2, the phosphorylation of Cyclin F by AKT leads to a significant increase in Cyclin F stability. Phosphorylation by AKT enhances Cyclin F assembly into SCF ligase complexes. Thus, phosphorylation contributes to the switch in Cyclin F, from being an APC/C substrate to being capable of targeting for Cdh1 degradation in late G1-phase [62]. The tight regulation of Cyclin F throughout the cell cycle, its substrates, phosphorylation by AKT, and regulation by other E3s, point to its critical role in cell cycle progression. Moreover, these results suggest that Cyclin F is a key regulatory node mediating the interaction between AKT-dependent growth factor signaling and the core cell cycle machinery.

3.4 FBXW7/FBW7/FBXO30

The SCF^{Fbxw7} ubiquitin ligase (also called SCF^{FBW7} or SCF^{FBXO30}) is the most tightly linked to cancer proliferation of all SCF-type E3s [63]. Fbxw7, is highly mutated in human cancers, and exhibits both truncating mutations throughout its gene body, as well as "hotspot" point mutations in its substrate binding motif. Interestingly, while Fbxw7 is generally considered a tumor suppressor [64], "hotspot" mutations are more commonly found in oncogenes, such as the common G12V mutation recurrently observed in oncogenic K-Ras in many human malignancies. SCF^{Fbxw7} promotes cell cycle progression by regulating the degradation of

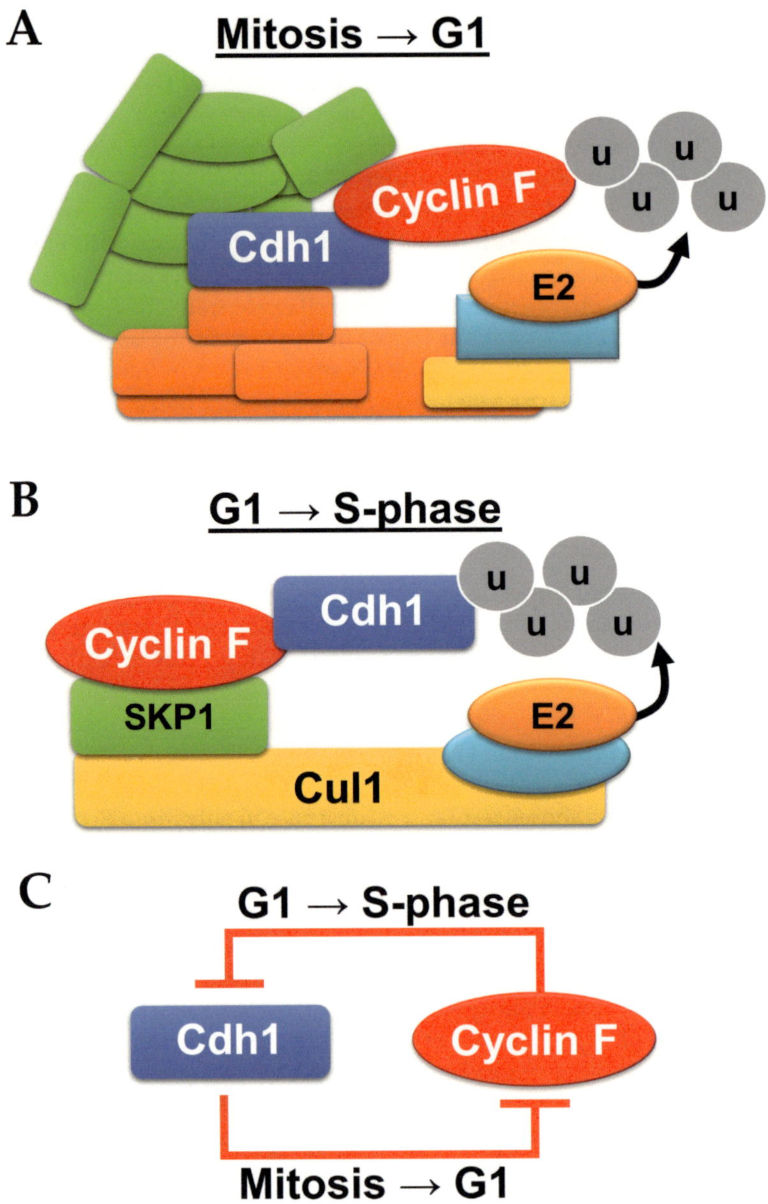

Figure 6.
$SCF^{Cyclin\ F}$ and APC/C constitute a double-negative feedback loop. (A) APC/C^{Cdh1} targets Cyclin F for degradation in late mitosis and early G1. (B) $SCF^{Cyclin\ F}$ targets Cdh1 for degradation in late G1 and S-phase. (C) Together, this suggests a temporally ordered, double negative feedback loop that promotes S-phase entry.

Cyclin E, the key activator of CDK2 at the G1/S boundary [63, 65–67]. In addition, Fbxw7 regulates the ubiquitination and destruction of numerous other pro-proliferative and cancer associated proteins, including Myc [68, 69], Notch [70, 71] and Jun [72].

Similar to other SCF ligases, the SCF^{Fbxw7} ligase recognizes substrates through phospho-degron motifs, with the most well characterized being that on Cyclin E. The phosphorylation of Cyclin E, by CDK2 or GSK3, can promote the degradation of Cyclin E by enhancing its binding to Fbxw7 [64, 66, 67, 73]. In addition, Fbxw7 homo-dimerizes, and this dimerization plays an important role in its ability to target substrates for degradation [74].

3.5 EMI1/Fbx05

Emi1 is a cell cycle regulated F-box domain-containing protein. However, Emi1 is unique among F-box proteins in that it has no known substrates, despite the fact that it binds tightly to the SCF adaptor Skp1. Emi1 is instead a key regulator of the cell cycle E3 ligase APC/C [75].

Many studies have demonstrated the potent and extensive role that Emi1 plays in inhibiting APC/C. Emi1 acts as a pseudo-substrate for APC/C, blocking the binding and ubiquitination of substrates [76]. In addition, Emi1 can alter the binding of the APC/C E2 ubiquitin conjugating enzymes, providing additional layers of regulation [77–79].

The association of Emi1 with S-phase entry is complex. Based largely on gain-of-function approaches, Emi1 was shown capable of inhibiting APC/C at the G1/S boundary and promoting S-phase entry [80]. This was fitting, since Emi1 abundance is controlled by the E2F family of transcription factors, which are activated in mid G1 and promote G1/S [80]. However, loss of Rca1, the fly version of Emi1, leads to an accumulation of cells in later stages of the cell cycle, not at G1/S [81]. Similarly, the loss of Emi1 in human cells was reported to induce the reactivation of APC/C during S and G2-phase, and to induce DNA re-replication as a result of the degradation of proteins which normally restrain licensing of replication origins [82, 83]. However, consistent with early gain-of-function studies, recent single cell approaches suggest that Emi1 contributes to the kinetics of APC/C inactivation at G1/S, and that Emi1 locks APC/C in an off state once S-phase begins [16]. Surprisingly, Emi1 might also be a substrate of the APC/C [84]. If Emi1 is a substrate of APC/C, this implies that Emi1 could be ubiquitinated by APC/C in early G1, and that it later accumulates as an inhibitor to inactivate APC/C and promote S-phase entry, much like Cyclin F [84]. This adds to our understanding of Emi1 degradation, wherein previous studies had shown it was degraded in mitosis by the SCF^{bTRCP} ubiquitin ligase [85, 86]. It will be important in the future to determine if altering the ubiquitination and degradation of Emi1 by APC/C accelerates progression through G1/S and to determine how this is coordinated with other SCF ligases that regulate G1/S.

4. Involvement of APC/C in G1/S

An extensive body of evidence has defined the role of APC/C^{Cdh1} in G1/S control [17]. Early studies in yeast showed that Cyclin proteolysis starts in late mitosis but then persists as cells continue through G1-phase [87]. In addition, yeast cells lacking Cdh1 are defective at arresting in G1-phase. Similar results have been observed across all eukaryotes in which loss of Cdh1 has been studied, including worms [88, 89], flies [90], chickens [91], mice [92] and humans [16, 61, 93]. The loss of Cdh1 accelerates progression through G0/G1 and promotes the start of S-phase. In addition, cells lacking Cdh1 are universally defective in G0/G1 arrest [17]. Accordingly, single allelic loss of Cdh1, the APC/C substrate receptor/coactivator in G0/G1-phase, is sufficient to cause tumors in mice [94]. Since the APC/C controls the stability of many dozens of substrates, it is unlikely that any one provides the basis for how cells enter S-phase in the absence of G1 APC/C function. Instead, it is more likely that the concerted upregulation of many cell cycle drivers together provides an explanation for the vital role of APC/C in restraining G1/S. Nevertheless, the APC/C is among a small group of key signaling molecules that prevent entry into S-phase of the cell cycle. These regulators include the retinoblastoma tumor suppressor and its related proteins p107 and p130, as well as the CDK inhibitors p21, p27 and p57.

Myriad mechanisms account for the inactivation of APC/C at the G1/S boundary, some of which were discussed above. This includes the degradation of Cdh1 by $SCF^{Cyclin\ F}$ and perhaps by the APC/C itself [61, 95]. The APC/C E2 enzymes, Ube2S and Ube2C, are unstable proteins and are also APC/C substrates [4, 96]. The substrate receptor Cdh1 is subject to CDK dependent phosphorylation, preventing its association with the APC/C and likely affecting its localization [89, 97–101]. Finally, accumulation of Emi1 is controlled by E2F, contributing to APC/C inhibition [16, 80, 84].

5. Distilling the complexity of ubiquitination in G1/S

The interconnected web of enzymes, substrates, and pathways discussed above paints a complicated picture of G1/S control. Remarkably, our understanding of the role of ubiquitin ligases in S-phase entry pales in comparison to studies performed on parallel kinase signaling cascades that converge on the E2F transcription factor. In quiescent and early G1 cells, E2F activity is repressed by the retinoblastoma tumor suppressor (RB), as well as the RB-like proteins P130 and P107. The phosphorylation of RB, first by Cyclin D-CDK4/6, and then by Cyclin E-Cdk2, inactivates RB and derepresses E2F. This derepression, in turn, triggers the transcriptional upregulation of many genes needed for S-phase entry.

How then do the pathways described above fit together with each other, and with the canonical CDK-RB-E2F pathway? We propose that multiple pathways act coordinately to promote the start of DNA replication. The most well-studied of these is the RB-E2F pathway, which promotes S-phase entry by promoting the expression of numerous cell cycle genes. In parallel, ubiquitin signaling pathways that control the degradation of numerous cell cycle proteins coordinate entry into S-phase. First, SCF^{Skp2} must be active and able to promote the degradation of CKI proteins. Second, SCF^{Fbw7} must be inactive or otherwise unable to ubiquitinate its substrates Cyclin E and Myc, which accumulate to promote cell cycle. Third, $SCF^{Cyclin\ F}$ must be available to trigger the degradation of Cdh1 and help promote the inactivation of APC/C. And finally, the APC/C must be inactivated, by Cyclin F and other pathways, allowing for the accumulation of cell cycle proteins (many of which are transcribed by E2F), to promote S-phase entry (**Figure 7**). It is notable that Cyclin F and Skp2, as well as many other cell cycle proteins, are downregulated by APC/C. Altogether, this suggests that aberrant APC/C inactivation could promote cancer cell cycles. Accordingly, single allelic loss of Cdh1 causes cancer in mice [94]. How APC/C might be inactivated in cancer remains an open question of significant importance that has only recently begun to be studied [17].

Upstream of these regulators are myriad kinase signaling cascades. These kinase cascades include, for example, the phosphorylation of RB by CDK4/6 and also CDK2; phosphorylation of Cyclin F and Skp2 by AKT; and, phosphorylation of Myc and Cyclin E, thereby marking them for degradation by Fbw7. Significantly, we hypothesize that these pathways control S-phase entry by globally remodeling the protein landscape either through changes in gene expression or protein degradation. The activity of CDK2, CDK4/6 and AKT is dysregulated in many cancers. This suggests that dysregulated cell cycle transcription, as well as dysregulated cell cycle ubiquitination, likely contributes to a weakening of the G1/S boundary and uncontrolled cancer cell cycles.

Testing this hypothesis and determining how these pathways are integrated remains an important question for future study. Determining the order of and

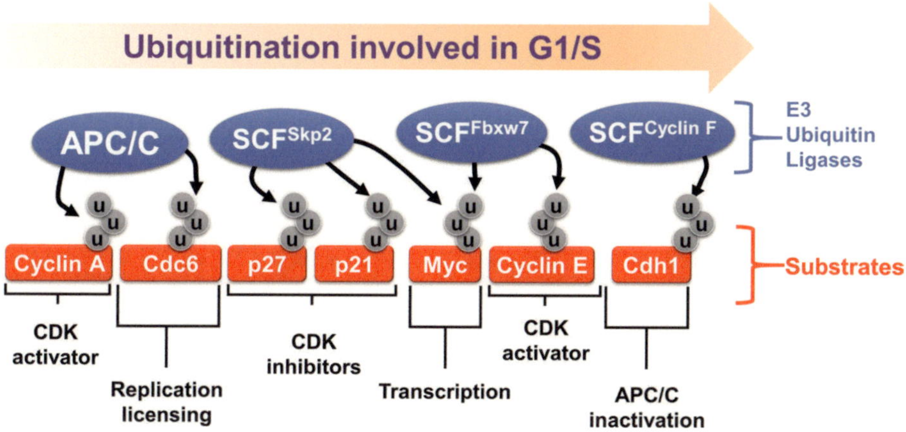

Figure 7.
Overview of ubiquitin signaling pathways involved in G1/S. A subset of substrates are shown. Note that the APC/C controls the stability of several dozen substrate proteins during late mitosis and early G1.

integration between these pathways is also critical. For example, recent live imaging studies demonstrated that CDK2 becomes active in mid-G1, several hours before APC/C is turned off. Moreover, these studies indicate that APC/C inactivation occurs at nearly the same time as DNA replication [16]. What is unclear is how Emi1, Cyclin F, and Skp2 accumulate at this time, as these proteins have never before been studied together in the same experimental system. In addition, the overwhelming majority of studies that have interrogated the kinetics of their accumulation have relied on bulk biochemical measurements (immunoblots) in synchronized cells. While informative, these studies would be better undertaken in asynchronous cells using either immunofluorescence or live cell reporters. Further, CDK2 activity begins to increase many hours before the inactivation of APC/C. It is therefore unknown how APC/C remains active into late G1-phase and is protected from CDK-dependent inactivation. Resolving these important questions will provide insight regarding how cells breach the G1/S boundary during the homeostatic cell cycles that occur during organismal development and growth, or in response to cell damage or wounding. Importantly, the G1/S boundary is universally dysfunctional in cancer and is the target of therapeutic interventions in the treatment of disease. Therefore, unraveling the complex pathways and mechanisms by which the ubiquitin system contributes to G1/S will shed light on both the etiology and treatment of cancer in the future.

Acknowledgements

Special thanks go to the Emanuele laboratory for feedback. Work in the Emanuele lab is supported by start-up funds from the University Cancer Research Fund and The National Institute of General Medical Sciences (NIH; R01GM120309).

Conflict of interest

The authors declare no competing conflicts of interest.

Author details

Michael James Emanuele* and Taylor Paige Enrico
The University of North Carolina at Chapel Hill, Chapel Hill, NC, USA

*Address all correspondence to: emanuele@email.unc.edu

References

[1] Williamson A, Werner A, Rape M. The colossus of ubiquitylation: Decrypting a cellular code. Molecular Cell. 2013;**49**:591-600

[2] Yau R, Rape M. The increasing complexity of the ubiquitin code. Nature Cell Biology. 2016;**18**:579-586

[3] Matsumoto ML, Wickliffe KE, Dong KC, Yu C, Bosanac I, Bustos D, et al. K11-linked polyubiquitination in cell cycle control revealed by a K11 linkage-specific antibody. Molecular Cell. 2010;**39**:477-484

[4] Williamson A, Wickliffe KE, Mellone BG, Song L, Karpen GH, Rape M. Identification of a physiological E2 module for the human anaphase-promoting complex. Proceedings of the National Academy of Sciences of the United States of America. 2009;**106**:18213-18218

[5] Yau RG, Doerner K, Castellanos ER, Haakonsen DL, Werner A, Wang N, et al. Assembly and function of heterotypic ubiquitin chains in cell-cycle and protein quality control. Cell. 2017;**0**:1-16

[6] Meyer H-J, Rape M. Enhanced protein degradation by branched ubiquitin chains. Cell. 2014;**157**:910-921

[7] Zhang J, Wan L, Dai X, Sun Y, Wei W. Functional characterization of anaphase promoting complex/Cyclosome (APC/C) E3 ubiquitin ligases in tumorigenesis. Biochimica et Biophysica Acta. 2014;**1845**:277-293

[8] Potapova TA, Daum JR, Pittman BD, Hudson JR, Jones TN, Satinover DL, et al. The reversibility of mitotic exit in vertebrate cells. Nature. 2006;**440**:954-958

[9] Sherr CJ, Bartek J. Cell cycle–targeted Cancer therapies. Annual Review of Cancer Biology. 2017;**1**:41-57

[10] Petroski MD, Deshaies RJ. Function and regulation of cullin-RING ubiquitin ligases. Nature Reviews. Molecular Cell Biology. 2005;**6**:9-20

[11] Bai C, Richman R, Elledge SJ. Human cyclin F. The EMBO Journal. 1994;**13**:6087-6098

[12] Bai C, Sen P, Hofmann K, Ma L, Goebl M, Harper JW, et al. SKP1 connects cell cycle regulators to the ubiquitin proteolysis machinery through a novel motif, the F-box. Cell. 1996;**86**:263-274

[13] Skowyra D, Craig KL, Tyers M, Elledge SJ, Harper JW. F-box proteins are receptors that recruit phosphorylated substrates to the SCF ubiquitin-ligase complex. Cell. 1997;**91**:209-219

[14] Feldman RM, Correll CC, Kaplan KB, Deshaies RJ. A complex of Cdc4p, Skp1p, and Cdc53p/cullin catalyzes ubiquitination of the phosphorylated CDK inhibitor Sic1p. Cell. 1997;**91**:221-230

[15] Brandeis M, Hunt T. The proteolysis of mitotic cyclins in mammalian cells persists from the end of mitosis until the onset of S phase. The EMBO Journal. 1996;**15**:5280-5289

[16] Cappell SD, Chung M, Jaimovich A, Spencer SL, Meyer T. Irreversible APCCdh1 inactivation underlies the point of No return for cell-cycle entry. Cell. 2016;**166**:167-180

[17] Kernan J, Bonacci T, Emanuele MJ. Who guards the guardian? Mechanisms that restrain APC/C during the cell cycle. Biochimica et Biophysica Acta (BBA)—Molecular Cell Research. 2018

[18] Murray AW, Solomon MJ, Kirschner MW. The role of cyclin synthesis and degradation in the control of maturation

promoting factor activity. Nature. 1989;**339**:280-286

[19] Sudakin V, Ganoth D, Dahan A, Heller H, Hershko J, Luca FC, et al. The cyclosome, a large complex containing cyclin-selective ubiquitin ligase activity, targets cyclins for destruction at the end of mitosis. Molecular Biology of the Cell. 1995;**6**:185-197

[20] King RW, Peters JM, Tugendreich S, Rolfe M, Hieter P, Kirschner MW. A 20S complex containing CDC27 and CDC16 catalyzes the mitosis-specific conjugation of ubiquitin to cyclin B. Cell. 1995;**81**:279-288

[21] Hartwell LH, Culotti J, Pringle JR, Reid BJ. Genetic control of the cell division cycle in yeast. Science. 1974;**183**:46-51

[22] Schwob E, Böhm T, Mendenhall MD, Nasmyth K. The B-type cyclin kinase inhibitor p40SIC1 controls the G1 to S transition in S cerevisiae. Cell. 1994;**79**:233-244

[23] Cappell SD, Baker R, Skowyra D, Dohlman HG. Systematic analysis of essential genes reveals important regulators of G protein signaling. Molecular Cell. 2010

[24] Chou S, Huang L, Liu H. Fus3-regulated Tec1 degradation through SCFCdc4 determines MAPK signaling specificity during mating in yeast. Cell. 2004

[25] Vendrell A, Martánez-Pastor M, González-Novo A, Pascual-Ahuir A, Sinclair DA, Proft M, et al. Sir2 histone deacetylase prevents programmed cell death caused by sustained activation of the Hog1 stress-activated protein kinase. EMBO Reports. 2011

[26] Hurst JH, Dohlman HG. Dynamic ubiquitination of the mitogen-activated protein kinase kinase (MAPKK) Ste7 determines mitogen-activated protein kinase (MAPK) specificity. The Journal of Biological Chemistry. 2013

[27] Drury LS, Perkins G, Diffley JFX. The Cdc4/34/53 pathway targets Cdc6p for proteolysis in budding yeast. The EMBO Journal. 1997

[28] Edenberg ER, Vashisht AA, Topacio BR, Wohlschlegel JA, Toczyski DP. Hst3 is turned over by a replication stress-responsive SCFCdc4 phospho-degron. Proceedings of the National Academy of Sciences of the United States of America. 2014;**111**:5962-5967

[29] Lyons NA, Morgan DO. Cdk1-dependent destruction of Eco1 prevents cohesion establishment after S phase. Molecular Cell. 2011

[30] Kishi T, Ikeda A, Nagao R, Koyama N. The SCFCdc4 ubiquitin ligase regulates calcineurin signaling through degradation of phosphorylated Rcn1, an inhibitor of calcineurin. Proceedings of the National Academy of Sciences of the United States of America. 2007

[31] Liu Q, Larsen B, Ricicova M, Orlicky S, Tekotte H, Tang X, et al. SCFCdc4 enables mating type switching in yeast by Cyclin-dependent kinase-mediated elimination of the Ash1 transcriptional repressor. Molecular and Cellular Biology. 2011

[32] Pagano M, Tam SW, Theodoras AM, Beer-Romero P, Del Sal G, Chau V, et al. Role of the ubiquitin-proteasome pathway in regulating abundance of the cyclin-dependent kinase inhibitor p27. Science. 1995;**269**:682-685

[33] Carrano AC, Eytan E, Hershko A, Pagano M. SKP2 is required for ubiquitin-mediated degradation of the CDK inhibitor p27. Nature Cell Biology. 1999;**1**:193-199

[34] Montagnoli A, Fiore F, Eytan E, Carrano AC, Draetta GF, Hershko A, et al. Ubiquitination of p27 is regulated

by Cdk-dependent phosphorylation and trimeric complex formation. Genes & Development. 1999

[35] Bornstein G, Bloom J, Sitry-Shevah D, Nakayama K, Pagano M, Hershko A. Role of the SCFSkp2 ubiquitin ligase in the degradation of p21Cip1 in S phase. The Journal of Biological Chemistry. 2003

[36] Kamura T, Hara T, Kotoshiba S, Yada M, Ishida N, Imaki H, et al. Degradation of p57Kip2 mediated by SCFSkp2-dependent ubiquitylation. Proceedings of the National Academy of Sciences of the United States of America. 2003

[37] Abbas T, Sivaprasad U, Terai K, Amador V, Pagano M, Dutta A. PCNA-dependent regulation of p21 ubiquitylation and degradation via the CRL4Cdt2 ubiquitin ligase complex. Genes & Development. 2008;**22**:2496-2506

[38] Tedesco D, Lukas J, Reed SI. The pRb-related protein p130 is regulated by phosphorylation-dependent proteolysis via the protein-ubiquitin ligase SCFSkp2. Genes & Development. 2002;**16**:2946-2957

[39] Bhattacharya S, Garriga J, Calbó J, Yong T, Haines DS, Graña X. SKP2 associates with p130 and accelerates p130 ubiquitylation and degradation in human cells. Oncogene. 2003;**22**:2443-2451

[40] Zheng N, Zhou Q, Wang Z, Wei W. Recent advances in SCF ubiquitin ligase complex: Clinical implications. Biochimica et Biophysica Acta. 2016;**1866**:12-22

[41] Chan C-H, Morrow JK, Li C-F, Gao Y, Jin G, Moten A, et al. Pharmacological inactivation of Skp2 SCF ubiquitin ligase restricts cancer stem cell traits and cancer progression. Cell. 2013;**154**:556-568

[42] Wu L, Grigoryan AV, Li Y, Hao B, Pagano M, Cardozo TJ. Specific small molecule inhibitors of Skp2-mediated p27 degradation. Chemistry & Biology. 2012;**19**:1515-1524

[43] Tansey WP. Mammalian MYC proteins and cancer. New Journal of Science. 2014;**2014**:1-27

[44] Von Der Lehr N, Johansson S, Wu S, Bahram F, Castell A, Cetinkaya C, et al. The F-box protein Skp2 participates in c-Myc proteosomal degradation and acts as a cofactor for c-Myc-regulated transcription. Molecular Cell. 2003;**11**:1189-1200

[45] Kim SY, Herbst A, Tworkowski KA, Salghetti SE, Tansey WP. Skp2 regulates Myc protein stability and activity. Molecular Cell. 2003;**11**:1177-1188

[46] Geng F, Wenzel S, Tansey WP. Ubiquitin and proteasomes in transcription. Annual Review of Biochemistry. 2012;**81**:177-201

[47] Jaenicke LA, von Eyss B, Carstensen A, Wolf E, Xu W, Greifenberg AK, et al. Ubiquitin-dependent turnover of MYC antagonizes MYC/PAF1C complex accumulation to drive transcriptional elongation. Molecular Cell. 2016;**61**:54-67

[48] Bashir T, Dorrello NV, Amador V, Guardavaccaro D, Pagano M. Control of the SCFSkp2–Cks1 ubiquitin ligase by the APC/CCdh1 ubiquitin ligase. Nature. 2004;**428**:190-193

[49] Wei W, Ayad NG, Wan Y, Zhang G-J, Kirschner MW, Kaelin WG. Degradation of the SCF component Skp2 in cell-cycle phase G1 by the anaphase-promoting complex. Nature. 2004;**428**:194-198

[50] Gao D, Inuzuka H, Tseng A, Chin RY, Toker A, Wei W. Phosphorylation by Akt1 promotes cytoplasmic localization of Skp2 and impairs APCCdh1-mediated

Skp2 destruction. Nature Cell Biology. 2009;**11**:397-408

[51] Liu P, Begley M, Michowski W, Inuzuka H, Ginzberg M, Gao D, et al. Cell-cycle-regulated activation of Akt kinase by phosphorylation at its carboxyl terminus. Nature. 2014

[52] Chan C-H, Li C-F, Yang W-L, Gao Y, Lee S-W, Feng Z, et al. The Skp2-SCF E3 ligase regulates Akt ubiquitination, glycolysis, herceptin sensitivity, and tumorigenesis. Cell. 2012;**149**:1098-1111

[53] D'Angiolella V, Esencay M, Pagano M. A cyclin without cyclin-dependent kinases: Cyclin F controls genome stability through ubiquitin-mediated proteolysis. Trends in Cell Biology. 2013;**23**:135-140

[54] Fischer M, Grossmann P, Padi M, DeCaprio JA. Integration of TP53, DREAM, MMB-FOXM1 and RB-E2F target gene analyses identifies cell cycle gene regulatory networks. Nucleic Acids Research. 2016;**44**:6070-6086

[55] Tetzlaff MT, Bai C, Finegold M, Harper JW, Mahon KA, Stephen J, et al. Cyclin F disruption compromises placental development and affects normal cell cycle execution. Molecular and Cellular Biology. 2004;**24**:2487-2498

[56] D'Angiolella V, Donato V, Vijayakumar S, Saraf A, Florens L, Washburn MP, et al. SCF Cyclin F controls centrosome homeostasis and mitotic fidelity through CP110 degradation. Nature. 2010;**466**:138-142

[57] Emanuele MJ, Elia AEH, Xu Q, Thoma CR, Izhar L, Leng Y, et al. Global identification of modular cullin-RING ligase substrates. Cell. 2011;**147**:459-474

[58] D'Angiolella V, Donato V, Forrester FM, Jeong Y, Pellacani C, Kudo Y, et al. Cyclin F-mediated degradation of ribonucleotide reductase M2 controls genome integrity and DNA repair. Cell. 2012;**149**:1023-1034

[59] Dankert JF, Rona G, Clijsters L, Geter P, Skaar JR, Bermudez-Hernandez K, et al. Cyclin F-mediated degradation of SLBP limits H2AX accumulation and apoptosis upon genotoxic stress in G2. Molecular Cell. 2016

[60] Walter D, Hoffmann S, Komseli E-S, Rappsilber J, Gorgoulis V, Sørensen CS. SCFCyclin F-dependent degradation of CDC6 suppresses DNA re-replication. Nature Communications. 2016;**7**:10530

[61] Choudhury R, Bonacci T, Arceci A, Lahiri D, Mills CA, Kernan JL, et al. APC/C and SCF(cyclin F) constitute a reciprocal feedback circuit controlling S-phase entry. Cell Reports. 2016;**16**:3359-3372

[62] Choudhury R, Bonacci T, Wang X, Truong A, Arceci A, Zhang Y, et al. The E3 ubiquitin ligase SCF(Cyclin F) transmits AKT signaling to the cell-cycle machinery. Cell Reports. 2017;**20**:3212-3222

[63] Welcker M, Clurman BE. FBW7 ubiquitin ligase: A tumour suppressor at the crossroads of cell division, growth and differentiation. Nature Reviews. Cancer. 2008;**8**:83-93

[64] Koepp DM, Schaefer LK, Ye X, Keyomarsi K, Chu C, Harper JW, et al. Phosphorylation-dependent ubiquitination of cyclin E by the SCFFbw7 ubiquitin ligase. Science. 2001;**294**:173-177

[65] Strohmaier H, Spruck CH, Kaiser P, Won KA, Strohmaier H, Reed SI. Archipelago regulates Cyclin E levels in Drosophila and is mutated in human cancer cell lines. Nature. 2001

[66] Strohmaier H, Spruck CH, Kaiser P, Won KA, Sangfelt O, Reed SI. Human F-box protein hCdc4 targets cyclin E for

proteolysis and is mutated in a breast cancer cell line. Nature. 2001

[67] Welcker M, Singer J, Loeb KR, Grim J, Bloecher A, Gurien-West M, et al. Multisite phosphorylation by Cdk2 and GSK3 controls cyclin E degradation. Molecular Cell. 2003

[68] Welcker M, Orian A, Jin J, Grim JE, Harper JW, Eisenman RN, et al. The FBW7 tumor supressor regulates glycogen synthase kinase 3 phosphorylation-dependent c-Myc protein degradation. PNAS. 2004

[69] Yada M, Hatakeyama S, Kamura T, Nishiyama M, Tsunematsu R, Imaki H, et al. Phosphorylation-dependent degradation of c-Myc is mediated by the F-box protein Fbw7. The EMBO Journal. 2004

[70] Wu G, Lyapina S, Das I, Li J, Gurney M, Pauley A, et al. SEL-10 is an inhibitor of notch signaling that targets notch for ubiquitin-mediated protein degradation. Molecular and Cellular Biology. 2001

[71] Gupta-Rossi N, Le Bail O, Gonen H, Brou C, Logeat F, Six E, et al. Functional interaction between SEL-10, an F-box protein, and the nuclear form of activated Notch1 receptor. The Journal of Biological Chemistry. 2001

[72] Nateri AS, Riera-Sans L, Da Costa C, Behrens A. The ubiquitin ligase SCFFbw7 antagonizes apoptotic JNK signaling. Science. 2004:80

[73] Won KA, Reed SI. Activation of cyclin E/CDK2 is coupled to site-specific autophosphorylation and ubiquitin-dependent degradation of cyclin E. The EMBO Journal. 1996

[74] Welcker M, Larimore EA, Swanger J, Bengoechea-Alonso MT, Grim JE, Ericsson J, et al. Fbw7 dimerization determines the specificity and robustness of substrate degradation. Genes & Development. 2013;**27**:2531-2536

[75] Reimann JDR, Freed E, Hsu JY, Kramer ER, Peters JM, Jackson PK. Emi1 is a mitotic regulator that interacts with Cdc20 and inhibits the anaphase promoting complex. Cell. 2001;**105**:645-655

[76] Miller JJ, Summers MK, Hansen DV, Nachury MV, Lehman NL, Loktev A, et al. Emi1 stably binds and inhibits the anaphase-promoting complex/cyclosome as a pseudosubstrate inhibitor. Genes & Development. 2006

[77] Frye JJ, Brown NG, Petzold G, Watson ER, Grace CRR, Nourse A, et al. Electron microscopy structure of human APC/C(CDH1)-EMI1 reveals multimodal mechanism of E3 ligase shutdown. Nature Structural & Molecular Biology. 2013;**20**:827-835

[78] Wang W, Kirschner MW. Emi1 preferentially inhibits ubiquitin chain elongation by the anaphase-promoting complex. Nature Cell Biology. 2013;**15**:797-806

[79] Chang L, Zhang Z, Yang J, McLaughlin SH, Barford D. Atomic structure of the APC/C and its mechanism of protein ubiquitination. Nature. 2015

[80] Hsu JY, Reimann JDR, Sørensen CS, Lukas J, Jackson PK. E2F-dependent accumulation of hEmi1 regulates S phase entry by inhibiting APC(Cdh1). Nature Cell Biology. 2002;**4**:358-366

[81] Grosskortenhaus R, Sprenger F. Rca1 inhibits APC-Cdh1(Fzr) and is required to prevent cyclin degradation in G2, Dev. Cell. 2002;**2**:29-40

[82] Machida YJ, Dutta A. The APC/C inhibitor, Emi1, is essential for

prevention of rereplication. Genes & Development. 2007;**21**:184-194

[83] DiFiore B, Pines J. Defining the role of Emi1 in the DNA replication-segregation cycle. Chromosoma. 2008;**117**:333-338

[84] Cappell SD, Mark KG, Garbett D, Pack LR, Rape M, Meyer T. EMI1 switches from being a substrate to an inhibitor of APC/CCDH1 to start the cell cycle. Nature. 2018

[85] Margottin-Goguet F, Hsu JY, Loktev A, Hsieh HM, Reimann JDR, Jackson PK. Prophase destruction of Emi1 by the SCFβTrCP/Slimbubiquitin ligase activates the anaphase promoting complex to allow progression beyond prometaphase. Developmental Cell. 2003

[86] Guardavaccaro D, Kudo Y, Boulaire J, Barchi M, Busino L, Donzelli M, et al. Control of meiotic and mitotic progression by the F box protein β-Trcp1 in vivo. Developmental Cell. 2003

[87] Amon A, Irniger S, Nasmyth K. Closing the cell cycle circle in yeast: G2 cyclin proteolysis initiated at mitosis persists until the activation of G1 cyclins in the next cycle. Cell. 1994;**77**:1037-1050

[88] Fay DS, Keenan S, Han M. fzr-1 and lin-35/Rb function redundantly to control cell proliferation in C elegans as revealed by a nonbiased synthetic screen. Genes & Development. 2002;**16**:503-517

[89] The I, Ruijtenberg S, Bouchet BP, Cristobal A, Prinsen MBW, van Mourik T, et al. Rb and FZR1/Cdh1 determine CDK4/6-cyclin D requirement in C elegans and human cancer cells. Nature Communications. 2015;**6**:5906

[90] Buttitta LA, Katzaroff AJ, Edgar BA. A robust cell cycle control mechanism limits E2F-induced proliferation of terminally differentiated cells in vivo. The Journal of Cell Biology. 2010;**189**:981-996

[91] Sudo T, Ota Y, Kotani S, Nakao M, Takami Y, Takeda S, et al. Activation of Cdh1-dependent APC is required for G1 cell cycle arrest and DNA damage-induced G2 checkpoint in vertebrate cells. The EMBO Journal. 2001;**20**:6499-6508

[92] Sigl R, Wandke C, Rauch V, Kirk J, Hunt T, Geley S. Loss of the mammalian APC/C activator FZR1 shortens G1 and lengthens S phase but has little effect on exit from mitosis. Journal of Cell Science. 2009;**122**:4208-4217

[93] Yuan X, Srividhya J, De Luca T, Lee J-HE, Pomerening JR. Uncovering the role of APC-Cdh1 in generating the dynamics of S-phase onset. Molecular Biology of the Cell. 2014;**25**:441-456

[94] García-Higuera I, Manchado E, Dubus P, Cañamero M, Méndez J, Moreno S, et al. Genomic stability and tumour suppression by the APC/C cofactor Cdh1. Nature Cell Biology. 2008;**10**:802-811

[95] Listovsky T, Oren YS, Yudkovsky Y, Mahbubani HM, Weiss AM, Lebendiker M, et al. Mammalian Cdh1/Fzr mediates its own degradation. The EMBO Journal. 2004;**23**:1619-1626

[96] Rape M, Kirschner MW. Autonomous regulation of the anaphase-promoting complex couples mitosis to S-phase entry. Nature. 2004;**432**:588-595

[97] Kramer ER, Scheuringer N, Podtelejnikov AV, Mann M, Peters JM. Mitotic regulation of the APC activator proteins CDC20 and CDH1. Molecular Biology of the Cell. 2000;**11**:1555-1569

[98] Hall MC, Warren EN, Borchers CH. Multi-kinase phosphorylation of the APC/C activator Cdh1 revealed

by mass spectrometry. Cell Cycle. 2004;**3**:1278-1284

[99] Höckner S, Neumann-arnold L, Seufert W. Dual control by Cdk1 phosphorylation of the budding yeast APC/C ubiquitin ligase activator Cdh1. Molecular Biology of the Cell. 2016;**27**:2198-2212

[100] Zachariae W, Schwab M, Nasmyth K, Seufert W. Control of cyclin ubiquitination by CDK-regulated binding of Hct1 to the anaphase promoting complex. Science. 1998;**282**:1721-1724

[101] Jaquenoud M, Van Drogen F, Peter M. Cell cycle-dependent nuclear export of Cdh1p may contribute to the inactivation of APC/CCdh1. The EMBO Journal. 2002;**21**:6515-6526

Chapter 2

Processes that Regulate the Ubiquitination of Chromatin and Chromatin-Associated Proteins

Alexander E. Hare and Jeffrey D. Parvin

Abstract

Ubiquitin is a post-translational modification important for many different processes in the cell, including antigen presentation and proteosomal degradation of proteins. It is heavily involved in the regulation of chromatin and the proteins that control chromatin-related processes. In this review, we will focus on ubiquitin-based chromatin regulation involved in four different processes. The first is DNA double strand break (DSB) repair and the role that ubiquitin plays in not just recruiting and stimulating DSB repair, but also the choice of pathway. The second is the PAF1 complex, which is involved in transcriptional elongation and interacts with RNAPII. The third is polycomb repressive complexes, specifically polycomb repressive complex 1, which utilizes ubiquitin to repress constitutively inactive genes. The last role of ubiquitin discussed is ubiquitin as a mitotic bookmark, which serves to provide a record of -active genes as cells transit mitosis. Each of these processes has independent pathways, but each is necessary for proper cellular function and organismal health.

Keywords: PRC1, RING1A, BMI1, bookmark ubiquitination

1. Introduction

Ubiquitin is most clearly associated with the process of targeted protein degradation, but it is involved in many cellular processes such as chromatin regulation, immune response, and antigen processing [1]. Proteosomal degradation is mediated through polyubiquitin chains linked via lysine-48 (K48) on the ubiquitin chain, interacting with the proteasome. Other processes utilize monoubiquitination or polymerization of ubiquitin molecules via another lysine. The ubiquitination system in humans is incredibly complex, with over 1000 known factors and over 10,000 known sites of ubiquitination, enabling its many and diverse roles in cellular biology.

In this review, we focus on four specific roles of ubiquitin in regulating chromatin: DNA repair, transcription elongation, epigenetic silencing via the polycomb repressive complex, and bookmark ubiquitination. In these processes, the ubiquitin moiety interfaces with many other epigenetic marks, such as acetylation, methylation, and histone modification to regulate a given process.

The process of ubiquitination (or ubiquitylation) is the attaching of one ubiquitin protein to a substrate and is performed by a cascade of three enzymes: E1 (activating), E2 (conjugating) and E3 (ligase) [2–5]. Substrates are proteins, and in the context of chromatin, histones are the most common class of substrate [6]. Histones form octamers containing two each of the four core histones (H2A, H2B, H3, H4), and when a histone octamer is wrapped with two turns of DNA, it is called a nucleosome. Each histone has a tail that extends outside the core of the nucleosome, where it is more accessible to the modifying enzymes. In addition to ubiquitination, other modifications occur, such as acetylation or methylation. The primary role of these marks is governing the localization on the genome of specific epigenetic marks as well as the compaction and decompaction of chromatin, which regulates accessibility of the transcription machinery to chromatin. Histone ubiquitination also serves as signaling molecules for other downstream regulators of transcription, which modulates transcription both directly and indirectly. Part of this concept includes histone cross-talk, where those regulators of transcription integrate signals of multiple distinct histone modifications on the same or nearby histones to generate a phenotype due to the composite signals [7]. Therefore, nucleosome modification has a fundamental function in silencing and activation of transcription. Most histone ubiquitination occurs as a monoubiquitination, but polyubiquitin chains have also been observed. There are a number of small ubiquitin-like modifier (SUMO) proteins that share structural resemblance to ubiquitin and play some similar roles [8]. SUMO proteins come in a variety of isoforms with varying capacity for chain formation and are conjugated to substrates in a similar manner as ubiquitination.

2. Ubiquitin in DNA damage repair

Survival of organisms and their cells depends on stability and integrity of DNA, but maintaining this integrity is a challenge for cells because they are constantly subjected to DNA damage from a variety of sources [9, 10]. DNA damage can cause disease and prevent faithful transfer of genetic information from one generation to the next. DNA double strand breaks (DSB) present a difficult problem to correct since there may be no template to guide error-free repair. In the DNA damage response, cells arrest the cell cycle and activate repair machinery. There are two primary methods eukaryotic cells use to repair DSBs: nonhomologous end joining (NHEJ) and homologous recombination (HR) [11]. The NHEJ pathway occurs throughout the cell cycle (except during mitosis) and is performed more commonly, but it is error-prone since it does not utilize a template [12]. By contrast, HR is only active during S and G2 phases of the cell cycle and because it uses the template in a sister chromatid, the repair has higher fidelity. The faulty repair observed in NHEJ can cause chromosomal rearrangements and mutations, leading to cancer susceptibility. In the following paragraphs, we highlight the roles of a variety of ubiquitin ligases and ubiquitin binding proteins to regulate the DNA damage response.

2.1 Initiation of DSB repair

When DSBs occur, ionizing radiation-induced foci form from genome-localized high concentrations of repair machinery with a host of bound factors necessary for DSB repair. To form these ionizing radiation-induced foci, histones near the damage site become modified with K63-linked polyubiquitin chains via the action of the E2 Mms2-Ubc13 and the E3 ligases RNF168 and RNF8 [13]. These K63 chains serve as markers for recruitment of downstream repair proteins and as transcriptional repressors to prevent propagation of problems caused by broken DNA strands.

RNF8 has a forkhead-associated domain that binds to ionizing radiation induced foci following a cascade of events starting with the MRE11-RAD50-NBS1 (MRN) complex binding to the DSB end, followed by ATM phosphorylation of a variant

H2A histone called H2AX. This phosphorylated H2AX-serine139 is known as γH2AX [14]. Sequentially, MDC1 (mediator of DNA damage checkpoint 1) binds, [15–17], and MDC1 serves as scaffold protein near sites of DNA damage, which it localizes to by using its BRCT (BRCA carboxyl terminus) domains to recruit RNF8. Following RNF8 recruitment, RNF8 ubiquitinates the linker histone H1, and thereby recruits RNF168 via ubiquitin binding domains (UBDs) binding to the ubiquitin mark [13]. RNF168, in turn, ubiquitinates histone H2A (**Figure 1A**).

Once RNF168 has been recruited by RNF8-mediated H1 ubiquitination, it can recognize its own H2A ubiquitination mark due to the UBD on its C-terminus, allowing self-propagation of the DSB repair response [18]. RNF168-mediated H2A is capable of monoubiquitination of H2AK13–15 [19]. Chain elongation by RNF168 is atypical: as mentioned, most DSB-related ubiquitination is comprised of polyubiquitin chains linked by K63, but RNF168, when overexpressed, creates K27 linked chains instead of linkages via K63 [20]. H2AK15ub is important for the recruitment of downstream factors, most importantly 53BP1, which promotes NHEJ [21]. RNF8 and RNF168 recruit more E3 ubiquitin ligases through direct interaction with HERC2 and via their ubiquitin ligase activity, but these other E3 ligases stimulate DSB repair as scaffolds, rather than as ubiquitin ligases [22]. For example, BRCA1 and BARD1 have E3 ligase function, but their role in DSB repair is independent of their ubiquitination activity [23, 24].

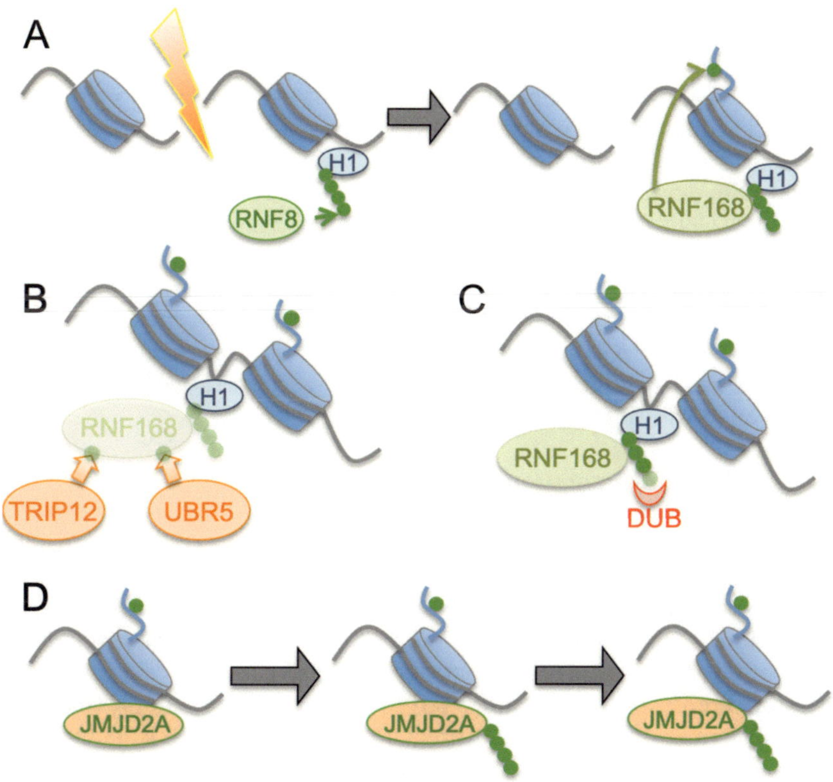

Figure 1.
Ubiquitination of chromatin regulates DSB repair. (A) Following RNF8 recruitment by ATM-phosphorylated MDC1, RNF8 ubiquitinates histone H1, which is necessary for RNF168 recruitment. (B) Ubiquitination regulates expression of DSB factors. UBR5 and TRIP12 are E3 ligases, which ubiquitinate RNF168 to target them for proteosomal degradation. (C) Deubiquitinases break down polyubiquitin chains, removing ubiquitin signals that recruit DSB repair factors. (D) Ubiquitination leads to factor removal from DSB sites. JMJD2A is recognized by 53BP1 and is involved in 53BP1 recruitment. When JMJD2A is ubiquitinated, segregase activity removes it from DSB sites.

In NHEJ, DNA broken strand ends are bound by the Ku70-Ku80 heterodimer, which in turns allows recruitment of the catalytic subunit of DNA-dependent protein kinase (DNA-PKcs) [25, 26]. Following phosphorylation of DNA-PKcs, the DNA ends are trimmed to make them ready for ligation. This trimming plays a major role in why NHEJ is error-prone and is more damaging to cells than HR. DNA ligase IV and its associated proteins are responsible for ligating these trimmed ends and finishing the NHEJ process.

HR is dependent on a series of posttranslational modifications, including ubiquitination, reviewed in [27]. These modifications control a carefully orchestrated system that recruits and displaces DNA repair factors at multiple different sites during the process of HR. The role of ubiquitin in regulating HR is predicated on the following well-established model of HR. When broken DNA strands are detected, the MRN complex begins end clipping via the endonuclease activity of MRE11, along with other proteins such as EXO1, CtIP, and DNA2, creating extensive ssDNA near the break site. In addition to end clipping, the MRN complex recruits ATM. Resection of these ends at the break site allows binding by single-strand DNA binding protein, replication protein A (RPA), which is then displaced by RAD51, which enables the important difference between HR and NHEJ. RAD51 searches for homology and locates a template with which to use to repair the damaged strand. Because HR depends on this template, it can only occur during the S and G2 phases of the cell cycle, when the damaged DNA has already been replicated and the copied DNA serves as a template strand. Choosing which repair pathway, NHEJ or HR, a cell uses for repair is an important determiner of genome stability and involves complex regulatory processes involving ubiquitin.

Regulation of DSB repair depends on a variety of ubiquitination writers, readers and erasers. The readers and erasers require specific UBDs to recognize their specific conformations of ubiquitination. There are more than 20 unique types UBDs that are found in mammalian proteins and a few of these are enriched in proteins associated with DNA damage repair: ubiquitin-interacting motif (UIM), ubiquitin-binding zinc finger (UBZ) and motif interacting with ubiquitin (MIU) [28]. These proteins possess multiple UBDs that each bind to the target cooperatively to increase specificity and affinity. In several ubiquitin ligases, including RNF168, RNF169, RAD18, and RAP80, specificity is increased further because they possess ligand-binding regions adjacent to their UBDs, allow cooperative binding at higher specificities and affinities than otherwise possible [29].

2.2 RNF8 and RNF168 regulation

RNF8 and RNF168 are important regulators of the entire DSB repair response, and require careful regulation themselves. Because these ubiquitin ligases can recognize the same mark they create, their action can cause over-recruitment and overproduction of ionizing radiation-induced foci without control mechanisms. Overproduction of these ionizing radiation-induced foci would cause widespread transcriptional repression across much larger portions of the genome than necessary. One mechanism of limitation is direct ubiquitination of RNF168 by TRIP12 and UBR5, ubiquitin ligases that recognize certain N-terminal domains and direct proteosomal protein degradation on those targets, causing a decrease in the amount of RNF168 in the cell (**Figure 1B**) [30].

Another method of RNF8 and RNF168 regulation occurs via deubiquitinating enzymes (DUBs) (**Figure 1C**). Ubiquitin-specific protease 3 (USP3) has been shown to increase genomic instability and lead to spontaneous tumors when depleted in mice. This finding was supported when it was shown that UPS3 depletion led to increased levels of H2A ubiquitination, indicating the role of properly regulated H2A ubiquitination in DSB repair [31]. USP3, USP16 and USP44 and their family members also deubiquitinate H2A and thus downregulate the DSB response. One of the significant differences between these DUBs is their affinity for different lysine chains. For example, USP3 is known to target the H2A protein K13 and K15 sites that RNF168 targets as well as the K119 and K120 monoubiquitination sites of PRC1. In contrast, PSMD14 deubiqutinates K63-linked poly-ubiquitin, a different RNF8-RNF168 mediated target [32, 33]. Another DUB, USP14, downregulates DSB repair by decreasing RNF168 ubiquitination and RNF168-mediated ubiquitin signals in the setting of inhibited autophagy [34].

RNF8 and RNF168-mediated DSB repair can also be downregulated by phosphorylation. During mitosis, chromatin structure undergoes massive changes and most nuclear processes pause. Phosphorylation of RNF8 and MDC1 (a scaffold protein) prevents DSB repair from occurring during mitosis by blocking their interaction with 53BP1.

2.3 Pathway choice

As there are two primary pathways of DSB repair that function through entirely different mechanisms, NHEJ versus HR, cells must decide which pathway to activate. HR requires a perfect homolog to use as a template across the DSB, and for this reason HR should only function following replication of the DNA during S phase or in G2. Cells must have built-in mechanisms to suppress HR during G1, since during this stage of the cell cycle HR would use inappropriate nonhomologous DNAs as template and thus be mutagenic [35]. In addition, during mitosis, NHEJ is repressed by phosphorylation and inactivation of 53BP1 and RNF8 by the cyclin-dependent kinase CDK1 [36]. This inactivation of DNA repair is protective against chromosomal fusions at telomeres that would lead to aneuploidy. The structure of the DSB is also a factor in the decision of cells to engage in which pathway. In general, more complex DSB structures cannot be repaired via NHEJ and require the more time-consuming HR pathway [37, 38].

The RNF8-RNF168 ubiquitination pathway plays an important role in determining which DSB pathway will predominate in a cell. BRCA1 stimulates HR but antagonizes NHEJ, and conversely 53BP1 antagonizes HR and promotes NHEJ [39, 40]. Ubiquitination via RNF8 and RNF168 leads to retention of 53BP1 and BRCA1 at DSB sites and the balance between these two proteins is the primary decision point between NHEJ and HR. 53BP1 functions to inhibit end resection that is necessary for HR, allowing only NHEJ to be performed. 53BP1 binds H2AK15ub (catalyzed by RNF168) through its own ubiquitin-dependent recruitment motif and also possesses Tudor domains, which recognize H4K20me2 [21, 41]. It is proposed that H4K20me2 is the signal that 53BP1 recognizes to promote its recruitment at DSB sites. RNF8–168 ubiquitinates other proteins that impact the pathway selection, including JMJD2A, JMJD3A, and L3MBTL1. When these three factors are ubiquitinated, they are released from H4K20me2. Through the action of JMJD2A, JMJD3A, and L3MBTL1 vacating H4K20me2, 53BP1 can bind freely without competition (**Figure 1D**) [42, 43]. H4K20me2 is another mechanism supporting the cell cycle dependent decision point between pathways. As S phase continues and more DNA is replicated, H4K20me2 becomes diluted between the two replicated DNA strands, reducing 53BP1 capacity for binding through its Tudor domains, and shifting the balance away from NHEJ to HR [44]. While 53BP1 is bound, RIF1 (and other factors) are recruited to 53BP1 and inhibit resection of the DNA ends. These factors are responsible for replacing BRCA1 at DSB sites, inhibiting HR. BRCA1, in turn, inhibits RIF1 binding at these sites, inhibiting NHEJ.

The antagonist of 53BP1 is BRCA1, which promotes HR over NHEJ by way of supporting RAD51 activity. BRCA1 is a scaffold with activity that also depends on

RAP80, which has a ubiquitin binding domain suspected to recognize RNF8-RNF168-mediated H2A ubiquitination and is a part of the BRCA1-A complex, which it targets to these sites of RNF8-RNF168 ubiquitination [45, 46]. However, RAP80 depletion does not lead to the expected abolishment of HR, but to increased HR activity. To explain this, it has been proposed that RAP80 is functioning to sequester BRCA1 away from DSB sites, so when RAP80 is removed, BRCA1 recruitment to DSB sites is unregulated, leading to the over activity of HR [47]. This would suggest an unknown regulator of BRCA1 recruitment to sites requiring HR activity. BRCA1 also antagonizes 53BP1 by recruiting phosphorylated UHRF1, an E3 ligase that ubiquitinates RIF1, which is bound to 53BP1 at DSB sites. Ubiquitinated RIF1 becomes displaced, reducing 53BP1 mediated repression of DNA end resection [48]. Cockayne syndrome B (CSB) protein has been proposed as fulfilling this role because it seems to antagonize 53BP1 support of NHEJ [49, 50]. In addition, when CSB is removed, DNA damage responses have been limited and CSB has been found accumulating at DSB sites.

One mechanism of HR regulation is the proteasome-mediated degradation of factors important to HR, such as CtIP [51]. The decision point depends on the resection of broken DNA ends by factors such as CtIP in HR. During S and G2, when HR is stimulated, CtIP is ubiquitinated by RNF138 to promote CtIP localization to DSB sites [52]. RNF138 also ubiquitinates the NHEJ factor Ku80 during S phase, causing the Ku70/80 heterodimer to dissociate from DSB sites, and thus suppressing NHEJ during S and G2 phases [53].

HR is also inhibited during G1 via ubiquitination of PALB2, a factor involved in HR along with BRCA1 [54]. This ubiquitination by the E3 ligase complex CRL3 is in the BRCA1-binding domain of PALB2 and sterically blocks the two proteins from binding. The ubiquitination is antagonized by USP11, a DUB that is degraded during G1. Thus, while the balance of CRL3 to USP11 is heavily in favor of CRL3 in G1, the BRCA1-binding site on PALB2 is ubiquitinated, and so the PALB2-BRCA1 interaction is blocked, preventing BRCA1 activity, and therefore, HR.

3. Ubiquitin and transcription elongation

A second key process regulated by ubiquitination of chromatin is transcription. As mRNA is transcribed, one protein complex that associates with the elongating RNA Polymerase II (RNAPII) is the Polymerase Associated Factor 1 Complex (PAF1) (**Figure 2A**) [55–57]. The PAF1 complex regulates RNAPII related transcription elongation and posttranscriptional events and is conserved across many species.

The PAF1 complex in humans is comprised of six protein subunits: PAF1, CDC73, CTR9, LEO1, RTF1, and WDR61 [58]. WDR61 is not present in yeast, although it is present in humans. Cells without PAF1 or CTR9 have a global decrease in protein levels and exhibit growth defects [59, 60]. The complex is found on active genes, at levels directly relating to transcription [61–63]. PAF1 binds directly to the carboxy terminal domain (CTD) of RNAPII via the Cdc73 subunit when the RNAPII CTD becomes phosphorylated via CDK9, and via Rtf1 binding along with the elongation factor Spt5 [64–66]. The localization and recruitment of PAF1C to specific sites on active genes is dependent upon many factors; in humans, PAF1C recruitment is highest at the transcription start site (TSS) or immediately (~2 nucleosomes) following the TSS [62, 67].

The PAF1 complex regulates transcription and the chromatin template to ensure its readiness for transcription. The impact of PAF1C on human chromatin was first established from its role in the ubiquitination of histone H2B at K120 (**Figure 2B**) [68]. H2Bub is an important epigenetic mark that is associated with both activating and deactivating transcription, though its primary effect on chromatin is

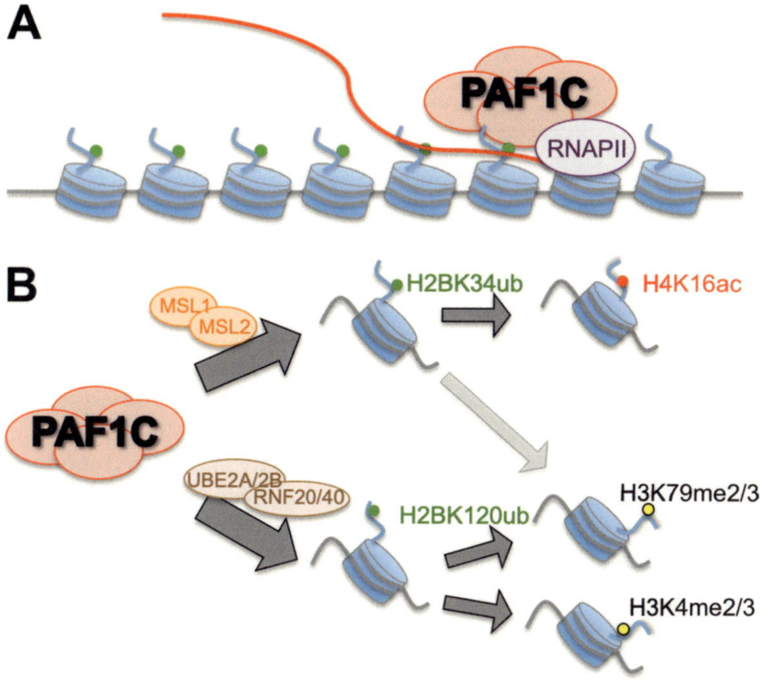

Figure 2.
The PAF1 complex ubiquitinates histone H2B during transcription elongation. (A) PAF1C ubiquitinates histones following transcription by RNAPII and elongation of mRNA. (B) The PAF1C controls multiple histone modifications, including non-ubiquitination events. The two histone ubiquitinations controlled by PAF1C are H2BK34ub and H2BK120ub. Ubiquitination of H2BK120ub is stimulated by PAF1C in concert with the UBE2A/2B and RNF20/40 heterodimers. H2BK120ub is necessary for H3K79me2/3 and H3K4me2/3, catalyzed by additional methyltransferases. The other histone ubiquitination, H2BK34ub, is created by PAF1C interaction with the MSL1/2 complex and promotes H4K16ac via MOF activity. The faded arrow represents crosstalk by which H2BK34ub regulates H3K4 and H3K79 methylation.

disrupting compaction [69]. The ubiquitination at H2BK120 is catalyzed by the E3 ligase complex containing RNF20/40, which interacts directly with the PAF1 complex, and is conjugated by the E2 UBE2A/2B. In addition to H2B ubiquitination at K120, monoubiquitination can also occur at K34 on H2B, a separate mark placed by the heterodimeric E3 ligase, MSL1/2 [70]. Both H2BK120ub and H2BK34ub stimulate histone methylation at H3K79 and H3K4, which has been demonstrated via decreases in both H2BK120ub and H2BK34ub following PAF1 depletion [71]. H2BK120ub is necessary for H3K4 and H3K79 trimethylation, while H2BK34ub functions through *trans*-tail crosstalk to regulate these methylations. The mechanism of this effect was revealed by experiments showing that depletion of PAF1 caused a decrease in RNF20/40 and MSL1/2 association to chromatin, indicating the role of PAF1C as promoting localization of these E3 ligases, the method by which PAF1C regulates H2Bub [71]. RNF20/40 and MSL1/2 each depend on the specific binding to chromatin by the other ligase and the corresponding histone mark, demonstrating how much interdependence exists between the two co-regulated ligases. This interaction has multiples sources. CDK9 is a kinase that promotes PAF1C association to chromatin, but is itself dependent on both PAF1C-mediated chromatin marks for its chromatin association [71].

Deregulation of this pathway and H2B monoubiquitination is commonly found in cancers. This can occur via multiple mechanisms, such as mutations in CDC73, one of the components of PAF1C, which has been observed in multiple cancers [72]. In addition, silencing of expression by methylation of the RNF20 promoter and RNF20 enhancers has also been observed in many breast cancers [73]. However, it has also been observed that decreased levels of H2Bub have also been shown to be associated with decreased tumor growth, an apparent contradiction to H2Bub as a cancer-causing mutation [73]. Deregulation of a mark can also occur from overactive removal; there are several DUBs responsible for H2Bub deubiquitination, including USP3, USP7, USP12, USP22, USP44, USP46, USP49 [74–78]. Upregulation of these DUBs can cause similar phenotypes as RNF20 depletion. Errors in H2B monoubiquitination lead to errors in chromatin structure on scales larger than the aberrantly ubiquitinated nucleosome [69].

Dysregulation of RNF20 and concomitant H2B ubiquitination has been linked to a wide variety of cancer pathways. One method for H2Bub depletion leading to cancer occurs via H2Bub regulated inflammation and the interaction with NF-κB [79]. Inflammation involves the production of cytokines and chemokines that promote oncogenic activity and NF-κB is a key regulator of the inflammatory system. Reduction in H2Bub has been shown to lead to activated NF-κB, and thus its downstream regulation targets, leading to active inflammation in mice. This was indeed shown to lead to increased colorectal cancer in these animals. Ovarian cancers also display H2Bub dysfunction. One study found that the majority of high grade serous ovarian cancers show global decreases in H2Bub [80]. The most deadly cancer worldwide is lung cancer, and one of the more common forms of lung cancer is lung adenocarcinoma. In human lung adenocarcinomas, H2Bub decreases have been associated with increased cancer burden and a less differentiated carcinoma, a marker of poor prognosis [79].

Mixed lineage leukemia is a classification of cancers that depend on the MLL1 gene, and rearrangements of MLL1 have been shown to be dependent on RNF20 and its role in chromatin regulation [81]. Cells lacking RNF20 showed decreased tumor growth [82]. This role of RNF20 allowing cancer progression is contrary to its role in protecting against the above cancers, but does serve to highlight the fundamental role that H2B ubiquitination plays in maintenance of chromatin.

4. Polycomb repressive complex

While the preceding section described how histone H2B is ubiquitinated at multiples sites as a part of active transcription process, this section describes the ubiquitination of histone H2A, which has an opposite impact on gene expression. Thc polycomb repressive complex 1 (PRC1) monoubiquitinates H2A at lysine 119. H2AK119ub is a repressive mark, associated with inactive transcription by condensing chromatin, making it less accessible by transcription factors and associated machinery [83]. This repressive mark is only the most common role of PRC1, as it has also been shown to have diverse effects that have the overall impact of permanently silencing chromatin as part of the differentiation process. The two polycomb group (PcG) complexes, PRC1 and PRC2, modify chromatin to repress transcription and lead to the methylation of the promoter DNA to stably repress transcription at targeted genes. PRC2 contains the methyltransferase EZH2, which methylates histone H3 on lysine 27, H3K27me3. This review will focus on PRC1. PRC2 is necessary for targeted recruitment of PRC1, as experiments have shown that knockdown of PRC2 components also decrease PRC1-mediated H2A ubiquitination. The ubiquitination function of PRC1 is antagonized by the last form of polycomb repressive complex, polycomb repressive deubiquitinase (PR-DUB), which deubiquitinates H3K119 [84, 85]. The complimentary actions of PRC1 and PR-DUB to regulate H2AK119ub suggests the fundamental role it plays in repressing transcription.

PRC1 complexes exist in a number of different forms that have the same general structure, with different proteins occupying each position. The core of each complex contains a RING protein and a polycomb group RING finger protein, which bind via their RING domains [86]. This core serves as the base for further PRC1 proteins to bind. There are two possible RING proteins, RING1A and RING1B, and six possible polycomb group ring finger (PCGF) proteins, PCGF1–6. All eight of these proteins possess a RAWUL (RING finger and WD40 Ubiquitin-like) domain somewhere in their structure, which bind additional proteins [87, 88]. These additional proteins include chromobox and human polyhomeotic homolog (HPH) proteins, which, when included in the PRC1, form what is known as canonical PRC1 [89]. It was previously assumed canonical PRC1 performed the H2Aub function that is associated with PRC1, but it is now known that noncanonical PRC1 also plays an important role in gene regulation [90]. Between the variability of chromobox, HPH, RING, and PCGF proteins, there are well over 100 unique combinations of canonical PRC1 complexes that can form. This diversity plays an important role in the diverse targeting and functions exhibited by PRC1. The PCGF member of the complex binds specifically to a variety of proteins, which are responsible for targeting and regulation of the PRC1 activity [91]. Accordingly, PCGF RAWUL domains exhibit more selective binding than their counterpart RAWUL domains on the RING1A or RING1B protein. The importance of the RING domains is that the RING proteins are E3 ubiquitin ligases, responsible for the primary activity of H2AK119 ubiquitination. However, PCGF-4 (also known as BMI-1) and RING1A both do not directly ubiquitinate H2A, as only RING1B directly ubiquitinates H2A. Instead complexes containing BMI-1 and RING1A serve to promote the RING1B E3 ligase activity [92].

4.1 PRC1 function

RING1B monoubiquitinates H2AK119 as part of the PRC1 activity following PRC2 methylation at H3K27 (**Figure 3**). PcG-regulated genes show aberrant transcriptional levels following removal of PRC1 via RING1B knockdown using shRNA [93]. PRC1-related ubiquitination and subsequent gene silencing is associated with multiple silencing contexts. PcG proteins are known to occupy and thus regulate, developmental genes, X-chromosome inactivation, and parent of origin imprinting. The most widely accepted model of the activity of PRC1-mediated inactivation of target genes is through chromatin compaction. Promoters of active genes become compacted in the setting of PRC1 action, preventing RNA polymerases from accessing the targeted gene, and therefore preventing transcription. This concept has been supported by *in vitro* experiments and *in vivo* experiments showing decreased nuclease digestion at genes with PRC1-mediated compaction of chromatin [94]. While the fact that PRC1-mediated ubiquitination of H2A leads to diminished transcription via chromatin compaction is indisputable, the mechanism is currently unclear. It has been shown that PRC1 does not have a role in regulating chromatin accessibility, only to nucleosome spacing and occupancy. Identifying the direct mechanism by which PRC1-mediated H2Aub inhibits transcription needs further elucidation.

Targeting of PcG complexes occurs via Polycomb Response Elements (PREs), which are DNA elements that cannot be recognized by any PcG protein because PcG proteins do not appear to possess any sequence specific DNA binding subunits [95]. While several proteins have been suggested to have a role in recognizing the PRE and enabling recruitment of PcG complexes, none have been confirmed to be sufficient to mediate PcG recruitment alone, suggesting the PcG recruitment is dependent on the interactions of several proteins coordinately creating a stable protein-DNA complex [96]. In addition to protein-DNA interactions to promote PcG recruitment, protein-protein interactions are important. PRC2 has histone methyltransferase function, methylating H3K27. PRC1 can directly recognize the H3K27me3 mark produced by PRC2 [97, 98]. Similarly, PRC2 can bind to H2Aub. This complementary interaction can serve to support preservation of PcG silencing

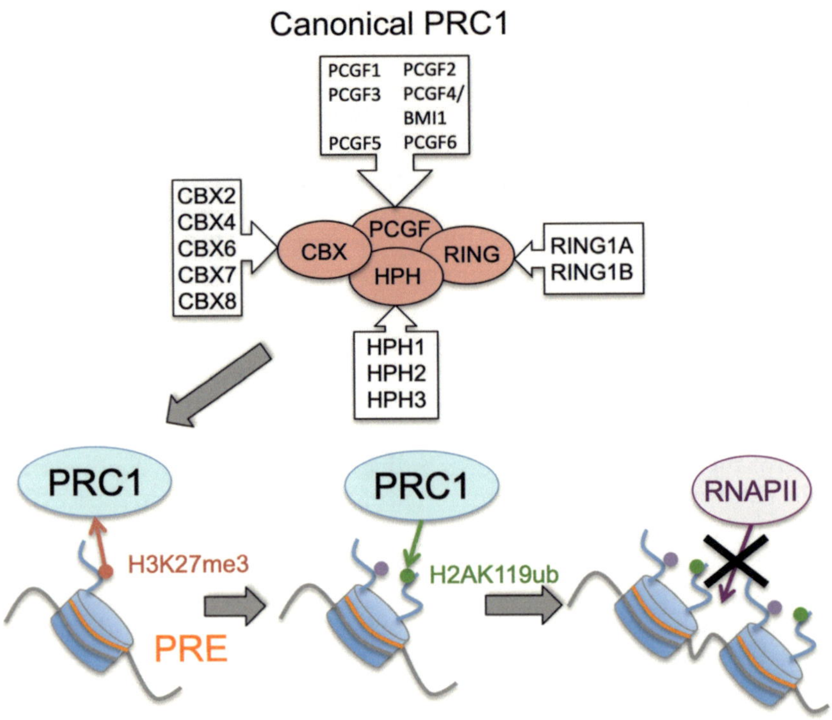

Figure 3.
Canonical PRC1 ubiquitinates H2AK119. Canonical PRC1 contains a RING protein and a PCGF protein, as defines PRC1, but is only called canonical in the presence of a Chromobox protein and a human polyhomeotic protein. This canonical PRC1 is responsible for the primary function of PRC1, ubiquitination of H2AK119. Recognition occurs at Polycomb Response Elements (PRE) containing specific DNA sequences that have been methylated by PRC2 at H3K27me3. PRC1-mediated H2AK119ub is a repressive mark, leading to decreased accessibility of targeted genes by transcription machinery, leading to inactivation of targeted genes.

across disruptive events to the genome, such as DNA synthesis, when histones are divided between the sister chromosomes [99]. Therefore, the most commonly accepted model of PcG recruitment is that PREs are recognized by adaptor proteins, which recruit PRC2 to promoters, which methylates H3K27. PRC1 recognizes H3K27me3 and is recruited to ubiquitinate H2AK119.

In addition to the repressive effect of PRC1, it has been shown to have activating effects on transcription. PcG proteins mediate their activity by regulating genome architecture [100]. It has been reported in mouse embryonic stem cells that RING1A and RING1B organize genes into three-dimensional interaction networks, which maintains interactions between promoters in the network. When PRC1 was removed, promoter-enhancer interactions were affected, leading to activation of affected promoters and increased transcription. This supports the compaction-based theory of PRC1 transcriptional repression and provides a mechanism for this activity. Deep sequencing of ChIP experiments against selected PRC1 proteins, including both RING1A and RING1B, has shown their enrichment at active transcriptional sites in human fibroblasts [101]. This experiment also showed cell-type specific binding of PRC1. RING1B, the primary ubiquitin ligase involved PRC1-mediated H2Aub, has been found associated with Aurora B kinase at active promoters in lymphocytes, while RING1B knockdown decreased transcription at these sites, suggesting an important activating function of RING1B [102]. Cells that have had conditionally-inactivated RING1A and RING1B, and thus inactivated PRC1, exhibit errors in DNA replication [103]. Slow elongation and even stalling of replication forks has been observed in these cells in specific pericentromeric regions.

These S phase errors were rescued by monoubiquitination events, suggesting the role of RING1A/B in S phase is dependent on their function as ubiquitin ligases. In breast cancer, RING1B has been found at oncogene promoters, playing an activating role and promoting cancer development and metastasis [104]. All these activating effects of PcG proteins suggest that there is much not known about the diverse array of proteins involved in PRC1 and that understanding PRC1 function may explain many previously unknown chromatin regulation events.

5. Ubiquitin as a mitotic bookmark

Mitotic bookmarks are a mechanism of dividing cells that maintain the epigenetic and transcriptional state despite the rigors demanded by the mitosis process [105]. Although epigenetic marks persist from mother cell to daughter cell, the compaction of the genome during mitosis requires many epigenetic marks to be temporarily erased. Every mitosis, the epigenome is bookmarked, erased, and reestablished as the cells reenter G1. Cells need a mechanism allowing them to reestablish cell specific chromatin marks after they have been erased during mitosis. The mechanism cells use for "remembering" chromatin architecture is mitotic bookmarking, whereby specific molecules or proteins are found on promoters of genes that enable memory of the chromatin state before mitosis. By definition, bookmarks must be deposited in association with active genes before or at the beginning of mitosis, persist throughout mitosis, and transmit gene expression memory to the cell after mitosis (**Figure 4**). These mitotic bookmarks involve multiple chromatin changes, including histone modifications and histone variants. Transcription factors also make up a large number of mitotic bookmarks. Many of those transcription factors bookmark specific subsets of genes. One example of a highly selective mitotic bookmark is Brd4, which is found only on the transcription start sites of genes that are expressed at the end of mitosis and beginning of G1 [106]. Mitotic bookmarks can also regulate a specific biological process, as in the case of GATA1, which occupies locations on key hematopoietic genes during mitosis [107]. Ubiquitin has also been found to play a role as a mitotic bookmark, but while many mitotic bookmarks are specific for certain genes or pathways, the mitotic bookmark ubiquitination appears to be generally acting at genes with high transcriptional activity.

This novel role of ubiquitin was first identified through a variant of ChIP-seq experiments that found ubiquitin present on certain sites during mitosis that were previously not described [108]. Those experiments showed that during interphase, ubiquitin was present on the chromatin of transcribed regions of transcriptionally active genes, consistent with the known function of PAF1C. The novel observation was that ubiquitinated chromatin associated proteins were bound to promoters during mitosis, contrasted to interphase, when ubiquitin localized to the promoter was absent. The fundamental difference between interphase and mitosis was a shift of the ubiquitin detected near promoters of the same genes that were previously ubiquitinated on their transcribed regions. For example [109], the GAPDH gene is heavily ubiquitinated over the gene body during G1 (**Figure 4A**, indicated in blue), while during mitosis that ubiquitination over the gene body is absent but ubiquitination is detected over the promoter (**Figure 4A**, gold) [108]. The ubiquitinated promoter sites were consistently ubiquitinated at 150 bp upstream of the transcription start sites, suggesting a specific function relating each of these promoters. The fact that this ubiquitin bookmark was identified on promoters of active genes further supported the conclusion that this novel finding of ubiquitin in mitosis was playing a role as a bookmark, not just an incidental observation. This conclusion was also supported by the fact that the promoter-associated, mitotic ubiquitin was found on the same genes as

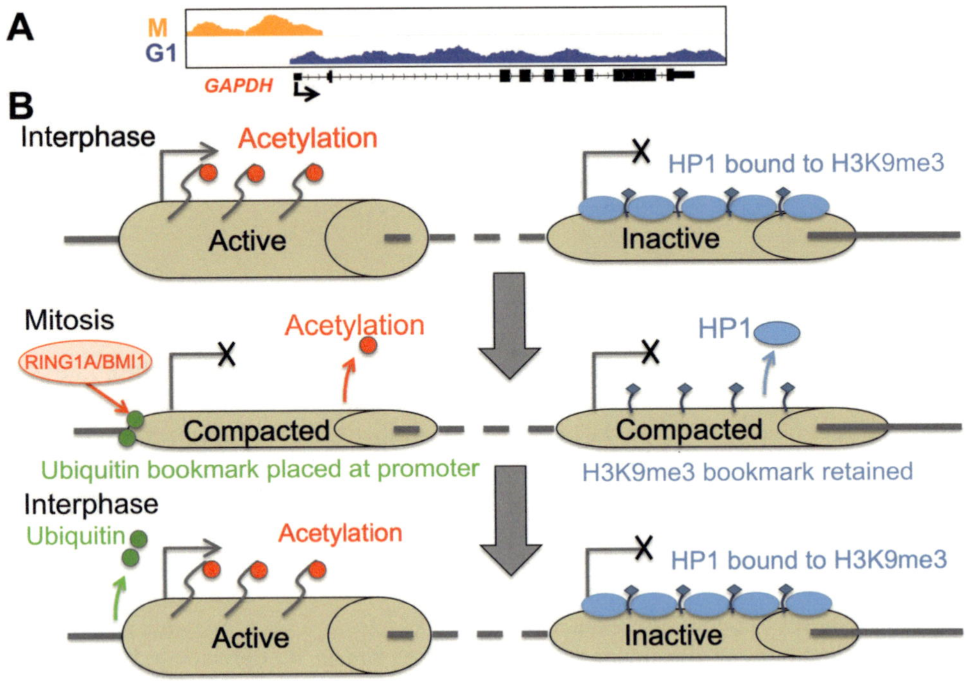

Figure 4.
A mitotic bookmark containing ubiquitin is necessary for maintaining the active chromatin state after completion of mitosis. (A) An example of measuring the ubiquitin density on an active gene (GAPDH) during mitosis (gold) and during G1 (blue) [108]. The localization of ubiquitin on the chromatin shift from over the gene body during G1 through G2 to over the promoter during mitosis. (B) Model for mitotic bookmarking. During interphase, active genes have active chromatin has associated epigenetic marks, such as acetylation, whereas repressive marks as heterochromatin protein 1 (HP1) are present on inactive genes. When cells transition into mitosis, all of those marks are removed. Instead, mitotic bookmarks are placed on the active genes to enable cells to "remember" which genes were active. Ubiquitin is found on promoters of a subset of active genes and is necessary to support transcription following completion of mitosis. This ubiquitin bookmark is dependent on the E3 ligases RING1A and BMI-1. The HP1 localization is bookmarked by H3K9me3.

those with PAF1C-associated transcriptional H2B-ubiquitin. The association between these two forms of ubiquitination suggests that the ubiquitin bookmark is dependent upon transcription, as is PAF1C-associated H2B-ubiquitin. However, the mechanisms underlying creation of these ubiquitination marks is different, suggesting that there is no direct relationship between these two transcription-associated marks.

The presence of the ubiquitin bookmark is dependent upon the E3 ligases RING1A and BMI-1, which are both parts of the polycomb repressive complex discussed previously [109]. Surprisingly, RING1B, the primary E3 ligase involved in the PRC1 primary function has no role in bookmark ubiquitination, suggesting that the role of RING1A and BMI-1 in creating the ubiquitin bookmark is independent of their role in the PRC1 complex. However, this remains untested and what factors interact with RING1A and BMI-1 when they are involved in bookmark ubiquitination is an open question. With the discovery of the ligases responsible for the ubiquitin bookmark, it was possible to test experimentally how the process is regulated. RING1A depletion caused a decrease in phosphorylated RNAPII at promoters; the phosphorylated RNAPII was used as a surrogate for transcriptional activity, indicating that the ubiquitin bookmark was necessary for the proper transcription of the bookmarked genes, one of the criteria for mitotic bookmark. So far, what is known about the ubiquitin bookmark is that is present during mitosis, responds to changes in gene expression, and that it impacts transcription. These are the basic requirements to satisfy the definition of a mitotic bookmark.

Beyond the basic outline of a bookmark, there are relatively few facts known about the ubiquitin bookmark. Its localization to promoters of active genes is notable, but only a subset of active genes is found to be bookmarked, suggesting the input of more factors than active transcription and ubiquitination via RNF20/40 with PAF1C. What these factors are, even what kind of signal they are, is still unknown. Only a little is also known about the mechanism by which bookmark ubiquitination affects transcription. H3K4me3 is a histone modification known to associate with sites of active transcription. H3K4me3 has been observed decreasing when RING1A has been depleted, suggesting that the lack of ubiquitin bookmarking has caused a decrease in transcription by decreasing H3K4me3 as a signal for transcription [109]. If this phenomenon is unique to H3K4me3 or if it is common to other histone modifications correlating to active transcription is unknown. Further studies to determine this mechanism will inform the importance of this bookmark and how broadly it affects cellular function and differentiation.

Currently the exact composition of the ubiquitin bookmark is undetermined. The ubiquitin bookmark must have a substrate protein that is directed to the sites identified to have ubiquitin bookmarks, and serves as the connection to chromatin. The signal has been detected via affinity-tagged ubiquitin molecules that do not discriminate between mono- or polyubiquitin, nor between the different lysine residues with which the polyubiquitin chain could be constructed. Ubiquitin is the only known component of the bookmark, but there must be other components yet to be identified. Given that most chromatin-associated proteins dissociate from the genome during mitosis, there are fewer candidates for the substrate than would be in interphase, though the possibility exists that a protein previously unknown to remain during mitosis exhibits that ability as part of the ubiquitin bookmark.

Another aspect for expanding our understanding of the ubiquitin bookmark is expanding the finding of the ubiquitin bookmark to other cell lines. Thus far, all the prior work done on the ubiquitin bookmark has been done in HeLa cells, a common model system. As the ubiquitin bookmark has not been demonstrated in any other cell lines, nor in tissue samples, questions of the ubiquity of the bookmark are raised. It is formally possible that the ubiquitin bookmark is unique to HeLa cells or just cancerous cell lines and is not apparent in tissues in organisms. Obviously, the role of the ubiquitin bookmark is only relevant in an actively dividing cell, although the majority of cells in living tissues are postmitotic. Detecting the presence or lack thereof of the ubiquitin bookmark in other cell lines should be one of the most pressing directions of current research. The significance of the ubiquitin bookmark as a relatively new and poorly understood process suggests a new field in epigenetics, or at least a significant evolution in our understanding of mitotic bookmarks as primarily transcription factors that control limited selections of genes to a much larger, potentially genome-wide scale.

6. Concluding remarks

Chromatin is dynamically modified as genes are silenced, as genes are expressed, as DNA damage is repaired, and as the genome is prepared for cell division. In this review, we highlighted the diverse roles of ubiquitin in each process. Understanding the complexity of the ubiquitin system is a monumental task of which the scientific community is only scratching the surface. Four important processes were reviewed here, and these processes are paramount to proper cellular functions and deregulation is generally implicated in cancers.

Conflict of interest

The authors have no conflicts of interest to declare.

Author details

Alexander E. Hare and Jeffrey D. Parvin*
Department of Biomedical Informatics, The OSU Comprehensive Cancer Center, Ohio State University, Columbus, OH, USA

*Address all correspondence to: jeffrey.parvin@osumc.edu

References

[1] Eytan E, Ganoth D, Armon T, Hershko A. ATP-dependent incorporation of 20S protease into the 26S complex that degrades proteins conjugated to ubiquitin. Proceedings of the National Academy of Sciences of the United States of America. 1989;**86**(20):7751-7755

[2] Schulman BA, Harper JW. Ubiquitin-like protein activation by E1 enzymes: The apex for downstream signalling pathways. Molecular Cell. 2009;**10**(5):319-331

[3] Ye Y. Building ubiquitin chains: E2 enzymes at work. Nature Reviews. Molecular Cell Biology. 2011;**10**(11):755-764

[4] Smit JJ, Sixma TK. RBR E3-ligases at work. EMBO Reports. 2014;**15**(2):142-154

[5] Deshaies RJ, Joazeiro CAP. RING domain E3 ubiquitin ligases. Annual Review of Biochemistry. 2009;**78**(1):399-434. Available from: http://www.annualreviews.org/doi/10.1146/annurev.biochem.78.101807.093809

[6] Swatek KN, Komander D. Ubiquitin modifications. Cell Research. 2016;**26**(4):399-422

[7] Lee J, Smith E, Shilatifard A. The language of histone crosstalk. Cell. 2013;**142**(5):682-685

[8] Jürgen Dohmen R. SUMO protein modification. Biochimica et Biophysica Acta (BBA)—Molecular Cell Research. 2004;**1695**(1-3, 113):31

[9] Lindahl T. Instability and decay of the primary structure of DNA. Nature. 1993;**363**:709-715

[10] Jackson SP, Bartek J. The DNA-damage response in human biology and disease. Nature. 2010;**461**(7267):1071-1078. Available from: http://www.pubmedcentral.nih.gov/articlerender.fcgi?artid=2906700&tool=pmcentrez&rendertype=abstract

[11] Kowalczykowski SC. An overview of the molecular mechanisms of recombinational DNA repair. Cold Spring Harbor Perspectives in Biology. 2015;**7**(11):a016410. Available from: http://cshperspectives.cshlp.org/lookup/doi/10.1101/cshperspect.a016410

[12] Beucher A, Birraux J, Tchouandong L, Barton O, Shibata A, Conrad S, et al. ATM and Artemis promote homologous recombination of radiation-induced DNA double-strand breaks in G2. The EMBO Journal. 2009;**28**(21):3413-3427

[13] Thorslund T, Ripplinger A, Hoffmann S, Wild T, Uckelmann M, Villumsen B, et al. Histone H1 couples initiation and amplification of ubiquitin signalling after DNA damage. Nature. 2015;**527**(7578):389-393

[14] Paull TT, Rogakou EP, Yamazaki V, Kirchgessner CU, Gellert M, Bonner WM. A critical role for histone H2AX in recruitment of repair factors to nuclear foci after DNA damage. Current Biology. 2000;**10**(15):886-895

[15] Mailand N, Bekker-Jensen S, Faustrup H, Melander F, Bartek J, Lukas C, et al. RNF8 ubiquitylates histones at DNA double-strand breaks and promotes assembly of repair proteins. Cell. 2007;**131**(5):887-900

[16] Huen MS, Grant R, Manke I, Minn K, Yu X, Yaff MB, et al. The E3 ubiquitin ligase RNF8 transduces the DNA damage signal via an ubiquitin-dependent signaling pathway. Cell. 2007;**131**(5):901914

[17] Kolas NK, Chapman JR, Nakada S, Ylanko J, Chahwan R, Sweeney FD,

et al. Orchestration of the DNA-damage response by the RNF8 ubiquitin ligase. Science. 2007;**318**(5856):1637-1640

[18] Stucki M, Clapperton JA, Mohammad D, Yaffe MB, Smerdon SJ, Jackson SP. MDC1 directly binds phosphorylated histone H2AX to regulate cellular responses to DNA double-strand breaks. Cell. 2005;**123**(7):1213-1226

[19] Mattiroli F, Vissers JHA, Van Dijk WJ, Ikpa P, Citterio E, Vermeulen W, et al. RNF168 ubiquitinates K13-15 on H2A/H2AX to drive DNA damage signaling. Cell. 2012;**150**(6):1182-1195. Available from: http://dx.doi.org/10.1016/j.cell.2012.08.005

[20] Gatti M, Pinato S, Maiolica A, Rocchio F, Prato MG, Aebersold R, et al. RNF168 promotes noncanonical K27ubiquitination to signal DNA damage. Cell Reports. 2015;**10**(2):226-238. Available from: http://dx.doi.org/10.1016/j.celrep.2014.12.021

[21] Fradet-Turcotte A, Canny MD, Escribano-Díaz C, Orthwein A, Leung CCY, Huang H, et al. 53BP1 is a reader of the DNA damage-induced H2A Lys15 ubiquitin mark. Nature. 2013;**499**(7456):50-54

[22] Bekker-Jensen S, Danielsen JR, Fugger K, Gromova I, Nerstedt A, Bartek J, et al. HERC2 coordinates ubiquitin-dependent assembly of DNA repair factors on damaged chromosomes. Nature Cell Biology. 2010;**12**(1):80-86

[23] Hashizume R, Fukuda M, Maeda I, Nishikawa H, Oyake D, Yabuki Y, et al. The RING heterodimer BRCA1-BARD1 is a ubiquitin ligase inactivated by a breast cancer-derived mutation. The Journal of Biological Chemistry. 2001;**276**(18):14537-14540

[24] Shakya R, Reid LJ, Reczek CR, Cole F, Egli D, Lin CS, et al. BRCA1 tumor suppression depends on BRCT phosphoprotein binding, but not its E3 ligase activity. Science. 2011;**334**(6055):525-528

[25] Fuentes L, Lebenkoff S, White K, Gerdts C, Hopkins K, Potter JE, et al. Phosphorylation of Ku dictates DNA double-strand break (DSB) repair pathway choice in S phase. Nucleic Acids Research. 2015;**93**(4):292-297

[26] Zhou Y, Lee J, Jiang W, Crowe J, Zha S, Paull T. Regulation of the DNA damage response by DNA-PKcs inhibitory phosphorylation of ATM. Molecular Cell. 2017;**65**(1):91-104

[27] Schwertman P, Bekker-Jensen S, Mailand N. Regulation of DNA double-strand break repair by ubiquitin and ubiquitin-like modifiers. Nature Reviews Molecular Cell Biology. 2016;**17**(6):379-394. Available from: http://dx.doi.org/10.1038/nrm.2016.58

[28] Dikic I, Wakatsuki S, Walters KJ. Ubiquitin-binding domains from structures to functions. Nature Reviews Molecular Cell Biology. 2009;**10**(10):659-671. Available from: http://dx.doi.org/10.1038/nrm2767

[29] Panier S, Ichijima Y, Fradet-Turcotte A, Leung CCY, Kaustov L, Arrowsmith CH, et al. Tandem protein interaction modules organize the ubiquitin-dependent response to DNA double-strand breaks. Molecular Cell. 2012;**47**(3):383-395

[30] Gudjonsson T, Altmeyer M, Savic V, Toledo L, Dinant C, Grøfte M, et al. TRIP12 and UBR5 suppress spreading of chromatin ubiquitylation at damaged chromosomes. Cell. 2012;**150**(4):697-709. Available from: http://dx.doi.org/10.1016/j.cell.2012.06.039

[31] Lancini C, van den Berk PCM, Vissers JHA, Gargiulo G, Song J-Y, Hulsman D, et al. Tight regulation of ubiquitin-mediated DNA damage

response by USP3 preserves the functional integrity of hematopoietic stem cells. Journal of Experimental Medicine. 2014;**211**(9):1759-1777. Available from: http://www.jem.org/lookup/doi/10.1084/jem.20131436

[32] Butler LR, Densham RM, Jia J, Garvin AJ, Stone HR, Shah V, et al. The proteasomal de-ubiquitinating enzyme POH1 promotes the double-strand DNA break response. EMBO Journal. 2012;**31**(19):3918-3934. Available from: http://dx.doi.org/10.1038/emboj.2012.232

[33] Kato K, Nakajima K, Ui A, Muto-Terao Y, Ogiwara H, Nakada S. Fine-tuning of DNA damage-dependent ubiquitination by OTUB2 supports the DNA repair pathway choice. Molecular Cell. 2014;**53**(4):617-630. Available from: http://dx.doi.org/10.1016/j.molcel.2014.01.030

[34] Sharma A, Alswillah T, Singh K, Chatterjee P, Willard B, Venere M, et al. USP14 regulates DNA damage repair by targeting RNF168-dependent ubiquitination. Autophagy. 2018;**14**(11):1976-1990. Available from: https://www.tandfonline.com/doi/full/10.1080/15548627.2018.1496877

[35] Karanam K, Kafri R, Loewer A, Lahav G. Quantitative live cell imaging reveals a gradual shift between DNA repair mechanisms and a maximal use of HR in mid S phase. Molecular Cell. 2012;**47**(2):320-329

[36] Orthwein A, Fradet-Turcotte A, Noordermeer SM, Canny MD, Brun CM, Strecker J, et al. Mitosis inhibits DNA double-strand break repair to guard against telomere fusions. Science. 2014;**344**(6180):189-193

[37] Reynolds P, Anderson JA, Harper JV, Hill MA, Botchway SW, Parker AW, et al. The dynamics of Ku70/80 and DNA-PKcs at DSBs induced by ionizing radiation is dependent on the complexity of damage. Nucleic Acids Research. 2012;**40**(21):10821-10831

[38] Shibata A, Conrad S, Birraux J, Geuting V, Barton O, Ismail A, et al. Factors determining DNA double-strand break repair pathway choice in G2 phase. The EMBO Journal. 2011;**30**(6):1079-1092

[39] Bouwman P, Aly A, Escandell JM, Pieterse M, Bartkova J, Van Der Gulden H, et al. 53BP1 loss rescues BRCA1 deficiency and is associated with triple-negative and BRCA-mutated breast cancers. Nature Structural & Molecular Biology. 2010;**17**(6):688-695

[40] Bunting SF, Callén E, Wong N, Chen H, Polato F, Gunn A, et al. 53BP1 inhibits homologous recombination in Brca1-deficient cells by blocking resection of DNA breaks. Cell. 2011;**141**(2):243-254

[41] Mathias JR, Dodd ME, Walters KB, Yoo SK, Erik A, Huttenlocher A. Structural basis for the methylation state-specific recognition of histone H4-K20 by 53BP1 and Crb2 in. DNA Repair. 2010;**33**(11):1212-1217

[42] Mallette FA, Mattiroli F, Cui G, Young LC, Hendzel MJ, Mer G, et al. RNF8- and RNF168-dependent degradation of KDM4A/JMJD2A triggers 53BP1 recruitment to DNA damage sites. The EMBO Journal. 2012;**31**(8):1865-1878

[43] Acs K, Luijsterburg MS, Ackermann L, Salomons FA, Hoppe T, Dantuma NP. The AAA-ATPase VCP/p97 promotes 53BP1 recruitment by removing L3MBTL1 from DNA double-strand breaks. Nature Structural & Molecular Biology. 2011;**18**(12):1345-1350. Available from: http://dx.doi.org/10.1038/nsmb.2188

[44] Pellegrino S, Michelena J, Teloni F, Imhof R, Altmeyer M. Replication-coupled dilution of H4K20me2 guides

53BP1 to pre-replicative chromatin. Cell Reports. 2017;**19**(9):1819-1831

[45] Hu Y, Scully R, Sobhian B, Xie A, Shestakova E, Livingston DM. RAP80-directed tuning of BRCA1 homologous recombination function at ionizing radiation-induced nuclear foci. Genes & Development. 2011;**25**(7):685-700

[46] Leung JWC, Makharashvili N, Agarwal P, Chiu LY, Pourpre R, Cammarata MB, et al. ZMYM3 regulates BRCA1 localization at damaged chromatin to promote DNA repair. Genes & Development. 2017;**31**(3):260-274

[47] Lombardi PM, Matunis MJ, Wolberger C. RAP80, ubiquitin and SUMO in the DNA damage response. Journal of Molecular Medicine. 2017;**95**(8):799-807

[48] Zhang H, Liu H, Chen Y, Yang X, Wang P, Liu T, et al. A cell cycle-dependent BRCA1-UHRF1 cascade regulates DNA double-strand break repair pathway choice. Nature Communications. 2016;**7**:1-14. Available from: http://dx.doi.org/10.1038/ncomms10201

[49] Batenburg NL, Thompson EL, Hendrickson EA, Zhu X-D. Cockayne syndrome group B protein regulates DNA double-strand break repair and checkpoint activation. EMBO Journal. 2015;**34**(10):1399-1416. Available from: http://emboj.embopress.org/cgi/doi/10.15252/embj.201490041

[50] Pascucci B, Fragale A, Marabitti V, Leuzzi G, Salvatore Calcagnile A, Parlanti E, et al. CSA and CSB play a role in the response to DNA breaks. Oncotarget. 2018;**9**(14):11581-11591. Available from: http://www.oncotarget.com/fulltext/24342

[51] Ferretti LP, Himmels SF, Trenner A, Walker C, Von Aesch C, Eggenschwiler A, et al. Cullin3-KLHL15 ubiquitin ligase mediates CtIP protein turnover to fine-tune DNA-end resection. Nature Communications. 2016;**7**:1-16. Available from: http://dx.doi.org/10.1038/ncomms12628

[52] Schmidt CK, Galanty Y, Sczaniecka-Clift M, Coates J, Jhujh S, Demir M, et al. Systematic E2 screening reveals a UBE2D-RNF138-CtIP axis promoting DNA repair. Nature Cell Biology. 2015;**17**(11):1458-1470

[53] Ismail IH, Gagné JP, Genois MM, Strickfaden H, Mcdonald D, Xu Z, et al. The RNF138 E3 ligase displaces Ku to promote DNA end resection and regulate DNA repair pathway choice. Nature Cell Biology. 2015;**17**(11):1446-1457

[54] Orthwein A, Noordermeer SM, Wilson MD, Landry S, Enchev RI, Sherker A, et al. A mechanism for the suppression of homologous recombination in G1 cells. 2016;**528**(7582):422-426

[55] Mueller CL, Jaehning JA. Ctr9 , Rtf1, and Leo1 are components of the Paf1/RNA polymerase II complex. Molecular and Cellular Biology. 2002;**22**(7):1971-1980

[56] Costa PJ, Arndt KM. Synthetic lethal interactions suggest a role for the *Saccharomyces cerevisiae* Rtf1 protein in transcription elongation. Genetics. 2000;**156**(2):535-547

[57] Squazzo SL, Costa PJ, Lindstrom DL, Kumer KE, Simic R, Jennings JL, et al. The Paf1 complex physically and functionally associates with transcription elongation factors in vivo. The EMBO Journal. 2002;**21**(7):1764-1774

[58] Zhu B, Mandal SS, Pham AD, Zheng Y, Erdjument-Bromage H, Batra SK, et al. The human PAF complex coordinates transcription with events downstream of RNA synthesis. Genes & Development. 2005;**19**(14):1668-1673

[59] Betz JL, Chang M, Washburn TM, Porter SE, Mueller CL, Jaehning JA. Phenotypic analysis of Paf1/RNA polymerase II complex mutations reveals connections to cell cycle regulation, protein synthesis, and lipid and nucleic acid metabolism. Molecular Genetics and Genomics. 2002;**268**(2):272-285

[60] Mueller CL, Porter SE, Hoffman MG, Jaehning JA. The Paf1 complex has functions independent of actively transcribing RNA polymerase II. Molecular Cell. 2004;**14**(4):447-456

[61] Chen FX, Woodfin AR, Gardini A, Rickels RA, Marshall SA, Smith ER, et al. PAF1, a molecular regulator of promoter-proximal pausing by RNA polymerase II. Cell. 2015;**162**(5):1003-1015. Available from: http://dx.doi.org/10.1016/j.cell.2015.07.042

[62] Van Oss SB, Shirra MK, Bataille AR, Wier AD, Yen K, Vinayachandran V, et al. The histone modification domain of Paf1 complex subunit Rtf1 directly stimulates H2B ubiquitylation through an interaction with Rad6. Molecular Cell. 2016;**64**(4):815-825

[63] Yu M, Yang W, Ni T, Tang Z, Nakadai T, Zhu J, et al. RNA polymerase II—Associated factor 1 regulates the release and phosphorylation of paused RNA polymerase II. Science. 2015;**350**(6266):1383-1386

[64] Zaborowska J, Egloff S, Murphy S. The pol II CTD: New twists in the tail. Nature Structural & Molecular Biology. 2016;**23**(9):771-777. Available from: http://dx.doi.org/10.1038/nsmb.3285

[65] Mbogning J, Nagy S, Pagé V, Schwer B, Shuman S, Fisher RP, et al. The PAF complex and Prf1/Rtf1 delineate distinct Cdk9-dependent pathways regulating transcription elongation in fission yeast. PLoS Genetics. 2013;**9**(12):29-31

[66] Liu Y, Warfield L, Zhang C, Luo J, Allen J, Lang WH, et al. Phosphorylation of the transcription elongation factor Spt5 by yeast Bur1 kinase stimulates recruitment of the PAF complex. Molecular and Cellular Biology. 2009;**29**(17):4852-4863. Available from: http://mcb.asm.org/cgi/doi/10.1128/MCB.00609-09

[67] Mayer A, Lidschreiber M, Siebert M, Leike K, Söding J, Cramer P. Uniform transitions of the general RNA polymerase II transcription complex. Nature Structural & Molecular Biology. 2010;**17**(10):1272-1278. Available from: http://www.nature.com/doifinder/10.1038/nsmb.1903

[68] Ng HH, Dole S, Struhl K. The Rtf1 component of the Paf1 transcriptional elongation complex is required for ubiquitination of histone H2B. The Journal of Biological Chemistry. 2003;**278**(36):33625-33628

[69] Fierz B, Chatterjee C, McGinty RK, Bar-Dagan M, Raleigh DP, Muir TW. Histone H2B ubiquitylation disrupts local and higher-order chromatin compaction. Nature Chemical Biology. 2011;**7**(2):113-119. Available from: http://dx.doi.org/10.1038/nchembio.501

[70] Wu L, Zee BM, Wang Y, Garcia BA, Dou Y. The RING finger protein MSL2 in the MOF complex is an E3 ubiquitin ligase for H2B K34 and is involved in crosstalk with H3 K4 and K79 methylation. Molecular Cell. 2011;**43**(1):132-144. Available from: http://dx.doi.org/10.1016/j.molcel.2011.05.015

[71] Wu L, Li L, Zhou B, Qin Z, Dou Y. H2B ubiquitylation promotes RNA pol II processivity via PAF1 and pTEFb. Molecular Cell. 2014;**54**(6):920-931

[72] Wang P, Tan M, Zhang C, Morreau H, Teh BT. HRPT2, a tumor suppressor gene for hyperparathyroidism–jaw tumor syndrome. Hormone and Metabolic Research. 2005;**37**(6):380-383

[73] Prenzel T, Begus-Nahrmann Y, Kramer F, Hennion M, Hsu C, Gorsler T, et al. Estrogen-dependent gene transcription in human breast cancer cells relies upon proteasome-dependent monoubiquitination of histone H2B. Cancer Research. 2011;**71**(17):5739-5753

[74] Nicassio F, Corrado N, Vissers JHA, Areces LB, Bergink S, Marteijn JA, et al. Human USP3 is a chromatin modifier required for S phase progression and genome stability. Current Biology. 2007;**17**(22):1972-1977

[75] Van Der Knaap JA, Kumar BRP, Moshkin YM, Langenberg K, Krijgsveld J, Heck AJR, et al. GMP synthetase stimulates histone H2B deubiquitylation by the epigenetic silencer USP7. Molecular Cell. 2005;**17**(5):695-707

[76] Joo HY, Jones A, Yang C, Zhai L, Smith AD IV, Zhang Z, et al. Regulation of histone H2A and H2B deubiquitination and xenopus development by USP12 and USP46. The Journal of Biological Chemistry. 2011;**286**(9):7190-7201

[77] Fuchs G, Shema E, Vesterman R, Kotler E, Wolchinsky Z, Wilder S, et al. RNF20 and USP44 regulate stem cell differentiation by modulating H2B monoubiquitylation. Molecular Cell [Internet]. 2012;**46**(5):662-673. Available from: http://dx.doi.org/10.1016/j.molcel.2012.05.023

[78] Zhang Z, Jones A, Joo H-Y, Cao Y, Chen S, Erdjument-Bromage H, et al. USP49 deubiquitinates histone H2B and regulates pre-mRNA intron splicing. Genes & Development. 2013;**27**:1581-1595

[79] Tarcic O, Pateras IS, Cooks T, Shema E, Kanterman J, Ashkenazi H, et al. RNF20 links histone H2B ubiquitylation with inflammation and inflammation-associated cancer. Cell Reports. 2016;**14**(6):1462-1476. Available from: http://dx.doi.org/10.1016/j.celrep.2016.01.020

[80] Dickson KA, Cole AJ, Gill AJ, Clarkson A, Gard GB, Chou A, et al. The RING finger domain E3 ubiquitin ligases BRCA1 and the RNF20/RNF40 complex in global loss of the chromatin mark histone H2B monoubiquitination (H2Bub1) in cell line models and primary high-grade serous ovarian cancer. Human Molecular Genetics. 2016;**25**(24):5460-5471

[81] Wang E, Kawaoka S, Yu M, Shi J, Ni T, Yang W, et al. Histone H2B ubiquitin ligase RNF20 is required for MLL-rearranged leukemia. Proceedings of theNational Academy of Sciences. 2013;**110**(10):3901-3906. Available from: http://www.pnas.org/cgi/doi/10.1073/pnas.1301045110

[82] Garrido Castro P, Van Roon EHJ, Pinhanços SS, Trentin L, Schneider P, Kerstjens M, et al. The HDAC inhibitor panobinostat (LBH589) exerts in vivo anti-leukaemic activity against MLL-rearranged acute lymphoblastic leukaemia and involves the RNF20/RNF40/WAC-H2B ubiquitination axis. Leukemia [Internet]. 2018;**32**(2):323-331. Available from: http://dx.doi.org/10.1038/leu.2017.216

[83] Aranda S, Mas G, Di Croce L. Regulation of gene transcription by polycomb proteins. Science Advances. 2015;**1**(11):1-15

[84] Bach SV, Hegde AN. The proteasome and epigenetics: Zooming in on histone modifications. Biomolecular Concepts. 2016;**7**(4):215-227

[85] Chittock EC, Latwiel S, Miller TCR, Müller CW. Molecular architecture of polycomb repressive complexes. Biochemical Society Transactions. 2017;**45**(1):193-205. Available from: http://biochemsoctrans.org/lookup/doi/10.1042/BST20160173

[86] Satijn DP, Otte AP. RING1 interacts with multiple polycomb-group proteins and displays tumorigenic activity. Molecular and Cellular Biology. 1999;**19**(1):57-68. Available from: http://www.ncbi.nlm.nih.gov/pubmed/9858531%5Cnhttp://www.pubmedcentral.nih.gov/articlerender.fcgi?artid=PMC83865

[87] Bezsonova I, Walker JR, Bacik JP, Duan S, Dhe-Paganon S, Arrowsmith CH. Ring1B contains a ubiquitin-like docking module for interaction with Cbx proteins. Biochemistry. 2009;**48**(44):10542-10548

[88] Sanchez-Pulido L, Devos D, Sung ZR, Calonje M. RAWUL: A new ubiquitin-like domain in PRC1 ring finger proteins that unveils putative plant and worm PRC1 orthologs. BMC Genomics. 2008;**9**:1-11

[89] Gao Z, Zhang J, Bonasio R, Strino F, Sawai A, Parisis F, et al. PCGF homologs, CBX proteins, and RYBP define functionally distinct PRC1 family complexes. Molecular Cell. 2012;**45**(3):344-356

[90] Morey L, Aloia L, Cozzuto L, Benitah SA, Di Croce L. RYBP and Cbx7 define specific biological functions of polycomb complexes in mouse embryonic stem cells. Cell Reports. 2013;**3**(1):60-69. Available from: http://dx.doi.org/10.1016/j.celrep.2012.11.026

[91] Junco S, Wang R, Gaipa J, Taylor A, Schirf V, Gearhart MD, et al. Structure of the polycomb group protein PCGF1 (NSPC1) in complex with BCOR reveals basis for binding selectivity of PCGF homologs. Structure. 2014;**21**(4):665-671

[92] Gray F, Cho HJ, Shukla S, He S, Harris A, Boytsov B, et al. BMI1 regulates PRC1 architecture and activity through homo- and hetero-oligomerization. Nature Communications. 2016;**7**:13343

[93] Yu M, Mazor T, Huang H, Huang HT, Kathrein KL, Woo AJ, et al. Direct recruitment of polycomb repressive complex 1 to chromatin by core binding transcription factors. Molecular Cell. 2012;**45**(3):330-343. Available from: http://dx.doi.org/10.1016/j.molcel.2011.11.032

[94] Bell O, Schwaiger M, Oakeley EJ, Lienert F, Beisel C, Stadler MB, et al. Accessibility of the Drosophila genome discriminates PcG repression, H4K16 acetylation and replication timing. Nature Structural & Molecular Biology. 2010;**17**(7):894-900

[95] Dorafshan E, Kahn TG, Schwartz YB. Hierarchical recruitment of polycomb complexes revisited. Nucleus. 2017;**8**(5):496-505. Available from: https://doi.org/10.1080/19491034.2017.1363136

[96] Kassis J, Brown J. Polycomb group response elements in Drosophila and vertebrates. Advances in Genetics. 2012;**81**:83-118. Available from: http://europepmc.org/abstract/MED/23419717

[97] Fischle W, Wang Y, Jacobs SA, Kim Y, Allis CD, Khorasanizadeh S. Molecular basis for the discrimination of repressive methyl-lysine marks in histone H3 by polycomb and HP1 chromodomains. Genes & Development. 2003;**17**(15):1870-1881

[98] Kalb R, Latwiel S, Baymaz HI, Jansen PWTC, Müller CW, Vermeulen M, et al. Histone H2A monoubiquitination promotes histone H3 methylation in polycomb repression. Nature Structural & Molecular Biology. 2014;**21**(6):569-571. Available from: http://dx.doi.org/10.1038/nsmb.2833

[99] Laprell F, Finkl K, Müller J. Propagation of polycomb-repressed chromatin requires sequence-specific recruitment to DNA. Science. 2017;**356**(6333):85-88

[100] Schoenfelder S, Sugar R, Dimond A, Javierre B, Furlan-magaril M, Segonds-pichon A, et al. Europe PMC funders group polycomb repressive complex PRC1 spatially constrains the mouse embryonic stem cell genome. 2016;**47**(10):1179-1186. Available from: https://www.ncbi.nlm.nih.gov/pmc/articles/PMC4847639/pdf/emss-67387.pdf

[101] Pemberton H, Anderton E, Patel H, Brookes S, Chandler H, Palermo R, et al. Genome-wide co-localization of polycomb orthologs and their effects on gene expression in human fibroblasts. Genome Biology. 2014;**15**(2):1-16

[102] Frangini A, Sjöberg M, Roman-Trufero M, Dharmalingam G, Haberle V, Bartke T, et al. The Aurora B kinase and the polycomb protein Ring1B combine to regulate active promoters in quiescent lymphocytes. Molecular Cell. 2013;**51**(5):647-661

[103] Bravo M, Nicolini F, Starowicz K, Barroso S, Cales C, Aguilera A, et al. Polycomb RING1A- and RING1B-dependent histone H2A monoubiquitylation at pericentromeric regions promotes S-phase progression. Journal of Cell Science. 2015;**128**(19):3660-3671. Available from: http://jcs.biologists.org/cgi/doi/10.1242/jcs.173021

[104] Chan HL, Beckedorff F, Zhang Y, Garcia-Huidobro J, Jiang H, Colaprico A, et al. Polycomb complexes associate with enhancers and promote oncogenic transcriptional programs in cancer through multiple mechanisms. Nature Communications. 2018;**9**(1):3377. Available from: http://www.nature.com/articles/s41467-018-05728-x

[105] Sarge K, Park-Sarge O. Mitotic bookmarking of formerly active genes: Keeping epigenetic memories from fading. Cell Cycle. 2009;**8**(6):818-823

[106] Dey A, Nishiyama A, Karpova T, McNally J, Ozato J. Brd4 marks select genes on mitotic chromatin and directs postmitotic transcription. Molecular Biology of the Cell. 2009;**20**(23):4899-4909

[107] Kadauke S, Udugama M, Pawlicki J, Achtman J, Jain D, Cheng Y, et al. Tissue-specific mitotic bookmarking by hematopoietic transcription factor GATA1. Cell. 2012;**150**(4):725-737

[108] Arora M, Zhang J, Heine GF, Ozer G, Liu HW, Huang K, et al. Promoters active in interphase are bookmarked during mitosis by ubiquitination. Nucleic Acids Research. 2012;**40**(20):10187-10202

[109] Arora M, Packard CZ, Banerjee T, Parvin JD. RING1A and BMI1 bookmark active genes via ubiquitination of chromatin-associated proteins. Nucleic Acids Research. 2015;**44**(5):2136-2144

Chapter 3

E3 Ubiquitin Ligases in Cancer and Their Pharmacological Targeting

Joseph Y. Ong and Jorge Z. Torres

Abstract

Ubiquitination plays many critical roles in protein function and regulation. Consequently, mutation and aberrant expression of E3 ubiquitin ligases can drive cancer progression. Identifying key ligase-substrate relationships is crucial to understanding the molecular basis and pathways behind cancer and toward identifying novel targets for cancer therapeutics. Here, we review the importance of E3 ligases in the regulating the hallmarks of cancer, discuss some of the key and novel E3 ubiquitin ligases that drive tumor formation and angiogenesis, and review the clinical development of inhibitors that antagonize their function. We conclude with perspectives on the field and future directions toward understanding ubiquitination and cancer progression.

Keywords: E3 ubiquitin ligase, cancer, pharmacological targeting

1. Introduction

The regulation and turnover of proteins is an essential aspect of cell homeostasis and one that is commonly disrupted in cancer cells [1]. Regulation of a protein's levels, activity, or localization is affected by ubiquitination, a posttranslational modification that involves the covalent attachment of a 76 amino acid ubiquitin molecule onto a substrate protein [2, 3]. Depending on the cellular context, ubiquitinated proteins can affect a myriad of cellular processes, including signaling [4], epigenetics [5], endosome trafficking [6], DNA repair [7] and protein stability via the 26S-proteasome [8].

The outcome of protein ubiquitination is affected primarily by two properties: what kind of ubiquitin linkage and how many ubiquitin molecules are present [2]. Ubiquitin is usually covalently attached to its substrate via a nucleophilic lysine residue on the substrate and the ubiquitin carboxy terminus. Ubiquitin itself can serve as a nucleophile via one of seven lysine residues (K6, K11, K27, K29, K33, K48, and K63) [9, 10] though K48- and K63-linkages seem to be the most abundant and are the most well-studied. In some cases, the N-terminal amide of the initiator methionine (M1) of the substrate can serve as the nucleophile [11, 12]. If one of the lysine residues or the initiator methionine of ubiquitin serves as the nucleophile for another ubiquitin molecule, a polyubiquitin chain is formed. A K48-linked polyubiquitin chain of four or more ubiquitin molecules is typically enough to target the substrate for 26S-proteasome mediated degradation [13]. Meanwhile, poly-K63 linkages are involved in many processes, including endocytic trafficking, inflammation, and DNA repair [5, 6, 14]. Other ubiquitin linkages [11], combinations of linkages (mixed or branched chains) [15–17], monoubiquitination [5, 18], and multi-monoubiquitination [19, 20] events have other diverse functions within the cell.

Ubiquitination occurs in three main steps [21, 22]. First, the E1 ubiquitin-activating enzyme (two in the human genome) covalently attaches to a ubiquitin molecule via a thioester bond in an ATP-dependent process. Next, the E1 enzyme transfers ubiquitin onto an E2 ubiquitin-conjugating enzyme (about 40 in the human genome). Finally, the E2 enzyme binds a substrate-bound E3 ligase (about 600 in the human genome) to transfer ubiquitin onto a lysine residue of the substrate. Repeating the cycle creates a polyubiquitin chain.

E3 ligases can function either as single peptides (like Parkin), simple complexes (e.g.: hetero/homodimers, like MDM2/MDMX or XIAP), or as large complexes (like Cullin-RING-ligase complexes or the anaphase promoting complex/cyclosome). There are two main classes of E3 ligases [23]: HECT (about 30 in the human genome) and RING ligases (including RING and RING-like ligases and their accessory proteins, about 600 in the human genome).

HECT ligases contain a C-terminus HECT domain that accepts the ubiquitin molecule from an E2 conjugating enzyme via a thioester bond before transferring the ubiquitin to the substrate [24]. RING ligases contain a zinc finger domain, and these proteins allow the E2 to transfer ubiquitin directly onto the substrate [25]. A subclass of RING ligases known as RING-between-RING (RBR) ligases contain two RING domains that have elements of both HECT and RING ligases: one RING domain binds the charged E2, while the other RING domain accepts the ubiquitin molecule before transferring it onto the substrate [26].

As E3 ligases ultimately determine the target of the ubiquitination machinery, they play a critical role in cell regulation. They regulate key players in processes like apoptosis (caspases), cell senescence and growth (p53, p21, p27; Hippo and

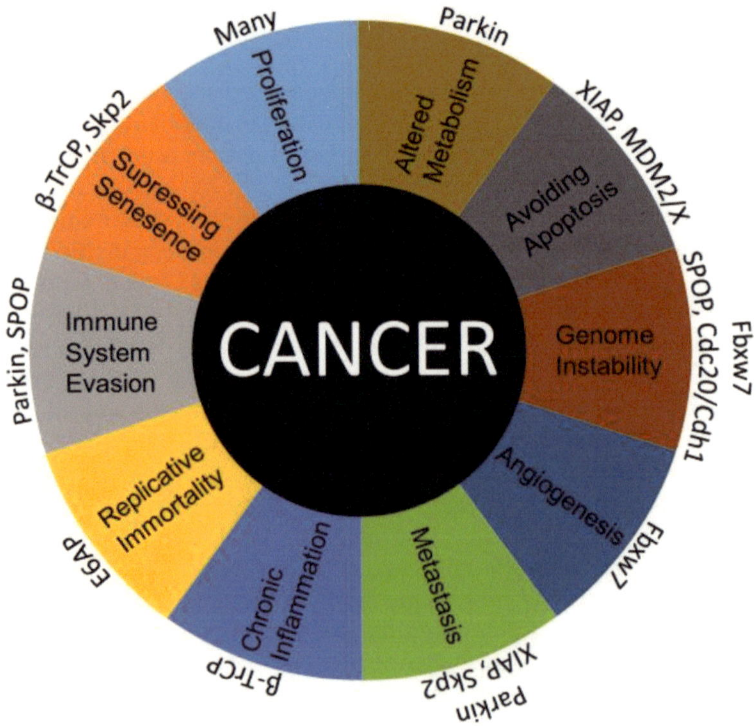

Figure 1.
E3 ubiquitin ligases (outer circle) regulate hallmarks of cancer (inner circle) to drive cancer progression.

Hedgehog signaling), proliferation and genomic stability (c-Myc, cyclins), immune system evasion (PD-L1), inflammation (NFκB), and metastasis and angiogenesis (Wnt signaling) (**Figure 1**). Misregulation or mutation of E3 ligases can lead to overexpression of oncogenes or downregulation of tumor suppressor genes, leading to cancer progression. Consequently, understanding the molecular targets and functions of E3 ligases serves as the basis for designing new cancer therapies.

Here, we describe some central and novel E3 ligases related to cancer development, pharmacological targeting of those ligases, and perspectives on understanding the role of E3 ligases in cancer progression.

2. E3 ligases and cancer progression

2.1 TP53

The tumor protein p53 (TP53) is a transcription factor that serves as one of the principal regulators of cell function and survival (reviewed in [27]), mediating cellular responses to proliferation, cell cycle control, DNA damage response pathways, and apoptosis. Consequently, it is mutated in approximately 50% of all cancer types. Thus, regulators of p53 serve as ideal candidates to understand and address cancer cell progression (**Table 1**).

E6AP (Ube3a) is a 100 kDa HECT domain ligase discovered for mediating the interaction between human papillomavirus protein E6 and p53 [28]. Neither E6AP nor E6 alone have a strong affinity for p53, but together, the E6/E6AP complex binds to p53 and changes the substrate specificity of E6AP [28], allowing E6AP to ubiquitinate p53 at the N-terminal DNA binding domain and target it for

E3 ligase		**Notable substrates and binding partners**	**Expression in cancer**	**Cancer types**
TP53	E6AP	p53	Gain of function via HPV E6	Cervical, breast [38, 166]
	MDM2/X	p53	Overexpressed	Many; liposarcomas [48, 167]
SCF	Skp2	p21, p27	Overexpressed	Many [95, 168]
	Fbxw7	Cyclin E, mTOR	Downregulated or dominant-negative mutant	Many; endometrial, cervical, blood [64, 67, 169]
	β-TrCP	IkB, β-catenin, Wee1, Cdc25a/b	Overexpressed (in some tissues)	Many [60, 168]
APC/C	Cdc20	Cyclin A/B, securin	Overexpressed	Pancreatic, lung, gastric [95, 168, 170]
	Cdh1	Cdc20, Plk1, Aurora kinase A/B	Underexpressed	Many [171]
Other	XIAP	Caspases 3, 7, 9	Overexpressed	Many [98, 99]
	Park2	Cyclin D/E, Cdc20/Cdh1, tubulin	Underexpressed	Breast, pancreatic, colorectal, ovarian [172]
	SPOP	PD-L1, androgen and estrogen receptor	Downregulated or dominant-negative mutant	Prostate, endometrial, kidney [139, 141, 150]

Table 1.
E3 ligases and cancer progression.

degradation [29]. Consequently, E6AP may play a role in HPV-mediated cervical cancers [30], particularly for those mediated by high-risk HPV16 strain, as E6 proteins from lower-risk strains of HPV lack the ability to degrade p53 [31].

The E6/E6AP complex plays other roles in cancer cell progression. Neither E6 nor E6AP alone can activate the hTERT promoter, but together, the E6/E6AP complex can activate the hTERT promoter, perhaps via interactions with c-Myc and NFX-1 to respectively activate and repress promoter activity [32]. The E6/E6AP complex has also been implicated in the ubiquitination of apoptosis-inducing proteins Bak [33], Fas [34], and TNFR1 [35]. Independent of E6 binding, endogenous E6AP targets include the tumor suppressor PML [36]; cell cycle regulators p27 [36], Cdk1, Cdk4; cell proliferation regulator MAPK1 [37];, and guanine nucleotide exchange factor ECT2 [38]. A published list of 130 likely substrates of E6AP includes β-catenin and PRMT5, proteins involved in cancer progression [37].

MDM2 is best known as a regulator of p53. MDM2 is a RING ligase [39] that forms stable heterodimers with a homolog, **MDMX** (MDM4), via their RING domains [40]. MDM2 localizes primarily in the nucleus bound to p300/CBP [41]. When complexed to p53, MDM2 inhibits p53 activity in two ways: first, MDM2 binds the N-terminal transactivation domain [42], inhibiting p53-mediated transcription [43]; secondly, MDM2 modulates p53 protein levels via ubiquitination near the C-terminus [44]. After MDM2 monoubiquitinates p53, p300 and CBP catalyze the polyubiquitination of p53, leading to p53 degradation [8, 41, 45]. Overexpression of MDM2 [46, 47], seen in many cancers where p53 is not mutated [48], leads to a loss of p53 activity.

During p53 activation, p53 is phosphorylated by multiple serine/threonine kinases at residues near the N-terminus, disrupting p53/MDM2 binding and stabilizing p53. For example, ATM kinase phosphorylates p53 at S15 [49] to promote p53-mediated transcription. Additionally, ATM phosphorylation of MDM2 on S395 disrupts the MDM2/p53 complex, allowing p53 to accumulate [50].

2.2 SCF complexes

The SCF complex is a multimeric ubiquitination complex with multiple roles in cell regulation (**Table 1**). The main scaffold of the SCF complex, Cullin 1 (Cul1), recruits the substrate to be ubiquitinated at the N-terminus and the charged ubiquitin at the C-terminus. Rather than bind the substrate directly, Cul1 uses two adaptor proteins: Cul1 binds directly to Skp1, which then binds to one of about 70 F-box proteins [51] that directly bind their substrates. At the C-terminus, Cul1 binds an adaptor protein, either Rbx1 or Rbx2 (also known as Roc1 or Roc2), that will bind a charged E2 ubiquitin conjugating enzyme [52, 53] .

Skp2 (Fbxl1) is a F-box protein that is most active during S-phase [54]. During S phase, Skp2 binds and ubiquitinates phosphorylated p27 [55] by binding the Cdk2-cyclin E complex [56]. Degradation of p27 frees inhibition of Cdk2-cyclinA/E complexes, allowing for progression into S-phase and entry into mitosis [57]. Other targets of Skp2 include p21 [58] and E-cadherin [59]. In some cases, Skp2 requires an accessory protein Cks1 to enhance binding to the substrate [60]. Skp2 both enhances c-Myc transcriptional activity and promotes c-Myc degradation [61]. Interestingly, p300-mediated acetylation of Skp2 changes the localization of Skp2 from nuclear to cytoplasmic, increasing cellular proliferation, motility, and tumorigenesis [59]. Skp2 is commonly overexpressed in a variety of cancers [62], including blood, colorectal, stomach, ovarian, and cervical cancers [60].

Fbxw7 (in yeast, Cdc4) contains a homodimerization domain, an F-box domain that binds Skp1, and eight WD40 repeats that form a beta-propeller structure to bind substrates [63]. Substrate binding is dependent on interaction between the arginine residues of the Fbxw7 WD40 domains and phosphorylated residues of the substrate in a recognition motif termed the Cdc4 phosphodegron (CPD) [63]. Mutations that disrupt substrate binding, especially point mutations of the arginine residues of the WD40 region, are commonly found in tumor samples [64]. Because Fbxw7 homodimerizes, these mutations may have a dominant-negative effect [65], as wild-type Fbxw7-mutant Fbxw7 dimers are able to effectively bind but not ubiquitinate their substrates [66]. Fbxw7 is deleted [67] or mutated in many cancers, with mutations being especially common in cancers of the bile duct and blood [68].

One well-characterized substrate of Fbxw7 is cyclin E [69]. The ubiquitination and degradation of cyclin E is dependent on phosphorylation of by Cdk2 and glycogen synthase kinase 3 (GSK3) [70]. Dimerization of Fbxw7 can also change its affinity for cyclin E as well as other substrates [71]. Other substrates of Fbxw7 include transcription factors c-Myc [72]; c-JUN, Notch 1; DNA-binding protein DEK [73]; and nutrient sensing protein mTOR [74]. Interestingly, the SV40 large T antigen contains a decoy CPD that can mislocalize Fbxw7 and inhibit Fbxw7-mediated degradation of cyclin E [75].

β-TrCP (BTRC), **Fbxw1a** (β-TrCP1) and **Fbxw11** (β-TrCP2) are protein homologs that appear to have redundant roles [76]. These F-box proteins can form homo- and hetereodimers with each other [76] and use WD40 domains to bind a DSG phosphodegron motif (such as DpSGXXpS) [60]. Overexpression of β-TrCP is seen in various types of cancers, including colorectal, pancreatic, breast, ovarian and melanomas [77].

β-TrCP plays an important role as a regulator of Cdk1. One substrate of β-TrCP is Wee1, a kinase that inhibits Cdk1 activity [78]. Phosphorylation of Wee1 at S53 and S123 by Plk1 and Cdk1 respectively allow β-TrCP to bind to and ubiquitinate Wee1, activating Cdk1 during G2 to promote rapid entry into mitosis. Similarly, in prophase, β-TrCP also ubiquitinates Emi1, an inhibitor of the APC/C [79]. Consequently, β-TrCP accelerates mitotic progression both by increasing Cdk1 activity and activating the APC/C. In the case of DNA damage, checkpoint proteins hyperphosphorylate Cdc25a [80], a phosphatase that activates Cdk1 by removing repressive phosphorylation events. β-TrCP binds to and ubiquitinates hyperphosphorylated Cdc25a, deactivating Cdk1 and delaying the cell cycle. β-TrCP also ubiquitinates Cdc25b [81], a phosphatase that activates Cdk2/cyclin A and Cdk1/cyclin B to progress through the G2/M transition [82]. Other β-TrCP substrates that are linked to cancer progression include the IkB family [83], β-catenin [76] and MDM2 [84].

2.3 APC/C

Proper cell cycling and successful mitotic events rely on the coordinated accumulation and destruction of cyclins [85]. Disruption of this coordination can lead to aberrant mitotic events, aneuploidy, and cancer [86] (**Table 1**). While entry into mitosis is mediated by activation of Cdk1/2, progression through and exit from mitosis is mediated principally by the anaphase promoting complex or cyclosome (**APC/C**).

The APC/C is a 1.2 megadalton complex whose activity is necessary for entry to and exit from mitosis [87]. The structure of the human APC/C was solved via cryoEM to 7.4 angstrom resolution, allowing for the identification of 20 subunits of the APC/C and a mechanistic understanding of its function [88]. APC/C ubiquitin ligase activity depends on two activating subunits, **Cdc20** or **Cdh1** (coded by gene FRZ1; not to be confused with the gene CDH1, which codes for E-cadherin), which are necessary for APC/C binding to substrate and subsequent degradation [89] via

K11 ubiquitin linkages [90]. In early mitosis, APC/C-Cdc20 degrades proteins such as cyclins A and B and Securin, the inhibitor of separase [91]. In later stages of mitosis and early G1, APC/C-Cdh1 degrades Cdc20, mitotic kinases like Plk1 and Aurora kinases A/B, and the contractile ring protein Anillin to ensure exit from mitosis and proper transition into G1 [92]. Binding of the substrate to APC/C is mediated by two main modalities [93]: for some substrates, Cdc20/Cdh1 binds the substrate through a KEN box motif; for others, both the APC/C subunit Apc10 and Cdc20/Cdh1 "sandwich" the substrate at the substrate's D box. Some substrates have both and/or additional motifs to bind the APC/C and Cdc20/Cdh1 [92].

Cdc20 is found overexpressed in many cancers, including lung, oral, liver, and colon cancers [94, 95]. Cdh1 is generally a tumor suppressor, as downregulation of Cdh1 is found in some aggressive cancer cell types [95], and loss of Cdh1 sensitizes cells to DNA damage [96].

2.4 Other

X-linked inhibitor of apoptosis protein (**XIAP**) is a IAP family E3 ligase characterized by three N-terminal baculovirus IAP repeat domains and a C-terminal RING domain [97]. Like other IAPs, XIAP plays a central role in mediating the cell's response to apoptosis. XIAP is overexpressed in many cancer cell lines, particularly in kidney and skin cancers [98, 99].

The linker region of XIAP between BIR1 and BIR2 binds to the active site and inhibits caspase 3 and caspase 7 [100]. The BIR3 domain of XIAP also binds to caspase 9, inhibiting caspase 9 dimerization and activity [101]. Moreover, XIAP ubiquitinates caspase 3 [102], caspase 9 [103], and caspase 7 [104] and targets them for degradation. As a final level of regulation, in addition to its ubiquitin E3 ligase role, XIAP can also function as a neddylation E3 ligase, neddylating and inhibiting the activity of caspases [105].

XIAP also plays important roles in cell motility. On one hand, XIAP degrades COMMD1 [106], a regulator of NFκB [107] and copper homeostasis. XIAP also binds to MAP3K7IP1, an event that activates kinase MAP3K7 to phosphorylate substrates leading to removal of NFκB inhibition [108]. XIAP also binds to survivin [109], activating NFκB signaling and encouraging cell metastasis by activating cell motility kinases Fadk1 and Src [110]. Conversely, XIAP has also been show to inhibit cell migration by binding to and ubiquitinating c-RAF to direct another ubiquitin ligase (CHIP) to degrade c-RAF [111]. Under non-stressed conditions, XIAP ubiquitinates and degrades MDM2, stabilizing p53 and inhibiting autophagy [112]. XIAP also binds to and monoubiquitinates TLE3, allowing β-catenin to activate Wnt-mediated transcription [113]. Finally, in addition to inflammation involving the NFκB pathway, XIAP suppresses TLR-based inflammation [114].

Park2 (PARKIN) is an RBR-E3 ligase with both RING and HECT ligase characteristics [115]. The Park2 locus is commonly deleted in cancers [116]. In mouse models, loss of Park2 causes spontaneous liver cancer [117] and contributes to colorectal cancer in mouse models [118]. Additionally, Park2 plays a central role in mitophagy [119], which may affect cell redox state [120], proliferation, and metastasis [121].

Park2 plays a prominent role in regulating cyclin levels. Park2 degrades cyclins D [122] and E [123] in a Cul1-dependent manner [124]. Park2 mutations found in cancer lead to stabilization of these G1/S-phase cyclins, an increase in the number of cells in S and G2/M phase [123, 124], and increased rates of cellular proliferation [122]. Moreover, Park2 associates with Cdc20 and Cdh1 during mitosis in an APC/C-independent manner and regulates the levels of many APC/C substrates including mitotic kinases and mitotic cyclins [125]. Park2 regulates microtubules and the mitotic spindle, cytokinetic bridge [126], cell motility [127], and invasion [128]. Park2 ubiquitinates and degrades HIF-1α to contribute to cell migration, and loss of Park2 leads to tumor metastasis in mouse models [129].

In Park2 knock-out mouse models, the resulting oxidative stress and the Warburg effect [130] caused an increase in the mRNA of Aim2, a protein involved in cytokine production [131]. In these mouse models, activation of Aim2 ultimately led to upregulation of PD-L1 in pancreatic tumors and lower rates of survival, an effect seen in human pancreatic tumors and patients [131]. Thus, Park2's roles in metabolism may affect the ability of the immune system to regulate cancer progression.

SPOP is a Cul3 substrate adaptor mutated in about 10% of prostate cancers and some kidney cancers [132]. SPOP has three basic domains: an N-terminal MATH domain for substrate recognition [133], a BTB domain for dimerization and interaction with Cul3 [134], and a BACK domain which assembles SPOP dimers into oligomers [134], a mechanism which increases SPOP binding to and ubiquitination of the substrate [135]. As SPOP regulates many proteins responsible for maintaining cell integrity, mutations in the MATH domain that disrupt binding to substrate encourage cancer progression [136].

SPOP plays a role in immunotherapy by ubiquitinating and degrading PD-L1 [137]. SPOP binding mutants cannot ubiquitinate PD-L1, resulting in larger tumor growth and fewer tumor-infiltrating lymphocytes compared to tumors harboring wild-type SPOP in mouse models [137]. Similarly, pancreatic cancer samples with mutant SPOP had higher levels of PD-L1, demonstrating a role for SPOP in immune system invasion [137].

Other notable SPOP substrates include the apoptotic protein Daxx [138, 139], deSUMOlyase SENP7 [140], c-Myc [141], HDAC6 [142], Cdc20 [143], proto-oncogene DEK [144], phosphatases PTEN and Dusp7 [139], hedgehog pathway proteins Gli2 and Gli3 [145, 146], and BET transcriptional coactivators BRD2–4 [147–149]. SPOP is also closely tied to hormone-activated pathways, as steroid receptor coactivator SRC-3 [150], androgen receptor (AR) [151], enhancer of AR-mediated transcriptional activity TRIM24 [144], and estrogen receptor α (ERα) [136] are all substrates of SPOP. Finally, wild-type, but not mutant SPOP degrades ERG [152]. Interestingly, in some prostate cancer samples, some tumors expressed a fused ERG protein due to genome rearrangements, a phenotype driven by SPOP mutation [153]. Unlike wild-type ERG, these ERG-fusions lack an SPOP binding site, contributing to cancer progression [154].

3. E3 ligases and their inhibitors

One ubiquitin-proteasome inhibitor has already found use in the treatment of cancer: Bortezomib is a 26S-proteasome inhibitor approved for treating certain types of myeloma and lymphoma that binds to and inhibits the proteasome from degrading other proteins [155]. Another compound still in clinical development is MLN4924 (Pevonedistat), an inhibitor of the Nedd8-activating enzyme and thus of Cullin RING ligase complexes [155]. As ubiquitination plays many important roles in cell regulation, these broad inhibitors can affect many cellular pathways, not just those that are therapeutically useful. As E3 ligases are specific for their substrates, E3 ligases serve as precise targets for therapeutic intervention (**Table 2**). Inhibition of E3 ligases will hopefully minimize off-target effects. Moreover, as some E3 ligases have many oncogenes as their substrates, targeting E3 ligases may serve to be more efficient than targeting individual substrates.

While most inhibitors have been identified via high throughput screens, the most clinically relevant inhibitors have been derived from structure–function analyses of E3 ligases complexed to their substrates. For example, the crystal structure of MDM2 bound to p53 allowed for the identification of the MDM2-p53 binding pocket and the design of small molecules [156] (like Nutlins and their derivatives) and stapled peptides [157] that bind to MDM2 and inhibit p53 binding. Similarly, the structure of the IAP family of E3 ligases and their endogenous inhibitors, the SMAC peptides, allowed for the development of higher affinity peptides [158] and peptidomimetics and the discovery of one small molecule inhibitor, Embelin [159]. Of the inhibitors mentioned here, MDM2 and XIAP inhibitors have advanced the farthest in clinical trials. A crystal structure of the SPOP substrate binding domain was also used to develop an SPOP inhibitor, suggesting that structural studies may greatly enhance development of small molecule inhibitors [160].

Most inhibitors disrupt E3 ligase-substrate binding by blocking the binding pocket of the E3 ligase. However, because HECT domains first transfer the ubiquitin molecule to themselves via a thioester bond [24], HECT ligases have an

E3 ligase		**Therapeutic**	**Mechanism**	**Model**			**In clinical trials**
				***In vitro* assay**	**Cell culture**	**Mouse model**	
TP53	E6AP	CM-11 peptides [161]	Binds HECT domain	X	X		
		Compound 9 [173]	Binds HPV E6	X	X		
	MDM2/X	Nutlins [156], RG7112 [174]	Binds p53 binding site	X	X	X	
		Idasanutlin (RG7388) [175]		X	X	X	X
		MI-888 [176], SAR405838 [151]	Binds p53 binding site	X	X	X	X
		AMG-232 [177]	Binds p53 binding site	X	X	X	X
		NVP-CGM097 [178], HDM201 [179]	Binds p53 binding site	X	X	X	X
		JNJ-26854165 (Serdemetan)	Assumed to bind to RING domain of MDM2 [180]		X	X	X
		ALRN-6924 [157]	Stapled peptide binds MDM2 and MDMX at p53 binding site	X	X	X	X
SCF	Skp2	Compound #25 [181]	Binds Skp1 binding site	X	X	X	
		C1, C2, C16, C20 [163, 182]	Presumed: Binds Skp2, Cks1 at p27 binding site	X	X		
		CpdA [165]	Inhibits Skp2-Skp1 binding	X	X		
		NSC689857, NSC681152 [164]	Inhibits Skp2-Cks1 binding	X			
	Fbxw7	Oridonin [183]	Stabilizes Fbxw7, increases the activity of kinase Gsk-3	X	X		

E3 ligase		Therapeutic	Mechanism	Model			In clinical trials
				In vitro assay	Cell culture	Mouse model	
	β-TrCP	Erioflorin [184]	Inhibits β-TrCP1 binding to substrate	X	X		
		GS143 [185]	Presumed: Inhibits binding of β-TrCP1 and p-IkBa	X	X		
APC/C	Cdc20	Apcin [186]	Binds to D-box binding site of Cdc20	X	X		
	Cdc20/ Cdh1	ProTAME [187]	Inhibits formation of APC/C-Cdc20, -Cdh1	X	X	X	
Other	XIAP	LCL161 [158]	Binds to BIR3 domain of XIAP [188]	X	X	X	X
		AEG 35156 [189]	XIAP antisense oligonucleotide		X		X
	SPOP	Palbociclib [137]	Cdk4 phosphorylates SPOP, destabilizes PD-L1	X	X	X	*
		Compound 6b [160]	Binds to substrate pocket	X	X	X	

**Palbociclib is clinically approved for treatment of breast cancer.*

Table 2.
E3 ligases and their inhibitors.

additional mode of pharmacological inhibition. The CM-11 peptides (E6AP inhibitors) are one such therapy that takes advantage of this step to inhibit or disrupt the HECT-Ubiquitin transthiolation reaction [161]. Future work may focus on designing small molecules that disrupt this function of the HECT domain.

To degrade its most clinically relevant targets p21 and p27, Skp2 functions with an adaptor protein, Cks1 [162]. At least two classes of inhibitors (NSC689857/NSC681152 [163] and the C1/2/16/20 compounds [164]) have been developed that disrupt the Skp2-Cks1 interaction. Similarly, the SCF ligase complex is only active upon the binding of an F-box protein to Skp1. CpdA inhibits Skp2-Skp1 binding [165]. These results suggest that another method of inhibitor design may focus on disrupting crucial activators and binding partners of E3 ligases instead of merely disrupting E3 ligase-substrate binding.

Upon phosphorylation by Cdk4, SPOP protein levels are stabilized, and PD-L1 expression levels decrease [137]. To improve the efficiency of anti-PD-L1 immunotherapies, mice treated with both Cdk4/6 inhibitors (to destabilize SPOP and thus stabilize PD-L1) and anti-PD-L1 immunotherapy showed improved survival when compared to untreated mice or mice with each individual treatment [137]. In this case, stabilization of an oncogenic protein led to improved efficacy of a complimentary therapy. Whether a similar combination of therapies can be used to improve the overall survival rate in other pathways remains to be seen.

4. Conclusions and perspectives

Recent research has highlighted the role of ubiquitination in cell regulation, division, and cancer cell progression. While much work has advanced the identification of E3 ubiquitin ligases and their substrates, untangling how these ligases act upon interconnected pathways remains a challenge in cancer cell biology. For example, understanding in which contexts certain E3 ligases are tumor-supportive or tumor-suppressive (like β-TrCP) is still not clear. Genome-wide analyses and advancements in systems biology have aided in and will continue to contribute to addressing these issues.

The tumor microenvironment has established itself as a central component in understanding and treating cancer progression. The macro-level questions of tumors—how cancers induce angiogenesis, interact with the immune system and cytokines, interact with the microbiome, and metastasize—are some questions that are best addressed with research in animal models, not human cell culture models. For example, the recent discoveries that both SPOP and Park2 play a role in mediating PD-L1 stability demonstrate the need to study the roles of E3 ligases in animal models. Given the recent success of immuno-oncology and CAR-T cell therapy, a further understanding how E3 ligases affect macro-level phenotypes like tumor sensitivity to immunotherapies may influence the design of clinical therapies.

While many E3 ligase inhibitors are being identified via high-throughput small molecule screens that assess inhibition of E3 ligase-substrate binding or ubiquitination activity, the most clinically advanced inhibitors have been refined from structural analysis of the E3 ligase binding pocket. The structures of many E3 ligases have already been determined (for example, all 11 ligases discussed here have at least a partial structure), so further pharmacological development may involve identifying binding pockets and designing inhibitors to perturb ligase function, and optimizing already identified inhibitors. On the other hand, E3 ligases are often redundant, so inhibition of one ligase may not completely stabilize a beneficial substrate. Nonetheless, the early clinical success of some E3 ligase inhibitors suggests that ubiquitin ligase inhibition is a promising venue for therapeutic intervention in cancer patients.

Author details

Joseph Y. Ong and Jorge Z. Torres*
Department of Chemistry and Biochemistry, University of California, Los Angeles, CA, USA

*Address all correspondence to: torres@chem.ucla.edu

References

[1] Hanahan D, Weinberg RA. Hallmarks of cancer: The next generation. Cell. 2011;**144**:646-674

[2] Komander D, Rape M. The ubiquitin code. Annual Review of Biochemistry. 2012;**81**:203-229

[3] Yau R, Rape M. The increasing complexity of the ubiquitin code. Nature Cell Biology. 2016;**18**:579-586

[4] Tao M et al. ITCH K63-ubiquitinates the NOD2 binding protein, RIP2, to influence inflammatory signaling pathways. Current Biology. 2009;**19**: 1255-1263

[5] Sigismund S, Polo S, Di Fiore PP. Signaling through monoubiquitination. Current Topics in Microbiology and Immunology. 2004;**286**:149-185

[6] Williams RL, Urbé S. The emerging shape of the ESCRT machinery. Nature Reviews. Molecular Cell Biology. 2007; **8**:355-368

[7] Hoege C, Pfander B, Moldovan G-L, Pyrowolakis G, Jentsch S. RAD6-dependent DNA repair is linked to modification of PCNA by ubiquitin and SUMO. Nature. 2002;**419**:135-141

[8] Haupt Y, Maya R, Kazaz A, Oren M. Mdm2 promotes the rapid degradation of p 53. Nature. 1997;**387**:296-299

[9] Xu P et al. Quantitative proteomics reveals the function of unconventional ubiquitin chains in proteasomal degradation. Cell. 2009;**137**:133-145

[10] Peng J et al. A proteomics approach to understanding protein ubiquitination. Nature Biotechnology. 2003;**21**:921-926

[11] Emmerich CH et al. Activation of the canonical IKK complex by K63/M1-linked hybrid ubiquitin chains. Proceedings of the National Academy of Sciences of the United States of America. 2013;**110**:15247-15252

[12] Kirisako T et al. A ubiquitin ligase complex assembles linear polyubiquitin chains. The EMBO Journal. 2006;**25**: 4877-4887

[13] Thrower JS, Hoffman L, Rechsteiner M, Pickart CM. Recognition of the polyubiquitin proteolytic signal. The EMBO Journal. 2000;**19**:94-102

[14] Chen ZJ, Sun LJ. Nonproteolytic functions of ubiquitin in cell signaling. Molecular Cell. 2009;**33**:275-286

[15] Ben-Saadon R, Zaaroor D, Ziv T, Ciechanover A. The polycomb protein Ring1B generates self atypical mixed ubiquitin chains required for its In vitro histone H2A ligase activity. Molecular Cell. 2006;**24**:701-711

[16] Liu C, Liu W, Ye Y, Li W. Ufd2p synthesizes branched ubiquitin chains to promote the degradation of substrates modified with atypical chains. Nature Communications. 2017;**8**:14274

[17] Ohtake F, Tsuchiya H, Saeki Y, Tanaka K. K63 ubiquitylation triggers proteasomal degradation by seeding branched ubiquitin chains. Proceedings of the National Academy of Sciences of the United States of America. 2018;**115**: E1401-E1408

[18] Alpi AF, Pace PE, Babu MM, Patel KJ. Mechanistic insight into site-restricted monoubiquitination of FANCD2 by Ube2t, FANCL, and FANCI. Molecular Cell. 2008;**32**:767-777

[19] Braten O et al. Numerous proteins with unique characteristics are degraded by the 26S proteasome following monoubiquitination. Proceedings of the National Academy of Sciences of the

United States of America. 2016;**113**: E4639-E4647

[20] Kravtsova-Ivantsiv Y, Cohen S, Ciechanover A. Modification by single ubiquitin moieties rather than polyubiquitination is sufficient for proteasomal processing of the p105 NF-κB precursor. Molecular Cell. 2009;**33**: 496-504

[21] George AJ, Hoffiz YC, Charles AJ, Zhu Y, Mabb AM. A comprehensive atlas of E3 ubiquitin ligase mutations in neurological disorders. Frontiers in Genetics. 2018;**9**:29

[22] Weissman AM, Shabek N, Ciechanover A. The predator becomes the prey: Regulating the ubiquitin system by ubiquitylation and degradation. Nature Reviews. Molecular Cell Biology. 2011;**12**:605-620

[23] Li W et al. Genome-wide and functional annotation of human E3 ubiquitin ligases identifies MULAN, a mitochondrial E3 that regulates the Organelle's dynamics and signaling. PLoS One. 2008;**3**:e1487

[24] Scheffner M, Kumar S. Mammalian HECT ubiquitin-protein ligases: Biological and pathophysiological aspects. Biochimica et Biophysica Acta (BBA)—Molecular Cell Research. 2014; **1843**:61-74

[25] Lipkowitz S, Weissman AM. RINGs of good and evil: RING finger ubiquitin ligases at the crossroads of tumour suppression and oncogenesis. Nature Reviews. Cancer. 2011;**11**: 629-643

[26] Dove KK, Stieglitz B, Duncan ED, Rittinger K, Klevit RE. Molecular insights into RBR E3 ligase ubiquitin transfer mechanisms. EMBO Reports. 2016;**17**:1221-1235

[27] Kastenhuber ER, Lowe SW. Putting p53 in context. Cell. 2017;**170**:1062-1078

[28] Scheffner M, Werness BA, Huibregtse JM, Levine AJ, Howley PM. The E6 oncoprotein encoded by human papillomavirus types 16 and 18 promotes the degradation of p53. Cell. 1990;**63**:1129-1136

[29] Martinez-Zapien D et al. Structure of the E6/E6AP/p53 complex required for HPV-mediated degradation of p53. Nature. 2016;**529**:541-545

[30] Alejo M et al. Contribution of human papillomavirus in neuroendocrine tumors from a series of 10,575 invasive cervical cancer cases. Papillomavirus Research. 2018;**5**: 134-142

[31] Elbel M, Carl S, Spaderna S, Iftner T. A comparative analysis of the interactions of the E6 proteins from cutaneous and genital papillomaviruses with p53 and E6AP in correlation to their transforming potential. Virology. 1997;**239**:132-149

[32] Liu X et al. HPV E6 protein interacts physically and functionally with the cellular telomerase complex. Proceedings of the National Academy of Sciences. 2009;**106**:18780-18785

[33] Thomas M, Banks L. Inhibition of Bak-induced apoptosis by HPV-18 E6. Oncogene. 1998;**17**:2943-2954

[34] Filippova M, Parkhurst L, Duerksen-Hughes PJ. The human papillomavirus 16 E6 protein binds to Fas-associated death domain and protects cells from Fas-triggered apoptosis. The Journal of Biological Chemistry. 2004;**279**:25729-25744

[35] Filippova M, Song H, Connolly JL, Dermody TS, Duerksen-Hughes PJ. The human papillomavirus 16 E6 protein binds to tumor necrosis factor (TNF) R1 and protects cells from TNF-induced apoptosis. The Journal of Biological Chemistry. 2002;**277**:21730-21739

[36] Raghu D et al. E6AP promotes prostate cancer by reducing p27 expression. Oncotarget. 2017;**8**: 42939-42948

[37] Wang Y et al. Identifying the ubiquitination targets of E6AP by orthogonal ubiquitin transfer. Nature Communications. 2017;**8**:2232

[38] Mansour M et al. The E3-ligase E6AP represses breast cancer metastasis via regulation of ECT2-rho signaling. Cancer Research. 2016;**76**:4236-4248

[39] Fang S, Jensen JP, Ludwig RL, Vousden KH, Weissman AM. Mdm2 is a RING finger-dependent ubiquitin protein ligase for itself and p53. The Journal of Biological Chemistry. 2000; **275**:8945-8951

[40] Wang X, Wang J, Jiang X. MdmX protein is essential for Mdm2 protein-mediated p53 polyubiquitination. The Journal of Biological Chemistry. 2011; **286**:23725-23734

[41] Ferreon JC et al. Cooperative regulation of p53 by modulation of ternary complex formation with CBP/p300 and HDM2. Proceedings of the National Academy of Sciences. 2009; **106**:6591-6596

[42] Kussie PH et al. Structure of the MDM2 oncoprotein bound to the p53 tumor suppressor transactivation domain. Science. 1996;**274**:948-953

[43] Oliner JD et al. Oncoprotein MDM2 conceals the activation domain of tumour suppressor p53. Nature. 1993; **362**:857-860

[44] Poyurovsky MV et al. The C terminus of p53 binds the N-terminal domain of MDM2. Nature Structural & Molecular Biology. 2010;**17**:982-989

[45] Grossman SR et al. Polyubiquitination of p53 by a ubiquitin ligase activity of p300. Science (80-). 2003;**300**:342-344

[46] Zack TI et al. Pan-cancer patterns of somatic copy number alteration. Nature Genetics. 2013;**45**:1134-1140

[47] Beroukhim R et al. The landscape of somatic copy-number alteration across human cancers. Nature. 2010;**463**: 899-905

[48] Oliner JD, Saiki AY, Caenepeel S. The role of MDM2 amplification and overexpression in tumorigenesis. Cold Spring Harbor Perspectives in Medicine. 2016;**6**:a026336

[49] Canman CE et al. Activation of the ATM kinase by ionizing radiation and phosphorylation of p53. Science. 1998; **281**:1677-1679

[50] Gannon HS, Woda BA, Jones SN. ATM phosphorylation of Mdm2 Ser394 regulates the amplitude and duration of the DNA damage response in mice. Cancer Cell. 2012;**21**:668-679

[51] Lydeard JR, Schulman BA, Harper JW. Building and remodelling Cullin-RING E3 ubiquitin ligases. EMBO Reports. 2013;**14**:1050-1061

[52] Wei D, Sun Y. Small RING finger proteins RBX1 and RBX2 of SCF E3 ubiquitin ligases: The role in cancer and as cancer targets. Genes & Cancer. 2010;**1**:700-707

[53] Lyapina SA, Correll CC, Kipreos ET, Deshaies RJ. Human CUL1 forms an evolutionarily conserved ubiquitin ligase complex (SCF) with SKP1 and an F-box protein. Proceedings of the National Academy of Sciences of the United States of America. 1998;**95**: 7451-7456

[54] Nakayama K et al. Skp2-mediated degradation of p27 regulates progression into mitosis. Developmental Cell. 2004; **6**:661-672

[55] Carrano AC, Eytan E, Hershko A, Pagano M. SKP2 is required for ubiquitin-mediated degradation of the CDK inhibitor p27. Nature Cell Biology. 1999;**1**:193-199

[56] Ungermannova D, Gao Y, Liu X. Ubiquitination of p27Kip1 requires physical interaction with cyclin E and probable phosphate recognition by SKP2. The Journal of Biological Chemistry. 2005;**280**:30301-30309

[57] Zhang H, Kobayashi R, Galaktionov K, Beach D. p19Skp1 and p45Skp2 are essential elements of the cyclin A-CDK2 S phase kinase. Cell. 1995;**82**:915-925

[58] Yu ZK, Gervais JL, Zhang H. Human CUL-1 associates with the SKP1/SKP2 complex and regulates p21(CIP1/WAF1) and cyclin D proteins. Proceedings of the National Academy of Sciences of the United States of America. 1998;**95**:11324-11329

[59] Inuzuka H et al. Acetylation-dependent regulation of Skp2 function. Cell. 2012;**150**:179-193

[60] Frescas D, Pagano M. Deregulated proteolysis by the F-box proteins SKP2 and beta-TrCP: Tipping the scales of cancer. Nature Reviews. Cancer. 2008;**8**: 438-449

[61] von der Lehr N et al. The F-box protein Skp2 participates in c-Myc proteosomal degradation and acts as a cofactor for c-Myc-regulated transcription. Molecular Cell. 2003;**11**: 1189-1200

[62] Wei Z et al. Downregulation of Skp2 inhibits the growth and metastasis of gastric cancer cells in vitro and in vivo. Tumor Biology. 2013;**34**:181-192

[63] Hao B, Oehlmann S, Sowa ME, Harper JW, Pavletich NP. Structure of a Fbw7-Skp1-Cyclin E complex: Multisite-phosphorylated substrate recognition by SCF ubiquitin ligases. Molecular Cell. 2007;**26**:131-143

[64] Davis RJ, Welcker M, Clurman BE. Tumor suppression by the Fbw7 ubiquitin ligase: Mechanisms and opportunities. Cancer Cell. 2014;**26**: 455-464

[65] Tang X et al. Suprafacial orientation of the SCFCdc4 dimer accommodates multiple geometries for substrate ubiquitination. Cell. 2007;**129**:1165-1176

[66] O'Neil J et al. FBW7 mutations in leukemic cells mediate NOTCH pathway activation and resistance to gamma-secretase inhibitors. The Journal of Experimental Medicine. 2007;**204**: 1813-1824

[67] Yeh C-H, Bellon M, Nicot C. FBXW7: A critical tumor suppressor of human cancers. Molecular Cancer. 2018; **17**:115

[68] Akhoondi S et al. *FBXW7/hCDC4* is a general tumor suppressor in human cancer. Cancer Research. 2007;**67**: 9006-9012

[69] Strohmaier H et al. Human F-box protein hCdc4 targets cyclin E for proteolysis and is mutated in a breast cancer cell line. Nature. 2001;**413**: 316-322

[70] Welcker M et al. Multisite phosphorylation by Cdk2 and GSK3 controls cyclin E degradation. Molecular Cell. 2003;**12**:381-392

[71] Welcker M et al. Fbw7 dimerization determines the specificity and robustness of substrate degradation. Genes & Development. 2013;**27**: 2531-2536

[72] Yada M et al. Phosphorylation-dependent degradation of c-Myc is mediated by the F-box protein Fbw7. The EMBO Journal. 2004;**23**:2116-2125

[73] Babaei-Jadidi R et al. FBXW7 influences murine intestinal homeostasis and cancer, targeting Notch, Jun, and DEK for degradation. The Journal of Experimental Medicine. 2011;**208**:295-312

[74] Mao J-H et al. FBXW7 targets mTOR for degradation and cooperates with PTEN in tumor suppression. Science. 2008;**321**:1499-1502

[75] Welcker M, Clurman BE. The SV40 large T antigen contains a decoy phosphodegron that mediates its interactions with Fbw7/hCdc4. The Journal of Biological Chemistry. 2005; **280**:7654-7658

[76] Nakayama K et al. Impaired degradation of inhibitory subunit of NF-kappa B (I kappa B) and beta-catenin as a result of targeted disruption of the beta-TrCP1 gene. Proceedings of the National Academy of Sciences of the United States of America. 2003;**100**: 8752-8757

[77] Zheng N, Zhou Q, Wang Z, Wei W. Recent advances in SCF ubiquitin ligase complex: Clinical implications. Biochimica et Biophysica Acta. 2016; **1866**:12-22

[78] Watanabe N et al. M-phase kinases induce phospho-dependent ubiquitination of somatic Wee1 by SCFbeta-TrCP. Proceedings of the National Academy of Sciences of the United States of America. 2004;**101**: 4419-4424

[79] Guardavaccaro D et al. Control of meiotic and mitotic progression by the F box protein beta-Trcp1 in vivo. Developmental Cell. 2003;**4**:799-812

[80] Busino L et al. Degradation of Cdc25A by β-TrCP during S phase and in response to DNA damage. Nature. 2003;**426**:87-91

[81] Kanemori Y, Uto K, Sagata N. Beta-TrCP recognizes a previously undescribed nonphosphorylated destruction motif in Cdc25A and Cdc25B phosphatases. Proceedings of the National Academy of Sciences of the United States of America. 2005;**102**: 6279-6284

[82] Lammer C et al. The cdc25B phosphatase is essential for the G2/M phase transition in human cells. Journal of Cell Science. 1998;**111**(Pt 16): 2445-2453

[83] Shirane M, Hatakeyama S, Hattori K, Nakayama K, Nakayama K. Common pathway for the ubiquitination of IkappaBalpha, IkappaBbeta, and IkappaBepsilon mediated by the F-box protein FWD1. The Journal of Biological Chemistry. 1999;**274**:28169-28174

[84] Inuzuka H et al. Phosphorylation by casein kinase I promotes the turnover of the Mdm2 oncoprotein via the SCF (beta-TRCP) ubiquitin ligase. Cancer Cell. 2010;**18**:147-159

[85] Satyanarayana A, Kaldis P. Mammalian cell-cycle regulation: Several Cdks, numerous cyclins and diverse compensatory mechanisms. Oncogene. 2009;**28**:2925-2939

[86] Malumbres M, Barbacid M. Cell cycle, CDKs and cancer: A changing paradigm. Nature Reviews. Cancer. 2009;**9**:153-166

[87] Zhang J, Wan L, Dai X, Sun Y, Wei W. Functional characterization of anaphase promoting complex/cyclosome (APC/C) E3 ubiquitin ligases in tumorigenesis. Biochimica et Biophysica Acta. 2014;**1845**:277-293

[88] Chang L, Zhang Z, Yang J, McLaughlin SH, Barford D. Atomic structure of the APC/C and its mechanism of protein ubiquitination. Nature. 2015;**522**:450-454

[89] Primorac I, Musacchio A. Panta rhei: The APC/C at steady state. The Journal of Cell Biology. 2013;**201**: 177-189

[90] Williamson A et al. Identification of a physiological E2 module for the human anaphase-promoting complex. Proceedings of the National Academy of Sciences of the United States of America. 2009;**106**: 18213-18218

[91] Pines J. Cubism and the cell cycle: The many faces of the APC/C. Nature Reviews. Molecular Cell Biology. 2011; **12**:427-438

[92] Davey NE, Morgan DO. Building a regulatory network with short linear sequence motifs: Lessons from the degrons of the anaphase-promoting complex. Molecular Cell. 2016;**64**:12-23

[93] He J et al. Insights into degron recognition by APC/C coactivators from the structure of an Acm1-Cdh1 complex. Molecular Cell. 2013;**50**:649-660

[94] Wu W et al. CDC20 overexpression predicts a poor prognosis for patients with colorectal cancer. Journal of Translational Medicine. 2013;**11**:142

[95] Lehman NL et al. Oncogenic regulators and substrates of the anaphase promoting complex/ cyclosome are frequently overexpressed in malignant tumors. The American Journal of Pathology. 2007;**170**: 1793-1805

[96] Ishizawa J et al. FZR1 loss increases sensitivity to DNA damage and consequently promotes murine and human B-cell acute leukemia. Blood. 2017;**129**:1958-1968

[97] Gyrd-Hansen M, Meier P. IAPs: From caspase inhibitors to modulators of NF-κB, inflammation and cancer. Nature Reviews. Cancer. 2010;**10**: 561-574

[98] Fong WG et al. Expression and genetic analysis of XIAP-associated factor 1 (XAF1) in cancer cell lines. Genomics. 2000;**70**:113-122

[99] Tamm I et al. Expression and prognostic significance of IAP-family genes in human cancers and myeloid leukemias. Clinical Cancer Research. 2000;**6**:1796-1803

[100] Paulsen M et al. Interaction with XIAP prevents full caspase-3/-7 activation in proliferating human tlymphocytes. European Journal of Immunology. 2008;**38**:1979-1987

[101] Shiozaki EN et al. Mechanism of XIAP-mediated inhibition of caspase-9. Molecular Cell. 2003;**11**:519-527

[102] Suzuki Y, Nakabayashi Y, Takahashi R. Ubiquitin-protein ligase activity of X-linked inhibitor of apoptosis protein promotes proteasomal degradation of caspase-3 and enhances its anti-apoptotic effect in Fas-induced cell death. Proceedings of the National Academy of Sciences. 2001;**98**: 8662-8667

[103] Morizane Y, Honda R, Fukami K, Yasuda H. X-linked inhibitor of apoptosis functions as ubiquitin ligase toward mature caspase-9 and cytosolic Smac/DIABLO. Journal of Biochemistry. 2005;**137**:125-132

[104] Creagh EM, Murphy BM, Duriez PJ, Duckett CS, Martin SJ. Smac/Diablo antagonizes ubiquitin ligase activity of inhibitor of apoptosis proteins. The Journal of Biological Chemistry. 2004; **279**:26906-26914

[105] Broemer M et al. Systematic In vivo RNAi analysis identifies IAPs as NEDD8-E3 ligases. Molecular Cell. 2010;**40**:810-822

[106] Burstein E et al. A novel role for XIAP in copper homeostasis through

regulation of MURR1. The EMBO Journal. 2004;**23**:244-254

[107] Ganesh L et al. The gene product Murr1 restricts HIV-1 replication in resting CD4+ lymphocytes. Nature. 2003;**426**:853-857

[108] Lu M et al. XIAP induces NF-κB activation via the BIR1/TAB1 interaction and BIR1 dimerization. Molecular Cell. 2007;**26**:689-702

[109] Arora V et al. Degradation of survivin by the X-linked inhibitor of apoptosis (XIAP)-XAF1 complex. The Journal of Biological Chemistry. 2007; **282**:26202-26209

[110] Mehrotra S et al. IAP regulation of metastasis. Cancer Cell. 2010;**17**:53-64

[111] Dogan T et al. X-linked and cellular IAPs modulate the stability of C-RAF kinase and cell motility. Nature Cell Biology. 2008;**10**:1447-1455

[112] Huang X, Wu Z, Mei Y, Wu M. XIAP inhibits autophagy via XIAP-Mdm2-p53 signalling. The EMBO Journal. 2013;**32**:2204-2216

[113] Hanson AJ et al. XIAP monoubiquitylates Groucho/TLE to promote canonical Wnt signaling. Molecular Cell. 2012;**45**:619-628

[114] Lawlor KE et al. XIAP loss triggers RIPK3- and caspase-8-driven IL-1β activation and cell death as a consequence of TLR-MyD88-induced cIAP1-TRAF2 degradation. Cell Reports. 2017;**20**:668-682

[115] Riley BE et al. Structure and function of Parkin E3 ubiquitin ligase reveals aspects of RING and HECT ligases. Nature Communications. 2013; **4**:1982

[116] Cesari R et al. Parkin, a gene implicated in autosomal recessive juvenile parkinsonism, is a candidate tumor suppressor gene on chromosome 6q25-q27. Proceedings of the National Academy of Sciences. 2003;**100**: 5956-5961

[117] Fujiwara M et al. Parkin as a tumor suppressor gene for hepatocellular carcinoma. Oncogene. 2008;**27**: 6002-6011

[118] Poulogiannis G et al. PARK2 deletions occur frequently in sporadic colorectal cancer and accelerate adenoma development in Apc mutant mice. Proceedings of the National Academy of Sciences of the United States of America. 2010;**107**:15145-15150

[119] Chourasia AH, Boland ML, Macleod KF. Mitophagy and cancer. Cancer & Metabolism. 2015;**3**:4

[120] Suen D-F, Narendra DP, Tanaka A, Manfredi G, Youle RJ. Parkin overexpression selects against a deleterious mtDNA mutation in heteroplasmic cybrid cells. Proceedings of the National Academy of Sciences. 2010;**107**:11835-11840

[121] Gupta A et al. PARK2 depletion connects energy and oxidative stress to PI3K/Akt activation via PTEN S-nitrosylation. Molecular Cell. 2017;**65**: 999-1013.e7

[122] Yeo CWS et al. Parkin pathway activation mitigates glioma cell proliferation and predicts patient survival. Cancer Research. 2012;**72**: 2543-2553

[123] Staropoli JF et al. Parkin is a component of an SCF-like ubiquitin ligase complex and protects postmitotic neurons from kainate excitotoxicity. Neuron. 2003;**37**:735-749

[124] Gong Y et al. Pan-cancer genetic analysis identifies PARK2 as a master regulator of G1/S cyclins. Nature Genetics. 2014;**46**:588-594

[125] Lee SB et al. Parkin regulates mitosis and genomic stability through Cdc20/Cdh1. Molecular Cell. 2015;**60**: 21-34

[126] Sun X et al. Parkin deficiency contributes to pancreatic tumorigenesis by inducing spindle multipolarity and misorientation. Cell Cycle. 2013;**12**: 1133-1141

[127] Tay S-P et al. Parkin enhances the expression of cyclin-dependent kinase 6 and negatively regulates the proliferation of breast cancer cells. The Journal of Biological Chemistry. 2010; **285**:29231-29238

[128] Wang H et al. PARK2 negatively regulates the metastasis and epithelial-mesenchymal transition of glioblastoma cells via ZEB1. Oncology Letters. 2017; **14**:2933-2939

[129] Liu J et al. Parkin targets HIF-1α for ubiquitination and degradation to inhibit breast tumor progression. Nature Communications. 2017;**8**:1823

[130] Zhang C et al. Parkin, a p53 target gene, mediates the role of p53 in glucose metabolism and the Warburg effect. Proceedings of the National Academy of Sciences of the United States of America. 2011;**108**:16259-16264

[131] Li C et al. PINK1 and PARK2 suppress pancreatic tumorigenesis through control of mitochondrial iron-mediated immunometabolism. Developmental Cell. 2018;**46**:441-455.e8

[132] Frank S, Nelson P, Vasioukhin V. Recent advances in prostate cancer research: Large-scale genomic analyses reveal novel driver mutations and DNA repair defects. F1000 Research. 2018;**7**: 1173

[133] Zhuang M et al. Structures of SPOP-substrate complexes: Insights into molecular architectures of BTB-Cul3 ubiquitin ligases. Molecular Cell. 2009; **36**:39-50

[134] van Geersdaele LK et al. Structural basis of high-order oligomerization of the cullin-3 adaptor SPOP. Acta Crystallographica. Section D, Biological Crystallography. 2013;**69**:1677-1684

[135] Pierce WK et al. Multiple weak linear motifs enhance recruitment and processivity in SPOP-mediated substrate ubiquitination. Journal of Molecular Biology. 2016;**428**:1256-1271

[136] Zhang P et al. Endometrial cancer-associated mutants of SPOP are defective in regulating estrogen receptor-α protein turnover. Cell Death & Disease. 2015;**6**:e1687

[137] Zhang J et al. Cyclin D-CDK4 kinase destabilizes PD-L1 via cullin 3-SPOP to control cancer immune surveillance. Nature. 2017;**553**:91-95

[138] Kwon JE et al. BTB domain-containing speckle-type POZ protein (SPOP) serves as an adaptor of Daxx for ubiquitination by Cul3-based ubiquitin ligase. The Journal of Biological Chemistry. 2006;**281**:12664-12672

[139] Li G et al. SPOP promotes tumorigenesis by acting as a key regulatory hub in kidney cancer. Cancer Cell. 2014;**25**:455-468

[140] Zhu H et al. SPOP E3 ubiquitin ligase adaptor promotes cellular senescence by degrading the SENP7 deSUMOylase. Cell Reports. 2015;**13**: 1183-1193

[141] Geng C et al. SPOP regulates prostate epithelial cell proliferation and promotes ubiquitination and turnover of c-MYC oncoprotein. Oncogene. 2017; **36**:4767-4777

[142] Tan Y et al. Cullin 3 SPOP ubiquitin E3 ligase promotes the poly-ubiquitination and degradation of

HDAC6. Oncotarget. 2017;**8**: 47890-47901

[143] Wu F et al. Prostate cancer-associated mutation in SPOP impairs its ability to target Cdc20 for poly-ubiquitination and degradation. Cancer Letters. 2017;**385**:207-214

[144] Theurillat J-PP et al. Ubiquitylome analysis identifies dysregulation of effector substrates in SPOP-mutant prostate cancer. Science. 2014;**346**(80): 85-89

[145] Chen M-H et al. Cilium-independent regulation of Gli protein function by Sufu in hedgehog signaling is evolutionarily conserved. Genes & Development. 2009;**23**:1910-1928

[146] Cai H, Liu A. Spop promotes skeletal development and homeostasis by positively regulating Ihh signaling. Proceedings of the National Academy of Sciences. 2016;**113**:14751-14756

[147] Dai X et al. Prostate cancer-associated SPOP mutations confer resistance to BET inhibitors through stabilization of BRD4. Nature Medicine. 2017;**23**: 1063-1071

[148] Zhang P et al. Intrinsic BET inhibitor resistance in SPOP-mutated prostate cancer is mediated by BET protein stabilization and AKT–mTORC1 activation. Nature Medicine. 2017;**23**: 1055-1062

[149] Janouskova H et al. Opposing effects of cancer-type-specific SPOP mutants on BET protein degradation and sensitivity to BET inhibitors. Nature Medicine. 2017;**23**:1046-1054

[150] Geng C et al. Prostate cancer-associated mutations in speckle-type POZ protein (SPOP) regulate steroid receptor coactivator 3 protein turnover. Proceedings of the National Academy of Sciences. 2013;**110**:6997-7002

[151] Wang S et al. SAR405838: An optimized inhibitor of MDM2-p53 interaction that induces complete and durable tumor regression. Cancer Research. 2014;**74**:5855-5865

[152] Gan W et al. SPOP promotes ubiquitination and degradation of the ERG oncoprotein to suppress prostate cancer progression. Molecular Cell. 2015;**59**:917-930

[153] Boysen G et al. SPOP mutation leads to genomic instability in prostate cancer. eLife. 2015;**4**:1-18. e09207

[154] An J et al. Truncated ERG oncoproteins from TMPRSS2-ERG fusions are resistant to SPOP-mediated proteasome degradation. Molecular Cell. 2015;**59**:904-916

[155] Zhao Y, Sun Y. Cullin-RING ligases as attractive anti-cancer targets. Current Pharmaceutical Design. 2013;**19**: 3215-3225

[156] Vassilev LT et al. In vivo activation of the p53 pathway by small-molecule antagonists of MDM2. Science (80-). 2004;**303**:844-848

[157] Carvajal LA et al. Dual inhibition of MDMX and MDM2 as a therapeutic strategy in leukemia. Science Translational Medicine. 2018;**10**: eaao3003

[158] Weisberg E et al. Smac mimetics: Implications for enhancement of targeted therapies in leukemia. Leukemia. 2010;**24**:2100-2109

[159] Nikolovska-Coleska Z et al. Discovery of embelin as a cell-permeable, small-molecular weight inhibitor of XIAP through structure-based computational screening of a traditional herbal medicine three-dimensional structure database. Journal of Medicinal Chemistry. 2004;**47**:2430-2440

[160] Guo Z-Q et al. Small-molecule targeting of E3 ligase adaptor SPOP in kidney cancer. Cancer Cell. 2016;**30**: 474-484

[161] Yamagishi Y et al. Natural product-like macrocyclic N-methyl-peptide inhibitors against a ubiquitin ligase uncovered from a ribosome-expressed de novo library. Chemistry & Biology. 2011;**18**:1562-1570

[162] Ganoth D et al. The cell-cycle regulatory protein Cks1 is required for SCFSkp2-mediated ubiquitinylation of p27. Nature Cell Biology. 2001;**3**:321-324

[163] Pavlides SC et al. Inhibitors of SCF-Skp2/Cks1 E3 ligase block estrogen-induced growth stimulation and degradation of nuclear p27 ^{kip1}: Therapeutic potential for endometrial cancer. Endocrinology. 2013;**154**: 4030-4045

[164] Ungermannova D et al. High-throughput screening AlphaScreen assay for identification of small-molecule inhibitors of ubiquitin E3 ligase $SCF^{Skp2-Cks1}$. Journal of Biomolecular Screening. 2013;**18**: 910-920

[165] Chen Q et al. Targeting the p27 E3 ligase SCF(Skp2) results in p27- and Skp2-mediated cell-cycle arrest and activation of autophagy. Blood. 2008; **111**:4690-4699

[166] Beaudenon S, Huibregtse JM. HPV E6, E6AP and cervical cancer. BMC Biochemistry. 2008;**9**:S4

[167] Wade M, Li Y-C, Wahl GM. MDM2, MDMX and p53 in oncogenesis and cancer therapy. Nature Reviews. Cancer. 2013;**13**:83-96

[168] Nakayama KI, Nakayama K. Ubiquitin ligases: Cell-cycle control and cancer. Nature Reviews. Cancer. 2006; **6**:369-381

[169] Lau AW, Fukushima H, Wei W. The Fbw7 and betaTRCP E3 ubiquitin ligases and their roles in tumorigenesis. Frontiers in Bioscience (Landmark Ed.). 2012;**17**:2197-2212

[170] Manchado E et al. Targeting mitotic exit leads to tumor regression In vivo: Modulation by Cdk1, Mastl, and the PP2A/B55α,δ phosphatase. Cancer Cell. 2010;**18**:641-654

[171] Wäsch R, Robbins JA, Cross FR. The emerging role of APC/CCdh1 in controlling differentiation, genomic stability and tumor suppression. Oncogene. 2010;**29**:1-10

[172] Xu L, Lin D, Yin D, Koeffler HP. An emerging role of PARK2 in cancer. Journal of Molecular Medicine. 2014;**92**: 31-42

[173] Baleja JD et al. Identification of inhibitors to papillomavirus typ. 16 E6 protein based on three-dimensional structures of interacting proteins. Antiviral Research. 2006;**72**:49-59

[174] Vu B et al. Discovery of RG7112: A small-molecule MDM2 inhibitor in clinical development. ACS Medicinal Chemistry Letters. 2013;**4**:466-469

[175] Ding Q et al. Discovery of RG7388, a potent and selective p53–MDM2 inhibitor in clinical development. Journal of Medicinal Chemistry. 2013; **56**:5979-5983

[176] Zhao Y et al. A potent small-molecule inhibitor of the MDM2–p53 interaction (MI-888) achieved complete and durable tumor regression in mice. Journal of Medicinal Chemistry. 2013; **56**:5553-5561

[177] Sun D et al. Discovery of AMG 232, a potent, selective, and orally bioavailable MDM2–p53 inhibitor in clinical development. Journal of Medicinal Chemistry. 2014;**57**: 1454-1472

[178] Jeay S et al. A distinct p53 target gene set predicts for response to the selective p53-HDM2 inhibitor NVP-CGM097. eLife. 2015;**4**:1-23. e06498

[179] Furet P et al. Discovery of a novel class of highly potent inhibitors of the p53–MDM2 interaction by structure-based design starting from a conformational argument. Bioorganic & Medicinal Chemistry Letters. 2016;**26**: 4837-4841

[180] Yuan Y, Liao Y-M, Hsueh C-T, Mirshahidi HR. Novel targeted therapeutics: Inhibitors of MDM2, ALK and PARP. Journal of Hematology & Oncology. 2011;**4**:16

[181] Chan C-H et al. Pharmacological inactivation of Skp2 SCF ubiquitin ligase restricts cancer stem cell traits and cancer progression. Cell. 2013;**154**: 556-568

[182] Wu L et al. Specific small molecule inhibitors of Skp2-mediated p27 degradation. Chemistry & Biology. 2012;**19**:1515-1524

[183] Huang H-L et al. Triggering Fbw7-mediated proteasomal degradation of c-Myc by oridonin induces cell growth inhibition and apoptosis. Molecular Cancer Therapeutics. 2012;**11**:1155-1165

[184] Blees JS et al. Erioflorin stabilizes the tumor suppressor Pdcd4 by inhibiting its interaction with the E3-ligase β-TrCP1. PLoS One. 2012;**7**: e46567

[185] Nakajima H, Fujiwara H, Furuichi Y, Tanaka K, Shimbara N. A novel small-molecule inhibitor of NF-κB signaling. Biochemical and Biophysical Research Communications. 2008;**368**: 1007-1013

[186] Sackton KL et al. Synergistic blockade of mitotic exit by two chemical inhibitors of the APC/C. Nature. 2014; **514**:646-649

[187] Zeng X et al. Pharmacologic inhibition of the anaphase-promoting complex induces a spindle checkpoint-dependent mitotic arrest in the absence of spindle damage. Cancer Cell. 2010;**18**: 382-395

[188] Sharma SK, Straub C, Zawel L. Development of peptidomimetics targeting IAPs. International Journal of Peptide Research and Therapeutics. 2006;**12**:21-32

[189] McManus DC et al. Loss of XIAP protein expression by RNAi and antisense approaches sensitizes cancer cells to functionally diverse chemotherapeutics. Oncogene. 2004;**23**: 8105-8117

Chapter 4

The Role of Lysine 63-Linked Ubiquitylation in Health and Disease

Paola Pontrelli, Francesca Conserva and Loreto Gesualdo

Abstract

A specific subfamily within the E2 protein family is involved in the synthesis of noncanonical poly-ubiquitin chains, linked through lysine 63 residues. The role of lysine 63-linked polyubiquitylation in diseases has emerged only recently. Under physiological conditions, this process does not seem to be involved in the classical protein degradation by the proteasome, but it is involved in the regulation of intracellular signaling, DNA damage response, cellular trafficking, and lysosomal targeting. The alteration of this process has been described in a number of pathological conditions, including immune disorders, diabetes, and cancer. In this chapter, we will describe the role of lysine 63-linked ubiquitylation in the regulation of diverse signaling pathways involved in cell behavior. We will also describe some pathological conditions in which altered lysine 63-linked ubiquitylation has been referred to play an important role.

Keywords: lysine 63-linked ubiquitylation, immune system, diabetes complications, autophagy, cancer

1. Introduction

The ubiquitin signaling system, often referred to as "ubiquitin code", is very complex, although the ubiquitin moieties engaged in protein ubiquitylation are always identical. In order to understand the complexity of the ubiquitin code, we need to remember that the ubiquitin moiety contains seven lysine residues, all of which can be potentially engaged in the formation of polyubiquitin chains, and the protein fate depends upon the specific lysine residue involved in the polyubiquitin link as well as the length of the polyubiquitin chain [1, 2].

Several types of polyubiquitin chains exist in cells and the type of chain defines how ubiquitinated proteins are regulated. For instance, we know that ubiquitin chains generated via lysine (K)48 of ubiquitin (K48 chains) function as a signal for proteolysis, while chains generated via K63 (K63 chains) are involved in nonproteolytic functions, such as DNA repair, protein kinase activation, and membrane trafficking [3].

The ubiquitin signaling starts with the activation of the ubiquitin moiety by an ubiquitin-activating enzyme E1, and the energy to initiate this process is provided by an ATP molecule. Exploiting its active site containing a cysteine, E1s attack the ubiquitin-AMP intermediate, forming a thioester bond. The subsequent reaction

IntechOpen

involves the transfer of the activated ubiquitin from the E1 to an E2 enzyme through a transthioesterification reaction; E2 enzymes also contain an active site that includes a cysteine residue. The final step is carried out by ubiquitin protein ligase E3 enzymes, which allows the transfer of ubiquitin to the lysine of the target protein [4, 5]. Although target specificity is given by E3 enzymes, not all E3s are able to ligate the ubiquitin molecule to their target directly, as this ability is dependent on the type of active site they are furnished with. There are two main types of catalytic domains on E3 enzymes. The E3s possessing a RING (really interesting new gene) domain in their active site do not contain a cysteine residue within the active-site; thus, they work by bringing the "charged" E2 in close proximity to the target protein, and ubiquitin is transferred by the E2. E3s that possess a HECT domain [homology to E6-AP carboxyl terminus] instead, contain a cysteine residue in their active site that allows the formation of a thioester intermediate and the subsequent transfer of ubiquitin to the target protein [6, 7].

In summary, the enzymatic cascade that leads to the formation of an isopeptide bond between ubiquitin and its target molecule involves E1, E2, and E3 enzymes, and the ubiquitin chain linkage specificity is generally conferred by E2s.

Among the E2s that participate in the specific formation of K63-linked ubiquitin chains, Ubc13 (also known as Ube2n) is probably the best characterized. To finalize the reaction, Ubc13 requires the concomitant presence of specific E2-like partner proteins, among which Ube2v1 (also known as Uev1A) is involved in the formation of K63-linked ubiquitin chains in the cytosol, while the protein Mms2 participates in the nuclear K63-linked chain formation. It was also shown that this enzymatic complex exerts its activity on previously mono-ubiquitinated substrates; thus, the priming of the substrate with the first ubiquitin molecule can likely be promoted by different E2s [8].

Overall, the complexity of the cellular responses elicited by ubiquitin is actually greater than previously foreseen; it was recently discovered that heterogeneous ubiquitin chains also exist, possessing both proteolytic and nonproteolytic functions. Also, ubiquitin itself can be modified through phosphorylation and/or acetylation [9, 10]. Finally, deubiquitinating enzymes (DUBs) exist which operate through either the editing or disassembly of ubiquitin chains, allowing the fine-tuning of the entire system [11].

Importantly, both the proteolytic and nonproteolytic functions of ubiquitin are crucial to regulate different intracellular signaling pathways involved in the modulation of immunity, inflammation, and cell survival [12]. Here we report an overview on the main pathways modulated by K63-linked ubiquitylation and the role of this post-translational modification in health and disease.

2. Physiological roles of lysine 63-linked ubiquitylation

2.1 Lysine 63-linked ubiquitylation in NF-κB signaling

NF-κB is a dimeric transcription factor that controls cytokine production and cell growth, ultimately modulating processes such as inflammation and immune responses. This protein complex is ubiquitously present in the cell cytoplasm in an inactive state but it can rapidly be processed into its active form by different regulatory mechanisms including ubiquitylation.

The canonical ubiquitylation trough K48-linked chains controls processing of the NF-κB precursor p100 and p105, as well as degradation of the NF-κB inhibitor IκB when it becomes phosphorylated by the IκB kinase (IKK) complex. The noncanonical ubiquitylation through K63-linked chains is instead involved in the activation of the IκB kinase complex IKK [13]. These events are mediated, among others, by a class of E3 enzymes known as the TNF receptor-associated factor (TRAF). TRAF proteins mediate NF-κB activation from a number of receptors such as the TNF receptor (TNFR), IL-1 receptor (IL-1R), and Toll-like receptor (TLR). TRAF6 transduces signals from IL1-R/TLR while TRAF2 transduces signals from TNFR. Both TRAF-2 and TRAF-6 are able to form K63-linked ubiquitin chains on their specific targets [14].

2.1.1 IL1-R/TLR-induced NF-κB signaling

In the NF-κB signaling triggered by activation of the IL1-R/TLR, the E2 protein complex consisting of Ubc13-Uev1A, in concert with the E3 enzyme TRAF6, leads to the formation of K63-linked ubiquitin chains on several target proteins including Interleukin 1 Receptor Associated Kinase 1 (IRAK1), NF-kappa-B essential modulator (NEMO), and TRAF6 itself [15]. Once ubiquitinated, TRAF6 can be recognized by a specific ubiquitin binding domain (UBD) within the protein TGF-β Activated Kinase 1 (MAP3K7) Binding Protein 2 (TAB2) and this interaction activates the TAB2-associated TAK1 kinase, which in turn phosphorylates and activates IKK. Active IKK promotes degradation of IκB and ultimately releases inhibition on NF-κB (**Figure 1**). It was shown that TAK1 can activate IKK only in the presence of the NF-κB essential modulator NEMO. Importantly, this protein contains a C-terminal domain that binds preferentially to K63 ubiquitin chains. Thus, this type of ubiquitylation might be useful to provide a scaffold that facilitates protein interactions [14, 16, 17].

2.1.2 TNFR-induced NF-κB signaling

NF-κB activation can also be triggered by molecules that bind TNF-R1. As for IL-1/TLR, TNF-R1 signal transduction involves a cascade of reactions that are regulated at different levels by protein ubiquitylation.

The model proposed for NF-κB activation suggests that upon exposure to TNF-α, TNF-R1 undergoes a conformational change that allows recruitment of the adaptor

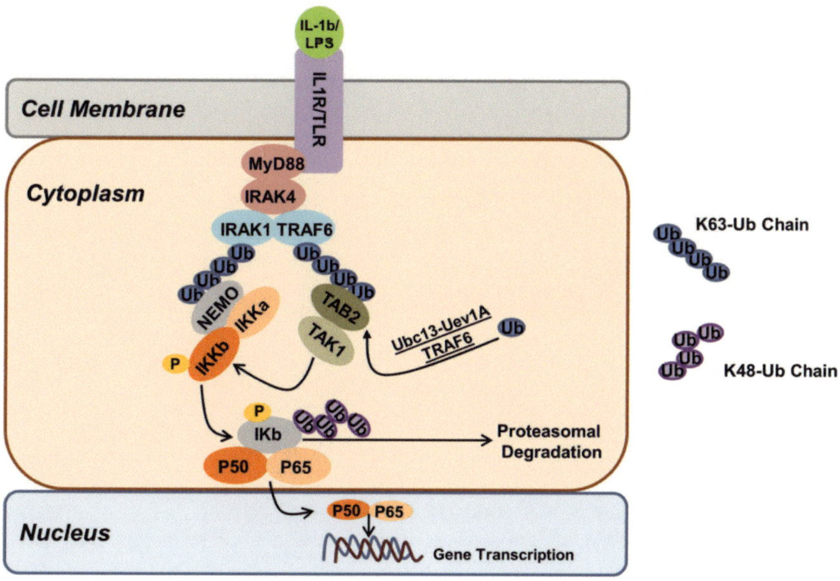

Figure 1.
IL1-R/TLR-induced NF-κB signaling.

tumor necrosis factor receptor type 1-associated death domain protein (TRADD) within the cytosol. This interaction elicits the enrollment of two additional proteins: TRAF2 and kinase receptor-interacting serine/threonine-protein kinase 1 (RIPK1). Unlike TRAF6, TRAF2 is unable to attach ubiquitin moieties to RIPK1 independently, but acts as a scaffold for the recruitment of two different E3 enzymes: cellular inhibitor of apoptosis protein-1 (c-IAP1) and c-IAP2. These proteins add, among others, K63-linked chains on RIPK1; thus, they are the actual effectors of RIPK1 ubiquitylation. The following events that lead to NF-κB activation involve the formation of a second E3 protein complex at the initial site of ubiquitylation known as linear ubiquitin chain assembly complex (LUBAC). In linear ubiquitin chains, the C-ter Gly76 of one ubiquitin is linked to the α-NH2 group of Met1 of another ubiquitin moiety. Thus, the LUBAC complex catalyzes the formation of linear (M1)-linked ubiquitin chains on RIPK1 proteins, c-IAP1 and c-IAP2. Once these linear chains are added, RIPK1 is structurally able to attract the kinase complexes TAK1 and IKK through their ubiquitin binding domain-containing subunits (TAB2/TAB3 and NEMO). This ultimately triggers IKK phosphorylation by TAK1 and NF-κB activation [18] (**Figure 2**). In summary, a variety of ubiquitin chain modifications seem to be required for NF-κB activation, and further studies will shed light on the many roles of each specific type of polyubiquitin chains.

2.2 Lysine 63-linked ubiquitylation in Wnt/β-catenin signaling

The Wnt/β-catenin signaling pathway is essential in the regulation of events such as cell proliferation, organized migration, self-renewal, and tissue polarity. Disruption of the Wnt/β-catenin signaling pathway has been linked with oncogenesis and other pathological conditions. Regulation of Wnt signaling is controlled by protein ubiquitylation at many levels, and K48- and K63-linked ubiquitin chains in particular have been shown to regulate this pathway through both proteolytic and nonproteolytic functions.

Within the canonical Wnt/β-catenin signaling, the absence of a Wnt ligand at the transmembrane receptor Frizzled (Fz) determines the rapid phosphorylation

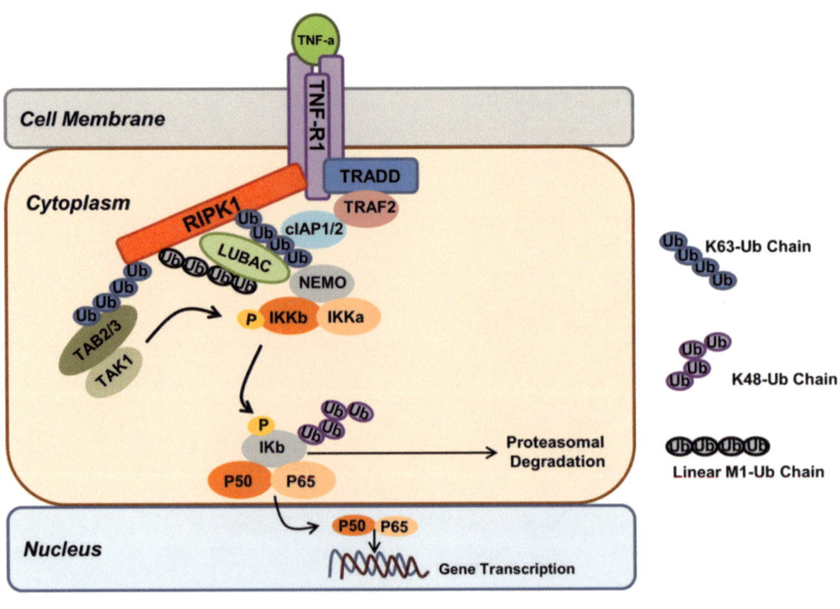

Figure 2.
TNFR1-induced NF-κB signaling.

of the free β-catenin in the cytosol. This reaction is catalyzed by a protein complex that includes the proteins axin, adenomatous polyposis coli (APC), casein kinase I (CK1), and glycogen synthase kinase 3 beta (GSK3β). When phosphorylated, β-catenin is recognized and ubiquitinated by a specific E3 ubiquitin ligase complex (SCF βTrCP), and degradation through the proteasome occurs (**Figure 3A**). Hence, one of the roles of the axin-APC complex is to maintain low cytosolic levels of β-catenin [19].

When a Wnt ligand binds a Fz receptor, in the presence of specific adapters known as low-density lipoprotein receptor-related proteins 5/6 (Lrp5/6), recruitment of the proteins Axin and Disheveled (DvI) occurs, and the formation of the protein complex that drives β-catenin degradation is inhibited. As a consequence, the levels of cytoplasmic β-catenin rise and the protein is transported into the nucleus where it forms a complex with the T cell factor (TCF) family of transcription factors, and activates transcription of its target genes [20] (**Figure 3B**).

A number of evidence demonstrate that K63-linked polyubiquitylation plays a role in the regulation of Wnt signaling although the precise molecular mechanisms that regulate these complex interactions have not been fully elucidated yet. We know that the formation of K63-linked polyubiquitin chains is promoted by the E2 protein complex Ubc13-Uev1a and recent evidence suggests that deletion of Ubc13 is associated with accumulation of β-catenin and increased transcription of Wnt target genes; however, the precise molecular mechanisms that drive these signaling events remain to be fully elucidated [21]. As largely known, deubiquitinating enzymes (DUBs) act in concert with ubiquitinating enzymes to precisely adjust the extent and duration of ubiquitin signals. Cellular experiments show that Trabid is a DUB protein able to reverse K63-linked hyperubiquitylation of the APC complex, thus acting as a positive regulator of the Wnt/β-catenin signaling [22].

Abnormal Wnt signaling underlies a wide range of pathological conditions in humans, including cancer. Hyperactivation of the Wnt pathway, for instance, is a characteristic of tumor cells from patients with cylindromatosis. These patients present mutations in the CYLD gene; this gene encodes for a DUB enzyme whose loss in human cells causes K63-linked hyperubiquitylation of the upstream Fz-binding effector protein Disheveled (Dvl) resulting in enhanced responsiveness to Wnt [23].

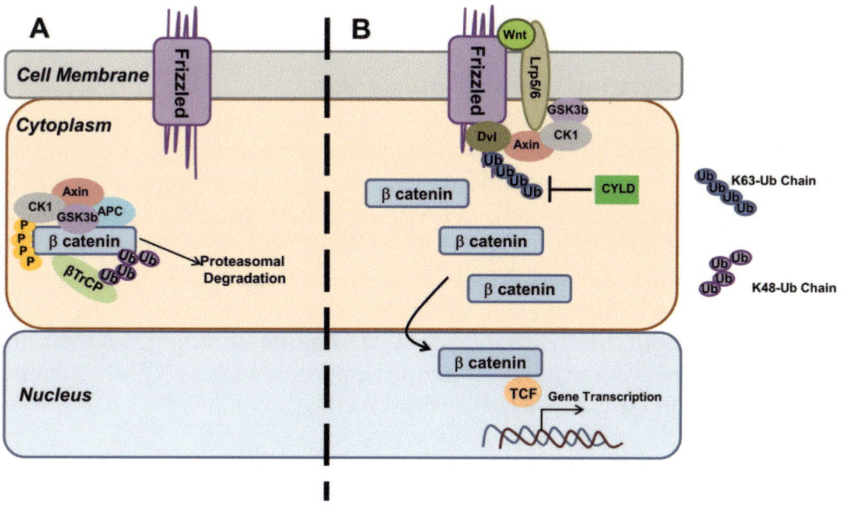

Figure 3.
Wnt/β-catenin signaling in the absence (A) or presence (B) of a Wnt ligand.

2.3 Lysine 63-linked ubiquitylation in membrane protein trafficking

Membrane proteins serve different functions: they allow cells to sense and/or interact with molecules in the extracellular space, confer a proper shape to the cell, regulate the osmotic pressure, and channel the passage of ions, endogenous compounds, xenobiotics, etc. Both the sorting and degradation of membrane-bound proteins are regulated at least in part by K63-linked protein ubiquitylation.

The synthesis of membrane protein takes place specifically on those ribosomes attached to the endoplasmic reticulum (ER). Once newly synthesized, proteins migrate from the inner lumen of the ER to the cis face of the Golgi apparatus (more proximal to the nucleus) where they undergo refinement and quality control to ensure proper folding. Within the Golgi apparatus, proteins are also sorted according to their predetermined cellular destination and finally secreted through the trans face (more distal to the nucleus). The migration of proteins through these cellular compartments is guided by lipid vesicles known as endosomes. Importantly, endosomes also coordinate the downregulation of cell-surface receptors through internalization of these proteins and subsequent degradation in the lysosomes.

A growing number of studies seem to suggest that K63-linked ubiquitin chains act as a signal for the internalization and intracellular sorting of integral membrane proteins [24]; more specifically, K63-linked ubiquitin chains have been shown to direct proteins to a specialized subclass of endosomes known as multivesicular bodies (MVBs). Once in the MVBs, proteins are either sent to the lysosomes for degradation, or secreted as exosomes via fusion with the plasma membrane; this process also serves to position membrane bound receptors to their specific location.

The first evidence that K63-linked ubiquitin chains could function as a signal to stimulate internalization of plasma membrane proteins through endocytosis and targeting into the lysosomal degradation pathway was acquired in *Saccharomyces cerevisiae* [25]. These preliminary observations led to the discovery of several mammalian proteins undergoing a similar regulatory mechanism, for example, the epidermal growth factor receptor (EGFR) [26], the human dopamine transporter (DAT) [27], the nerve growth factor receptor tyrosine receptor kinase A (TrkA) [28], major histocompatibility complex class I molecules [29], and the prolactin receptor [30] and possibly the low-density lipoprotein receptor (LDLR) [31].

3. Lysine 63-linked ubiquitylation in diseases

3.1 Lysine 63 ubiquitylation in immune disorders

Protein ubiquitylation has emerged as a key mechanism in the modulation of the immune system development and intensity of the immune responses [32, 33]. K63-linked ubiquitylation is involved in immune cell development since germline ablation of Ubc13 induces embryonic lethality [33]. Moreover, through the regulation of different intracellular signaling pathways, K63 ubiquitylation has emerged as critical for T cell differentiation [34]. K63 ubiquitylation chains have been described as fundamental for both the innate and adaptive immune systems given their involvement in master pathways controlling immune responses, such as NF-κB signaling [35] and MAPK activation [36], as previously described.

3.1.1 Lysine 63 ubiquitylation and adaptive immune response

The adaptive immune response, also called acquired immunity, refers to antigen-specific immune response; thus, antigen presentation induces the development of effective T- and B-cell responses. B-cells can be divided into memory B cells that express membrane-bound antibodies, or plasma B cells that can secrete antibodies to identify free pathogens circulating into the body. All B cells express a B cell receptor involved in antigen binding, internalization, and processing of antigens, other than activation of intracellular signaling pathways. T cells instead mature into the thymus where they start to express T cell receptors (TCRs) and CD4 and CD8 receptors. T cell receptors, assisted by CD4 or CD8 receptors, recognize antigens bound to certain major histocompatibility complex class 1 (MHCI) and class 2 (MHCII), expressed by antigen presenting cells such as macrophages and dendritic cells. Mature T cells can be mainly divided into Helper T cells, CD4+ cells involved in the activation of other immune cells, cytotoxic T cells, CD8+ that removes pathogens and infected cells, and T regulatory cells. T regulatory cells (Treg) play a central role in the regulation of the adaptive immune response and represent a T lymphocyte subpopulation that maintains tolerance to self-antigens and prevents autoimmune disease [37]. Tregs are produced in the thymus as a subpopulation of T cells and express a transcription factor (Forkhead box protein 3) involved in Treg development and function [38]. Tregs can also be induced from naive T cells in the periphery in the presence of transforming growth factor b (TGF-β).

K63 ubiquitylation is specifically involved in the suppressive function of Treg cells, and it has been described as fundamental for the immunosuppressive function of Tregs in murine models *in vivo* [39]. It is also well known that NF-κB signaling can modulate Treg cell differentiation [40]. Chang et al. demonstrated that Ubc13 deficiency in Treg, with the subsequent reduction in K63 ubiquitylation, impaired the *in vivo* suppressive function of these cells. Ubc13 deficiency in Ubc13$^{Treg-KO}$ mice, in fact, was able to influence the IKK signaling axis normally required for the expression of specific Treg functional factors, such as IL-10 and SOCS1 [39]. Both IL-10 and SOCS1 can specifically regulate Treg stability and inhibition activity.

Defects in Treg cells have been described in several human immune disorders including systemic lupus erythematosus (SLE). Treg disorders in SLE patients are characterized by abnormal peripheral tolerance that has been linked to a deficiency in the E3 ubiquitin ligase Cbl-b [41], involved in the regulation of T cell receptor signaling, during the induction of peripheral tolerance. Interestingly SLE patients were also characterized by an altered pattern of K63 ubiquitinated proteins in Tregs, with a decreased expression of K63 ubiquitinated proteins, related to increased pSTAT-3 expression [42]. These processes could be responsible for the loss of Treg suppressive capacity in SLE patients.

K63 ubiquitylation can also influence B cell receptor, T cell receptor, and IL-1 receptor (IL-1R)-mediated immune responses. Murine Ubc13-deficient T cells showed altered proliferation in response to diverse stimuli and impaired intracellular signaling altering the activation of both NF-κB and MAP kinases into T cells [43]. Also, murine Ubc13-deficient (Ubc13−/−) B cells showed impaired activation of the B cell receptor and CD40-induced activation, as well as Toll-like receptor mediated activation [44]. All this evidence underlies the meaning of this process in the mammalian immune response, thus indicating the importance of investigating K63 ubiquitylation in immune disorders.

3.1.2 Lysine 63 ubiquitylation and innate immune response

Innate immune response is triggered upon infections with pathogens such as bacteria, parasites, and viruses. It includes several mechanisms consisting in the physical and chemical barrier to infectious agents, in the activation of the complement cascade, and in the recruitment to the sites of infection of immune cells [such as macrophages and neutrophils] able to produce cytokines, thus inducing the inflammatory response. Innate immune response can also influence the adaptive immune system through antigen presentation.

It has been demonstrated that K63 ubiquitylation facilitates the innate immune signaling activated by diverse receptors such as Toll-like receptors, able to recognize pathogen components (lipopolysaccharide—LPS—from Gram-negative bacteria or lipoteichoic acid from Gram-positive bacteria), or cytokine receptors [45]. Also, in the regulation of the innate immune response, such as the adaptive immune response, the IKK-NF-κB pathway and MAPK activation plays a central role, since they are activated by the engagement of Toll-like receptors (TLRs).

An important component of the innate immune system is represented by natural killer (NK) cells, involved in the direct elimination of infected or transformed cells and able to secrete diverse cytokines, including IFN-γ, thus increasing the inflammatory response and the recruitment of immune cells. Also, IFN-γ production in NK cells is regulated by K63 ubiquitylation through its involvement in the NF-κB pathway [46], underlying once again the importance of this mechanism in the modulation of the immune response.

Ubiquitylation signaling has been described as specifically involved in the anti-viral innate immune response. In fact, the importance of the ubiquitylation pathway in the innate immune response has been validated by the discovery of some viruses encoding deubiquitinating proteases. Deubiquitinating enzymes (DUBs) catalyze the removal of ubiquitin from different cellular substrates, thus influencing several intracellular processes [47]. These deubiquitinating proteases produced by viruses, can lead to the suppression of the anti-viral immune response in order to promote viral replication [48].

Herpes Simplex Virus 1 (HSV1), a dsDNA virus belonging to the alpha-herpesvirus subfamily, can cause humans gingivostomatitis, cold sores, and herpetic keratitis. HSV1 dsDNA induces NF-κB signaling activation that promotes the anti-viral immune-response. However, to evade the innate immune system, these viruses encode for a DUB domain, called UL36USP, which is also similar to an open reading frame encoded by other viruses such as the human cytomegalovirus (HCMV) [49]. HCMV is a member of the beta-herpes virus subfamily, whose infection, normally asymptomatic, once reactivated can cause severe disease in immune-compromised and immune-suppressed individuals. Interestingly the UL36USP deubiquitinase activity inhibits NF-κB activation, by deubiquitinating IκBα, thus blocking its degradation and, consequently, finally quenching IFN production [50], a cytokine important for the anti-viral immune response.

Also, retroviral infections, including HIV-1, are mediated by modulation of the ubiquitylation system. In particular, the retrovirus factor TRIM5 (tripartite motif-containing protein 5) promotes innate immune signaling by activating, through K63 ubiquitin, MAP3K7 kinase complex with the subsequent stimulation of AP-1 and NF-κB signaling [51].

K63 ubiquitylation is also involved in bacterial cytoplasmic infections. These infections induce the cytosolic exposure of peptidoglycans and are characterized by the activation of the nuclear oligomerization domain 2 (NOD2) intracellular signaling. Polymorphisms in NOD2 have been associated with 15–30% of genetic Crohn's disease [52], an inflammatory bowel disease that may affect any part of the gastrointestinal tract, from mouth to anus, characterized by a dysfunctional immune response to normal microbiota [53]. It has been demonstrated that the activation by NOD2 of the K63-specific E3 ubiquitin ligase TRAF6, leading to NF-κB stimulation, is seriously compromised in Crohn's disease-patients with the NOD2 allele L1007insC [54]. In addition, NOD2 (nucleotide oligomerization domain 2) regulates the formation of K63-linked polyubiquitin chains on the I kappa kinase (IKK) scaffolding protein, NEMO.

Thus, ubiquitin-mediated regulation of the innate immune response could represent an important node in the management of pathogen infection and could symbolize a novel target for future therapies.

3.2 Lysine 63 ubiquitylation in diabetes and diabetic nephropathy

Diabetes mellitus is a metabolic disorder characterized by the reduction and altered function of pancreatic insulin-producing β-cells, and by organ damage [55]. Type 1 diabetes is induced by autoimmune destruction of β-cells responsible for insulin insufficiency, while type 2 diabetes is due to peripheral insulin resistance and subsequent β-cell expansion and hyperinsulinemia [56]. The number of diabetic patients is increasing decade by decade, and this high prevalence is registered worldwide with a projection of more than 438 million of diabetic patients with 7.8% prevalence by 2030 all over the world.

Chronic hyperglycemia and oxidative stress have been described as pro-apoptotic signals for pancreatic β-cell, thus influencing the metabolic state of these cells and the cell fate decisions [56]. Several intracellular signaling pathways can contribute to modulate β-cell function. Among these, also post-translational modification has been recognized to play a role. SUMOylation, a post-translational modification consisting in covalent attachment to target proteins of the small ubiquitin-like modifier (SUMO) peptides, has been recently described as a key event regulating β-cell survival and function [57]. Different cytokines can induce autoimmune destruction of pancreatic β-cell through the modulation of several intracellular signaling pathways characterized by the activation of phosphorylation and ubiquitylation events, including K63 ubiquitylation, in the cells. In fact, one of the factors involved into cytokine-mediated apoptosis of β-cell is represented by the mixed lineage kinase MLK3 [58], a pro-apoptotic factor involved in a cascade of events ultimately leading to mitochondrial outer membrane permeabilization, thus compromising mitochondrial integrity. Humphrey et al. demonstrated that IL-1β, one of the cytokines involved in the autoimmune destruction of pancreatic β-cells, stimulates K63-linked ubiquitylation of MLK3, thus promoting its activity and finally influencing the progression toward β-cell death [59]. Thus, K63 ubiquitylation could represent a potential target for therapeutic intervention in promoting β cell survival in diabetic patients.

Diabetes is also responsible for the insurgence of different complications such as retinopathy, cardiovascular diseases, and renal diseases. In the last years, the importance of the ubiquitylation pathway in diabetes and diabetic complications, such as cardiac diseases [60] and diabetic nephropathy [61–64], has emerged significantly [65].

Hyperglycemia, hypertension, and other hemodynamic changes intensify the filtration and reabsorption processes and this can lead to kidney failure progressing toward end-stage renal disease (ESRD). The incidence of ESRD due to diabetes varies among countries between 15 and 45%, with a mean value of 33%, which means that 33% of patients are starting renal replacement therapy.

Renal damage in type 2 diabetic patients can be characterized by different patterns including diabetic glomerulosclerosis, vascular and ischemic glomerular changes, and other glomerulonephritis in the presence or absence of diabetic lesions [66]. Pure diabetic nephropathy is characterized by mesangial proliferation, podocyte loss, glomerular basal membrane thickening, and nodular extracellular matrix accumulation with the classical Kimmelstiel-Wilson lesions.

The specific role of K63 ubiquitylation in diabetic nephropathy has been recently described [67]. Hyperglycemic conditions induce in tubular cells an increased expression of specifically K63-ubiquitinated proteins. Also, kidney biopsies from diabetic nephropathy patients are characterized by increased K63 ubiquitylation at tubular cells when compared to diabetic patients without renal damage or patients with other nephritides such as membranous nephropathy. Interestingly, increased K63 ubiquitylation in glucose-stimulated tubular cells was able to promote epithelial to mesenchymal transition, a process already described as involved in diabetic nephropathy dysfunction [68]. Epithelial to mesenchymal transition represents a potential source of myofibroblasts involved in the progression of kidney fibrosis. Also, in *in vivo* kidney biopsies of diabetic nephropathy patients, tubular cells characterized by increased accumulation of K63 ubiquitinated proteins were also characterized by expression of mesenchymal markers [67], thus underlying the importance of K63 ubiquitylation in the progression of renal fibrosis in diabetic patients.

Other than in epithelial-to-mesenchymal transition, hyperglycemia-induced K63 ubiquitylation is also involved in the apoptotic death of tubular cells through the deregulation of autophagy. Autophagy is an intracellular process involved in degradation of damaged proteins/organelle or in the intracellular response to nutrient deprivation, stress, and extracellular environmental changes. In human glomerulopathies, changes in the ubiquitin-proteasome system have been correlated with autophagy [69]. Impaired autophagy has also been described as a characteristic feature of diabetics [70], and recently, the molecular mechanisms responsible for this alteration have been correlated to hyperglycemia-induced K63 ubiquitylation [71]. In diabetic nephropathy patients *in vivo*, those tubules characterized by increased expression of the autophagic factor LC3 were also characterized by increase in K63 ubiquitinated protein accumulation. Interestingly, accumulation of autophagic particles into tubular cells, due to K63 ubiquitinated protein accumulation, could be responsible for increased apoptosis of these cells, as observed both *in vitro* and *in vivo* in kidney biopsies from diabetic nephropathy patients [70]. Taken together, all this evidence support the role of K63 ubiquitylation in the progression of tubular damage in diabetic nephropathy patients, which could be responsible for the progression of kidney fibrosis and for the induction of apoptosis of tubular cells with the consequent reduction of renal function.

3.3 Lysine 63 ubiquitylation and neurodegenerative disorders

Neurodegenerative disorders like Parkinson's disease or dementia or Alzheimer's disease are, in some cases, characterized by the presence of insoluble deposits in neurons containing components of the ubiquitin-proteasome system. It has been reported, in fact, that the ubiquitylation pathways play an important role in the pathogenesis of these diseases.

The most common cause of familial Parkinson's disease is characterized by mutations in Parkin [72], a ubiquitin ligase whose loss of function leads to both toxic accumulation of its substrates [73], and impaired formation of Lewy bodies, fibrillary masses of molecules implicated in the formation and degradation of alpha-synuclein aggregates [74]. Alpha-synuclein is a protein expressed in neurons involved in the formation of synaptic vesicles in presynaptic terminals and in the release of the dopamine, one of the brain's neurotransmitters. It has been demonstrated that K63 ubiquitylation plays an important role in the generation of these aggregates. Lim et al. observed that parkin-mediated ubiquitylation of proteins within Lewy-body-like inclusions was augmented by K63 ubiquitylation and occurs mainly through K63 linkages [75]. Interestingly, it has been demonstrated that the ubiquitin hydrolase UCHL1 is able to promote also K63-linked ubiquitylation of alpha-synuclein [76]. Thus, K63 ubiquitylation could represent a mechanism by which protein inclusion can occur and by which proteins are stabilized [77] forming aggregates observed in neurodegenerative disorders including Parkinson disease and dementia.

Interestingly, K63 ubiquitylation can contribute not only to inclusion biogenesis, but also to the clearance of inclusions. Intracellular alpha-synuclein can be ubiquitinated in K63 residues also by the E3 ubiquitin ligase Nedd4-1 [neural precursor cell expressed developmentally down-regulated protein 4-1], thus inducing its degradation by the endo-lysosomal pathway [78]. The first evidence of the role of K63 ubiquitylation in modulating autophagy clearance of aggregates in neurodegenerative disorders was described by Tan et al. in 2008 [79]. These authors demonstrated that K63 polyubiquitin chains linked to protein inclusions represent a target, driving aggregates to the clearance by autophagy.

Poly-ubiquitinated proteins can also be accumulated into the mitochondria during proteolytic stress and can be responsible for mitochondria-mediated cell death during proteasomal dysfunction, as demonstrated in an *in vitro* model of dopaminergic degeneration [80]. Interestingly, also monoamine oxidases (MAOs), located on the outer mitochondrial membrane and involved in the control of the neurotransmitters levels in the brain, can induce K63 ubiquitylation of mitochondrial proteins and promote autophagy of damaged organelles in neuroblastoma cells [81].

Taken together, these results evidenced the importance of K63 ubiquitylation in neurodegenerative disorders and open novel scenarios for the treatment of these diseases.

3.4 Lysine 63 ubiquitylation in cancer

The involvement of ubiquitin proteasome system dysregulation in the degradations of apoptotic proteins and subsequent induction of tumor formation has been well established, whereas, the finding that nonproteolytic ubiquitylation has a role in cancer and metastasis is recent.

With respect to K63-linked ubiquitylation, it was recently shown that both Ube2v1 and its partner Ubc13 are overexpressed in breast cancer, ovarian cancer, prostate cancer, and colorectal cancer, as well as in lymphoma [82, 83]. In breast cancer in particular, Ubc13 was identified as a key protein for metastasis spreading to the lung. This action appears to be mediated through TGF-β-induced activation of the TAK1-p38 MAP Kinase cascade. Notably, it was also demonstrated that *in vivo* inhibition of UBE2v1 through RNA interference can prevent breast tumor growth and metastasis formation [84, 85].

The role of Ubc13-Uev1A in tumorigenesis was also suggested by Pulvino et al., in accordance with the finding that small-molecule inhibitor of Ubc13-Uev1A interaction, known as NSC697923, can inhibit proliferation and survival of diffuse large B-cell lymphoma cells via inhibition of the Nf-KB signaling in these cells [86].

Additional proof that K63-linked ubiquitylation is involved in the regulation of tumorigenesis comes from the observation that several E3s and DUBs involved in this particular type of ubiquitylation/deubiquitylation are important regulators of proteins that guide cell cycle, DNA damage, and cell death [87–90]. For a complete review on the role of ubiquitylation on tumor formation and metastasis, refer to the work by Gallo et al. [83].

4. Conclusions

The role of post-translational modification in the regulation of cell behavior in response to diverse stimuli is emerging more and more and provides additional information regarding layers of regulation in cells. In this scenario, K63 ubiquitylation is starting to play an increasingly important role, since several crucial mechanisms involved in cell-signaling are regulated by this type of post-translational modification.

A vast group of human disorders such as Alzheimer's disease, Parkinson's disease, Type II Diabetes, and cancer have been investigated. In this chapter, we focused on the description of the main intracellular signaling pathways regulated by post-translational modification with respect to K63-linked ubiquitylation. The goal of this overview was to summarize the main findings regarding the main regulatory mechanisms that contribute to the disease pathogenesis or progression.

A better understanding of the different layers of regulation within molecular pathways, including the ubiquitin code, will indeed clear the path to a more precise manipulation of those aberrant post-translational signaling events that cause disease, marking a new era in therapeutic management and personalized medicine.

The specific targets of these modifications, influencing intracellular signaling pathways and cellular behavior, represent the future of target therapy; thus, the investigation of these mechanisms should be further analyzed in depth.

Acknowledgements

The personnel working on this project have received funding from the Innovative Medicines Initiative 2 Joint Undertaking under grant agreement No 115974. This Joint Undertaking receives support from the European Union's Horizon 2020 research and innovation program and EFPIA with JDRF.

Conflict of interest

None.

Author details

Paola Pontrelli*, Francesca Conserva and Loreto Gesualdo
Department of Emergency and Organ Transplantation, Nephrology, Dialysis and Transplantation Unit, University of Bari Aldo Moro, Bari, Italy

*Address all correspondence to: paola.pontrelli@uniba.it

References

[1] Ikeda F, Dikic I. Atypical ubiquitin chains: New molecular signals. 'Protein Modifications: Beyond the Usual Suspects' review series. EMBO Reports. 2008;**9**(6):536-542

[2] Akutsu M, Dikic I, Bremm A. Ubiquitin chain diversity at a glance. Journal of Cell Science. 2016;**129**(5):875-880

[3] Chen ZJ, Sun LJ. Nonproteolytic functions of ubiquitin in cell signaling. Molecular Cell. 2009;**33**(3):275-286

[4] Ye Y, Rape M. Building ubiquitin chains: E2 enzymes at work. Nature Reviews. Molecular Cell Biology. 2009;**10**(11):755-764

[5] Schulman BA, Harper JW. Ubiquitin-like protein activation by E1 enzymes: The apex for downstream signalling pathways. Nature Reviews. Molecular Cell Biology. 2009;**10**(5):319-331

[6] Komander D, Rape M. The ubiquitin code. Annual Review of Biochemistry. 2012;**81**:203-229

[7] Deshaies RJ, Joazeiro CA. RING domain E3 ubiquitin ligases. Annual Review of Biochemistry. 2009;**78**:399-434

[8] Hodge CD, Spyracopoulos L, Glover JN. Ubc13: The Lys63 ubiquitin chain building machine. Oncotarget. 2016;7(39):64471-64504

[9] Swatek KN, Komander D. Ubiquitin modifications. Cell Research. 2016;**26**(4):399-422

[10] Swaney DL, Rodriguez-Mias RA, Villen J. Phosphorylation of ubiquitin at Ser65 affects its polymerization, targets, and proteome-wide turnover. EMBO Reports. 2015;**16**(9): 1131-1144

[11] Amerik AY, Hochstrasser M. Mechanism and function of deubiquitinating enzymes. Biochimica et Biophysica Acta. 2004;**1695**(1-3):189-207

[12] Hayden MS, Ghosh S. Shared principles in NF-kappaB signaling. Cell. 2008;**132**(3):344-362

[13] Oeckinghaus A, Ghosh S. The NF-kappaB family of transcription factors and its regulation. Cold Spring Harbor Perspectives in Biology. 2009;**1**(4):a000034

[14] Chen ZJ. Ubiquitin signalling in the NF-kappaB pathway. Nature Cell Biology. 2005;7(8):758-765

[15] Keating SE, Bowie AG. Role of non-degradative ubiquitination in interleukin-1 and toll-like receptor signaling. The Journal of Biological Chemistry. 2009;**284**(13): 8211-8215

[16] Deng L, Wang C, Spencer E, Yang L, Braun A, You J, et al. Activation of the IkappaB kinase complex by TRAF6 requires a dimeric ubiquitin-conjugating enzyme complex and a unique polyubiquitin chain. Cell. 2000;**103**(2):351-361

[17] Wang C, Deng L, Hong M, Akkaraju GR, Inoue J, Chen ZJ. TAK1 is a ubiquitin-dependent kinase of MKK and IKK. Nature. 2001;**412**(6844):346-351

[18] Wajant H, Scheurich P. TNFR1-induced activation of the classical NF-kappaB pathway. The FEBS Journal. 2011;**278**(6):862-876

[19] Tauriello DV, Maurice MM. The various roles of ubiquitin in Wnt pathway regulation. Cell Cycle. 2010;**9**(18): 3700-3709

[20] Clevers H. Wnt/beta-catenin signaling in development and disease. Cell. 2006;**127**(3):469-480

[21] Wu X, Yamamoto M, Akira S, Sun SC. Regulation of hematopoiesis by the K63-specific ubiquitin-conjugating enzyme Ubc13. Proceedings of the National Academy of Sciences of the United States of America. 2009;**106**(49):20836-20841

[22] Tran H, Hamada F, Schwarz-Romond T, Bienz M. Trabid, a new positive regulator of Wnt-induced transcription with preference for binding and cleaving K63-linked ubiquitin chains. Genes & Development. 2008;**22**(4):528-542

[23] Tauriello DV, Haegebarth A, Kuper I, Edelmann MJ, Henraat M, Canninga-van Dijk MR, et al. Loss of the tumor suppressor CYLD enhances Wnt/beta-catenin signaling through K63-linked ubiquitination of Dvl. Molecular Cell. 2010;**37**(5):607-619

[24] Mukhopadhyay D, Riezman H. Proteasome-independent functions of ubiquitin in endocytosis and signaling. Science. 2007;**315**(5809):201-205

[25] Galan JM, Haguenauer-Tsapis R. Ubiquitin lys63 is involved in ubiquitination of a yeast plasma membrane protein. The EMBO Journal. 1997;**16**(19):5847-5854

[26] Huang F, Zeng X, Kim W, Balasubramani M, Fortian A, Gygi SP, et al. Lysine 63-linked polyubiquitination is required for EGF receptor degradation. Proceedings of the National Academy of Sciences of the United States of America. 2013;**110**(39):15722-15727

[27] Vina-Vilaseca A, Sorkin A. Lysine 63-linked polyubiquitination of the dopamine transporter requires WW3 and WW4 domains of Nedd4-2 and UBE2D ubiquitin-conjugating enzymes. The Journal of Biological Chemistry. 2010;**285**(10):7645-7656

[28] Geetha T, Jiang J, Wooten MW. Lysine 63 polyubiquitination of the nerve growth factor receptor TrkA directs internalization and signaling. Molecular Cell. 2005;**20**(2): 301-312

[29] Duncan LM, Piper S, Dodd RB, Saville MK, Sanderson CM, Luzio JP, et al. Lysine-63-linked ubiquitination is required for endolysosomal degradation of class I molecules. The EMBO Journal. 2006;**25**(8):1635-1645

[30] Varghese B, Barriere H, Carbone CJ, Banerjee A, Swaminathan G, Plotnikov A, et al. Polyubiquitination of prolactin receptor stimulates its internalization, postinternalization sorting, and degradation via the lysosomal pathway. Molecular and Cellular Biology. 2008;**28**(17):5275-5287

[31] Zhang L, Xu M, Scotti E, Chen ZJ, Tontonoz P. Both K63 and K48 ubiquitin linkages signal lysosomal degradation of the LDL receptor. Journal of Lipid Research. 2013;**54**(5):1410-1420

[32] Bhoj VG, Chen ZJ. Ubiquitylation in innate and adaptive immunity. Nature. 2009;**458**(7237):430-437

[33] Wu X, Karin M. Emerging roles of Lys63-linked polyubiquitylation in immune responses. Immunological Reviews. 2015;**266**(1):161-174

[34] Chen Z, Luo X, Lu Y, Zhu T, Wang J, Tsun A, et al. Ubiquitination signals critical to regulatory T cell development and function. International Immunopharmacology. 2013;**16**(3):348-352

[35] Ajibade AA, Wang HY, Wang RF. Cell type-specific function of TAK1 in innate immune signaling. Trends in Immunology. 2013;**34**(7): 307-316

[36] Matsuzawa A, Tseng PH, Vallabhapurapu S, Luo JL, Zhang W, Wang H, et al. Essential cytoplasmic translocation of a cytokine receptor-assembled signaling complex. Science. 2008;**321**(5889):663-668

[37] Schmidt A, Oberle N, Krammer PH. Molecular mechanisms of treg-mediated T cell suppression. Frontiers in Immunology. 2012;**3**:51

[38] Sakaguchi S, Yamaguchi T, Nomura T, Ono M. Regulatory T cells and immune tolerance. Cell. 2008;**133**(5):775-787

[39] Chang JH, Xiao Y, Hu H, Jin J, Yu J, Zhou X, et al. Ubc13 maintains the suppressive function of regulatory T cells and prevents their conversion into effector-like T cells. Nature Immunology. 2012;**13**(5):481-490

[40] Long M, Park SG, Strickland I, Hayden MS, Ghosh S. Nuclear factor-kappaB modulates regulatory T cell development by directly regulating expression of Foxp3 transcription factor. Immunity. 2009;**31**(6):921-931

[41] Gomez-Martin D, Ibarra-Sanchez M, Romo-Tena J, Cruz-Ruiz J, Esparza-Lopez J, Galindo-Campos M, et al. Casitas B lineage lymphoma b is a key regulator of peripheral tolerance in systemic lupus erythematosus. Arthritis and Rheumatism. 2013;**65**(4):1032-1042

[42] Romo-Tena J, Rajme-Lopez S, Aparicio-Vera L, Alcocer-Varela J, Gomez-Martin D. Lys63-polyubiquitination by the E3 ligase casitas B-lineage lymphoma-b (Cbl-b) modulates peripheral regulatory T cell tolerance in patients with systemic lupus erythematosus. Clinical and Experimental Immunology. 2018;**191**(1):42-49

[43] Yamamoto M, Sato S, Saitoh T, Sakurai H, Uematsu S, Kawai T, et al. Cutting edge: Pivotal function of Ubc13 in thymocyte TCR signaling. Journal of Immunology. 2006;**177**(11):7520-7524

[44] Yamamoto M, Okamoto T, Takeda K, Sato S, Sanjo H, Uematsu S, et al. Key function for the Ubc13 E2 ubiquitin-conjugating enzyme in immune receptor signaling. Nature Immunology. 2006;**7**(9):962-970

[45] Fiil BK, Gyrd-Hansen M. Met1-linked ubiquitination in immune signalling. The FEBS Journal. 2014;**281**(19):4337-4350

[46] Chen J, Hao L, Li C, Ye B, Du Y, Zhang H, et al. The endoplasmic reticulum adaptor protein ERAdP initiates NK cell activation via the Ubc13-mediated NF-kappaB pathway. Journal of Immunology. 2015;**194**(3):1292-1303

[47] Komander D, Clague MJ, Urbe S. Breaking the chains: Structure and function of the deubiquitinases. Nature Reviews. Molecular Cell Biology. 2009;**10**(8):550-563

[48] Bailey-Elkin BA, Knaap RCM, Kikkert M, Mark BL. Structure and function of viral deubiquitinating enzymes. Journal of Molecular Biology. 2017;**429**(22):3441-3470

[49] Kattenhorn LM, Korbel GA, Kessler BM, Spooner E, Ploegh HL. A deubiquitinating enzyme encoded by HSV-1 belongs to a family of cysteine proteases that is conserved across the family Herpesviridae. Molecular Cell. 2005;**19**(4):547-557

[50] Ye R, Su C, Xu H, Zheng C. Herpes simplex virus 1 ubiquitin-specific protease UL36 abrogates NF-kappaB activation in DNA sensing signal pathway. Journal of Virology. 2017;**91**(5)

[51] Pertel T, Hausmann S, Morger D, Zuger S, Guerra J, Lascano J, et al. TRIM5 is an innate immune sensor for

the retrovirus capsid lattice. Nature. 2011;**472**(7343):361-365

[52] Fritz JH, Ferrero RL, Philpott DJ, Girardin SE. Nod-like proteins in immunity, inflammation and disease. Nature Immunology. 2006;**7**(12):1250-1257

[53] Podolsky DK. Inflammatory bowel disease. The New England Journal of Medicine. 2002;**347**(6):417-429

[54] Abbott DW, Yang Y, Hutti JE, Madhavarapu S, Kelliher MA, Cantley LC. Coordinated regulation of Toll-like receptor and NOD2 signaling by K63-linked polyubiquitin chains. Molecular and Cellular Biology. 2007;**27**(17):6012-6025

[55] Weir GC, Bonner-Weir S. Islet beta cell mass in diabetes and how it relates to function, birth, and death. Annals of the New York Academy of Sciences. 2013;**1281**:92-105

[56] Remedi MS, Emfinger C. Pancreatic beta-cell identity in diabetes. Diabetes, Obesity & Metabolism. 2016;**18 Suppl 1**: 110-116

[57] MacDonald PE. A post-translational balancing act: The good and the bad of SUMOylation in pancreatic islets. Diabetologia. 2018;**61**(4):775-779

[58] Humphrey RK, Newcomb CJ, Yu SM, Hao E, Yu D, Krajewski S, et al. Mixed lineage kinase-3 stabilizes and functionally cooperates with TRIBBLES-3 to compromise mitochondrial integrity in cytokine-induced death of pancreatic beta cells. The Journal of Biological Chemistry. 2010;**285**(29):22426-22436

[59] Humphrey RK, Yu SM, Bellary A, Gonuguntla S, Yebra M, Jhala US. Lysine 63-linked ubiquitination modulates mixed lineage kinase-3 interaction with JIP1 scaffold protein in cytokine-induced pancreatic beta cell death. The Journal of Biological Chemistry. 2013;**288**(4):2428-2440

[60] Portbury AL, Ronnebaum SM, Zungu M, Patterson C, Willis MS. Back to your heart: Ubiquitin proteasome system-regulated signal transduction. Journal of Molecular and Cellular Cardiology. 2012;**52**(3):526-537

[61] Pontrelli P, Oranger A, Barozzino M, Conserva F, Papale M, Gesualdo L. The pathological role of the ubiquitination pathway in diabetic nephropathy. Minerva Medica. 2018;**109**(1):53-67

[62] Goru SK, Kadakol A, Gaikwad AB. Hidden targets of ubiquitin proteasome system: To prevent diabetic nephropathy. Pharmacological Research. 2017;**120**:170-179

[63] Yang XD, Xiang DX, Yang YY. Role of E3 ubiquitin ligases in insulin resistance. Diabetes, Obesity & Metabolism. 2016;**18**(8):747-754

[64] Gao C, Huang W, Kanasaki K, Xu Y. The role of ubiquitination and sumoylation in diabetic nephropathy. BioMed Research International. 2014;**2014**:160692

[65] Aghdam SY, Gurel Z, Ghaffarieh A, Sorenson CM, Sheibani N. High glucose and diabetes modulate cellular proteasome function: Implications in the pathogenesis of diabetes complications. Biochemical and Biophysical Research Communications. 2013;**432**(2):339-344

[66] Mazzucco G, Bertani T, Fortunato M, Bernardi M, Leutner M, Boldorini R, et al. Different patterns of renal damage in type 2 diabetes mellitus: A multicentric study on 393 biopsies. American Journal of Kidney Diseases. 2002;**39**(4):713-720

[67] Pontrelli P, Conserva F, Papale M, Oranger A, Barozzino M, Vocino G, et al. Lysine 63 ubiquitination is

involved in the progression of tubular damage in diabetic nephropathy. The FASEB Journal. 2017;**31**(1):308-319

[68] Zhao L, Chi L, Zhao J, Wang X, Chen Z, Meng L, et al. Serum response factor provokes epithelial-mesenchymal transition in renal tubular epithelial cells of diabetic nephropathy. Physiological Genomics. 2016;**48**(8):580-588

[69] Cybulsky AV. The intersecting roles of endoplasmic reticulum stress, ubiquitin-proteasome system, and autophagy in the pathogenesis of proteinuric kidney disease. Kidney International. 2013;**84**(1):25-33

[70] Yamahara K, Kume S, Koya D, Tanaka Y, Morita Y, Chin-Kanasaki M, et al. Obesity-mediated autophagy insufficiency exacerbates proteinuria-induced tubulointerstitial lesions. Journal of the American Society of Nephrology. 2013;**24**(11):1769-1781

[71] Pontrelli P, Oranger A, Barozzino M, Divella C, Conserva F, Fiore MG, et al. Deregulation of autophagy under hyperglycemic conditions is dependent on increased lysine 63 ubiquitination: A candidate mechanism in the progression of diabetic nephropathy. Journal of Molecular Medicine. 2018;**96**(7):645-659

[72] Lucking CB, Durr A, Bonifati V, Vaughan J, De Michele G, Gasser T, et al. Association between early-onset Parkinson's disease and mutations in the parkin gene. The New England Journal of Medicine. 2000;**342**(21):1560-1567

[73] Dawson TM. New animal models for Parkinson's disease. Cell. 2000;**101**(2):115-118

[74] Wakabayashi K, Tanji K, Mori F, Takahashi H. The Lewy body in Parkinson's disease: Molecules implicated in the formation and degradation of alpha-synuclein aggregates. Neuropathology. 2007;**27**(5):494-506

[75] Lim KL, Chew KC, Tan JM, Wang C, Chung KK, Zhang Y, et al. Parkin mediates nonclassical, proteasomal-independent ubiquitination of synphilin-1: Implications for Lewy body formation. The Journal of Neuroscience. 2005;**25**(8):2002-2009

[76] Liu Y, Fallon L, Lashuel HA, Liu Z, Lansbury PT Jr. The UCH-L1 gene encodes two opposing enzymatic activities that affect alpha-synuclein degradation and Parkinson's disease susceptibility. Cell. 2002;**111**(2): 209-218

[77] Lim KL, Dawson VL, Dawson TM. Parkin-mediated lysine 63-linked polyubiquitination: A link to protein inclusions formation in Parkinson's and other conformational diseases? Neurobiology of Aging. 2006;**27**(4):524-529

[78] Sugeno N, Hasegawa T, Tanaka N, Fukuda M, Wakabayashi K, Oshima R, et al. Lys-63-linked ubiquitination by E3 ubiquitin ligase Nedd4-1 facilitates endosomal sequestration of internalized alpha-synuclein. The Journal of Biological Chemistry. 2014;**289**(26):18137-18151

[79] Tan JM, Wong ES, Kirkpatrick DS, Pletnikova O, Ko HS, Tay SP, et al. Lysine 63-linked ubiquitination promotes the formation and autophagic clearance of protein inclusions associated with neurodegenerative diseases. Human Molecular Genetics. 2008;**17**(3):431-439

[80] Sun F, Kanthasamy A, Anantharam V, Kanthasamy AG. Mitochondrial accumulation of polyubiquitinated proteins and differential regulation of apoptosis by polyubiquitination sites Lys-48 and -63. Journal of Cellular and Molecular Medicine. 2009;**13**(8B):1632-1643

[81] Ugun-Klusek A, Theodosi TS, Fitzgerald JC, Burte F, Ufer C, Boocock DJ, et al. Monoamine oxidase-A promotes protective autophagy in human SH-SY5Y neuroblastoma cells through Bcl-2 phosphorylation. Redox Biology. 2018;**20**:167-181

[82] Ma L, Broomfield S, Lavery C, Lin SL, Xiao W, Bacchetti S. Up-regulation of CIR1/CROC1 expression upon cell immortalization and in tumor-derived human cell lines. Oncogene. 1998;**17**(10):1321-1326

[83] Gallo LH, Ko J, Donoghue DJ. The importance of regulatory ubiquitination in cancer and metastasis. Cell Cycle. 2017;**16**(7):634-648

[84] Wu X, Zhang W, Font-Burgada J, Palmer T, Hamil AS, Biswas SK, et al. Ubiquitin-conjugating enzyme Ubc13 controls breast cancer metastasis through a TAK1-p38 MAP kinase cascade. Proceedings of the National Academy of Sciences of the United States of America. 2014;**111**(38):13870-13875

[85] Wu Z, Shen S, Zhang Z, Zhang W, Xiao W. Ubiquitin-conjugating enzyme complex Uev1A-Ubc13 promotes breast cancer metastasis through nuclear factor-small ka, CyrillicB mediated matrix metalloproteinase-1 gene regulation. Breast Cancer Research. 2014;**16**(4):R75

[86] Pulvino M, Liang Y, Oleksyn D, DeRan M, Van Pelt E, Shapiro J, et al. Inhibition of proliferation and survival of diffuse large B-cell lymphoma cells by a small-molecule inhibitor of the ubiquitin-conjugating enzyme Ubc13-Uev1A. Blood. 2012;**120**(8):1668-1677

[87] Wu HT, Kuo YC, Hung JJ, Huang CH, Chen WY, Chou TY, et al. K63-polyubiquitinated HAUSP deubiquitinates HIF-1alpha and dictates H3K56 acetylation promoting hypoxia-induced tumour progression. Nature Communications. 2016;**7**:13644

[88] Ng HM, Wei L, Lan L, Huen MS. The Lys63-deubiquitylating enzyme BRCC36 limits DNA break processing and repair. The Journal of Biological Chemistry. 2016;**291**(31):16197-16207

[89] Wang B, Jie Z, Joo D, Ordureau A, Liu P, Gan W, et al. TRAF2 and OTUD7B govern a ubiquitin-dependent switch that regulates mTORC2 signalling. Nature. 2017;**545**(7654):365-369

[90] Yao F, Zhou Z, Kim J, Hang Q, Xiao Z, Ton BN, et al. SKP2- and OTUD1-regulated non-proteolytic ubiquitination of YAP promotes YAP nuclear localization and activity. Nature Communications. 2018;**9**(1):2269

Chapter 5

Regulation of Selective Proteolysis in Cancer

Pai-Sheng Chen

Abstract

Proteins are the fundamental building blocks of cells for diverse cellular and physiological functions. The dynamic equilibrium of protein turnover is balanced by protein synthesis and proteolysis. The newly synthesized proteins undergo proper folding into the three-dimensional conformations for executing biological functions and constructing cellular components like organelles. On the other hand, ubiquitin-proteasome system (UPS) and lysosome are two major proteolytic systems by which the unneeded, misfolded, or damaged proteins are selectively sent for clearance to maintain the quality and quantity of cellular proteins. Loss of the ability to maintain cellular proteolysis in control has been known to contribute as disease-causing factors. In this chapter, the function, regulation, and pathological roles of dysregulated proteolysis will be described in a concise view, focusing on the link between cancer and UPS.

Keywords: ubiquitin-proteasome system, proteolysis, cancer

1. Introduction

Protein ubiquitination is a multistep process. It is initiated by an ATP-required activation and covalent binding of E1 ubiquitin-activating enzyme (E1) with ubiquitin [1]. The E1 then passes the ubiquitin to E2 ubiquitin-conjugating enzyme (E2) followed by forming complex with the E3 ubiquitin ligase (E3), which specifically recognizes substrate protein and catalyzes the ubiquitin transfer. Theoretically, E3s can function as oncogenes or tumor suppressors depending on the specificities on substrate proteins they targeted in cancer (**Figure 1**). For instance, MDM2 is oncogenic since it is the E3 for tumor-suppressive p53, while von Hippel–Lindau (VHL) disease tumor suppressor is tumor suppressive since it is the E3 for oncogenic HIF-1alpha (HIF-1α). However, alternative functions of E3 are also observed since multiple targets with diverse roles may be regulated by a common E3. Here, the selective ubiquitin-proteasome system (UPS) for p53, HIF-1α, and other cancer-related proteins are exemplified.

Modification of substrate proteins by ubiquitination is the major way for selective proteolysis by proteasome. Ubiquitination is a reversible process controlled by the balance of ubiquitination and deubiquitination systems. This balance of ubiquitination is regulated by E3 ubiquitin ligases (E3s) [2] and deubiquitylating enzymes (DUBs) [3]. In addition to UPS-mediated protein degradation, ubiquitination is also involved in diverse non-proteolytic molecular and cellular functions, such as protein trafficking, activation, DNA repair, and apoptosis [4]. For example, K63-linked chains regulate DNA repair and NF-κB activation [5–7]. The TNF-α

IntechOpen

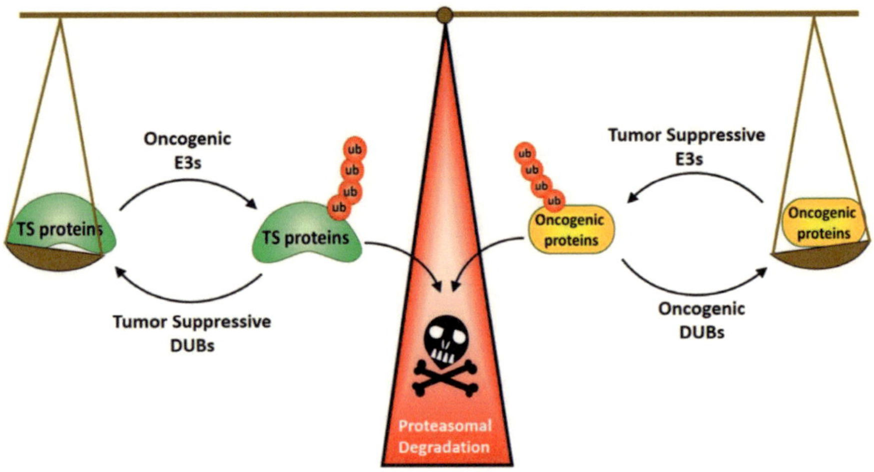

Figure 1.
Roles of E3s and DUBs in cancer.

receptor–associated factor 6 (TRAF6) mediated K63-linked polyubiquitination of NF-κB essential modulator (NEMO) for IκB kinase (IKK) activation [8]. These studies indicate the versatile function of ubiquitination machinery. In this section, we focus on the role of ubiquitination in proteasomal degradation. There are seven lysine residues at positions 6, 11, 27, 29, 33, 48, and 63 of ubiquitin, which are utilized for polyubiquitination. These lysine residues serve as acceptors for other ubiquitins. Different types (mono, multi, poly) and links (K6, K11, K27, K29, K33, K48, K63) of ubiquitination determine the fate of tagged substrates [9–11]. For proteasome degradation, K48- and K11-linked polyubiquitination is the canonical signal that tags substrate proteins [12–16]. Recent studies showed that branched K48- and K11-linked chains enhanced proteasomal degradation, whereas homotypic K11 linkages prevent substrate protein recognition by proteasome [17]. K29-linked polyubiquitin is the most abundant atypical linkage in mammalian cells [18]. But little is known about its cellular function. The use of K29-linked chains as a degradation signal is also unclear as these chains may accumulate as a consequence of proteosomal stress induced by proteasome inhibition rather than via the accumulation of K29-linkage-modified proteasome substrates. The K63 linkage, while it can be recognized by the proteasome [19], is widely regarded as a non-degradative signaling modification that is known to regulate signal transduction and endocytosis [20, 21]. In addition, efficient proteasome activity has also been found to rely on the presence of K6-linked ubiquitination [22].

2. UPS-mediated proteolysis in cancer

2.1 Regulatory network for p53 degradation

Tumor protein p53 is a well-known tumor suppressor [23]. As a guardian of genome, p53 can sense DNA damages, activate repair systems, pause cell growth, or initiate apoptosis when necessary [23]. These functions establish a cellular protective machinery, thus loss of expression or tumor-suppressive activities of p53 are observed as a hallmark in cancer. Deregulation of p53 is orchestrated by multiple pathways, such as gene mutation and enhanced proteasomal degradation. As an E3 for p53, overexpression of MDM2 in human cancers has been linked to p53 degradation and tumorigenesis [24]. The transactivation domain of p53 is recognized by MDM2 and followed by MDM2-dependent ubiquitination and further proteasome degradation [25]. ATM serine/threonine kinase (ATM) is a DNA damage sensor participating in multiple mechanisms for p53 regulation. ATM-mediated phosphorylation of p53 at Ser15 is induced by genotoxic stress and therefore causes its escape from MDM recognition to further trigger cells to initiate DNA repair system through p53 [26–28]. Also, the MDM2-mediated p53 degradation is diminished when ATM-mediated MDM2 phosphorylation is induced by DNA damage [29]. Moreover, there is a negative feedback loop in which p53 activates the transcription of MDM2 [30]. Several inhibitors, such as Nutlin-3 and RG7112, were developed to disrupt the interaction between p53 and MDM2 and are currently undergoing clinical trials [31]. On the other hand, herpesvirus-associated ubiquitin-specific protease (HAUSP) is a deubiquitinase for p53. It removes ubiquitination and stabilizes p53 even in the presence of MDM2 [32]. Moreover, MDM2 is also stabilized by HAUSP through a p53-independent pathway [33, 34], suggesting a feedback regulatory loop between p53 and MDM2. The ATM-mediated phosphorylation, nuclear translocation, and stabilization of USP10 synergistically help nuclear HASUP stabilize p53 in the presence of DNA damage [35]. In addition, the constitutive photomorphogenesis protein 1 (COP1) forms an E3 ubiquitin ligase complex with cullin 4 (CUL4), DNA damage-binding protein 1 (DDB1), de-etiolated 1 (DET1), and ring-box 1 (RBX1) to target p53 [36]. Under genotoxic stress, ATM phosphorylates COP1 at Ser387 for degradation and subsequent p53 induction. Since p53 is targeted by COP1 for proteasomal degradation, downregulation or inactivation of COP1 subsequently activates p53 in cancer. Like MDM2, a transcriptional activation of COP1 by p53 forms a negative feedback loop [37]. Overexpression of COP1 is correlated with reduced p53 and has been observed in ovarian, breast, and liver cancers. P53-induced RING-H2 (Pirh2, also known as RCHY1) is another E3 ubiquitin ligase belonging to the RING finger family. Like MDM2, Pirh2 is considered as an oncogene to facilitate p53 protein degradation by UPS through a MDM2-independent manner [38, 39]. Notably, similar to the p53-MDM2 and p53-COP1 feedback loop, Pirh2 is also upregulated transcriptionally by p53. Interestingly, several researches suggested that Pirh2, but not MDM2, plays a major role in DNA damage-induced p53 degradation [38]. Moreover, in contrast to MDM2, Pirh2 can still recognize the p53 with Ser15 phosphorylation for UPS [40]. Elevated Pirh2 level has been found in human cancers and is correlated with unfavorable prognosis of cancer patients [41, 42]. The regulatory network for p53 degradation is illustrated in **Figure 2**.

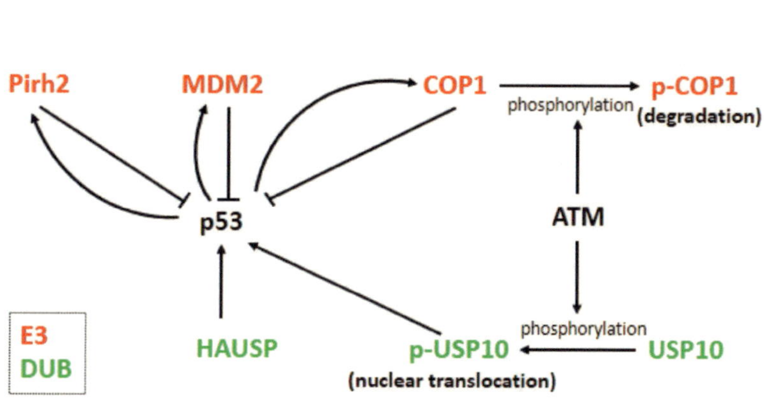

Figure 2.
Regulatory network for p53 degradation.

2.2 Ubiquitination system of HIF-1α

During tumorigenesis, the increased tumor mass leads to the reduction of available intratumoral oxygen, which is theoretically a survival stress to normal cells. However, cancer cells develop several mechanisms to face this stressful condition, such as the activation of hypoxia-inducible factor 1α (HIF-1α). Through transcriptional regulation of downstream genes, accumulation of HIF-1α is not only observed on facilitating angiogenesis at the initiation of rapid tumor growth (also called angiogenic switch) but also enhances metastasis and malignant progression of cancer [43]. Expression of HIF-1α is tightly controlled by ubiquitination in coordination with hypoxia (**Figure 3**). Inactivation of Von Hippel-Lindau (VHL) in familial kidney cancer syndrome contributes to oncogenic effects [44]. At the molecular level, VHL interacts with cullin 2, elongin B, elongin C, and Rbx [45–48]. This complex then targets HIF-1α for ubiquitination and proteasomal degradation [48, 49]. Under normoxia, prolyl hydroxylase (PHD) hydroxylates HIF-1α and facilitates its binding through N-TAD domain with VHL complex, leading to sustained ubiquitination and subsequent degradation of HIF-1α. The PHD-mediated post-translational modification (PTM) is abolished when cells encounter hypoxia during tumor growth. The stabilized HIF-1α is then accumulated in cancer cells and translocated to nucleus in complex with HIF-1β and other cofactors, resulting in transcriptional activation of downstream genes [50]. The transactivation activity of HIF-1α is also regulated by factor inhibiting HIF-1 (FIH-1). Through interaction with the ID and C-TAD domains, FIH-1 represses HIF-1α-mediated transactivation in association with histone deacetylase [51]. The HIF-1α-independent functions of VHL have also been pointed out. Besides HIF-1α, growing evidence has identified several targets of VHL with oncogenic properties in cancer [52]. It is well known that both downregulation of VHL and accumulation of HIF-1α are associated with cancer progression [53]. In addition to VHL, the chaperone-dependent E3 carboxy

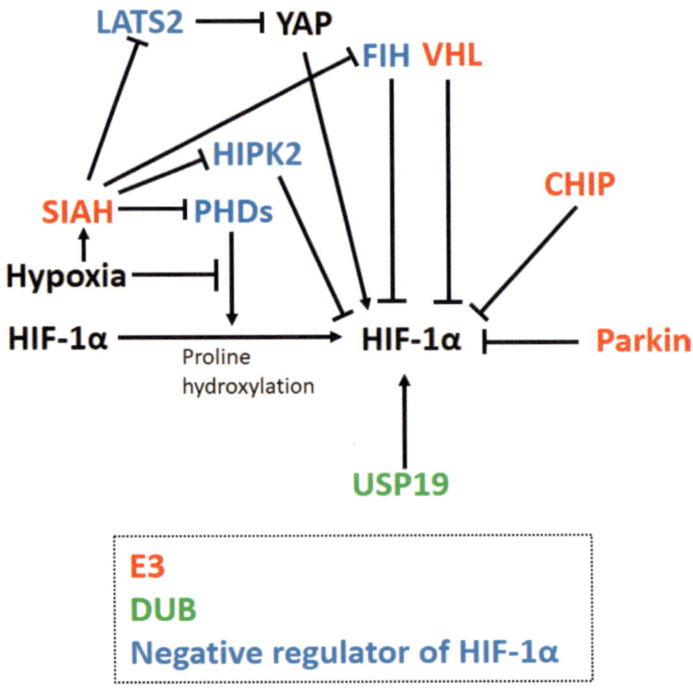

Figure 3.
Regulation of HIF-1α ubiquitination.

terminus of Hsp70-interacting protein (CHIP) is also identified to ubiquitinate HIF-1α for protein degradation [54]. Cellular response to hypoxia is also modulated by the E3s seven in absentia homolog (Siah) family proteins [55]. As another layer for HIF-1α regulation, Siah proteins are accumulated by transcriptional regulation and post-translational modification (PTM) under hypoxia [55]. The increased Siah proteins subsequently activate the degradation of PHDs and factors inhibiting HIF-1 (FIH) reduce prolyl hydroxylation of HIF-1α and consequently prevent VHL-mediated degradation [51, 55]. In addition to this regulation, there are several mechanisms known to cooperatively activate HIF-1α. For example, HIF-1α is stabilized when its ubiquitination is removed by ubiquitin-specific protease-19 (USP19) [56]. Siah proteins ubiquitinate the HIF-1α inhibitor, homeodomain-interacting protein kinase 2 (HIPK2), for degradation and thus enhance HIF-1α activity [57]. Siah2 also enhances the ubiquitination and degradation of large tumor suppressor kinase 2 (LATS2) resulting in suppressed HIPPO pathway and activated Yes-associated protein 1 (YAP1) that subsequently stabilizes HIF-1α [58]. Parkin is a recently-identified E3 for HIF-1α [59]. It facilitates HIF-1α polyubiquitination at K477 for proteasomal degradation through the interaction with HIF-1α. Alternatively, under the stimulations by hypoxia or growth factors, the induced HIF-1α brings Parkin and Dicer together, following by ubiquitination and autophagic degradation of Dicer, and eventually enhances cancer metastasis [60]. The findings exemplify the dual role of E3, which in this case, the target substrate (HIF-1α or Dicer) determines the fate of cellular function (**Figure 4**).

2.3 Cellular signaling regulated by UPS

Networks of signaling pathways coordinately orchestrate the cellular functions. Dysregulation of signal transduction pathways, especially those controlling oncogenic behaviors, is tightly regulated and also controlled by UPS. E3s play as modulators through regulating the proteolysis of key proteins in signaling networks. Several E3s can mediate substrate degradation to modulate PI3K/Akt/mTOR and RAS/MAPK, which are two central pathways, coordinately to control a broad range of tumor-promoting functions.

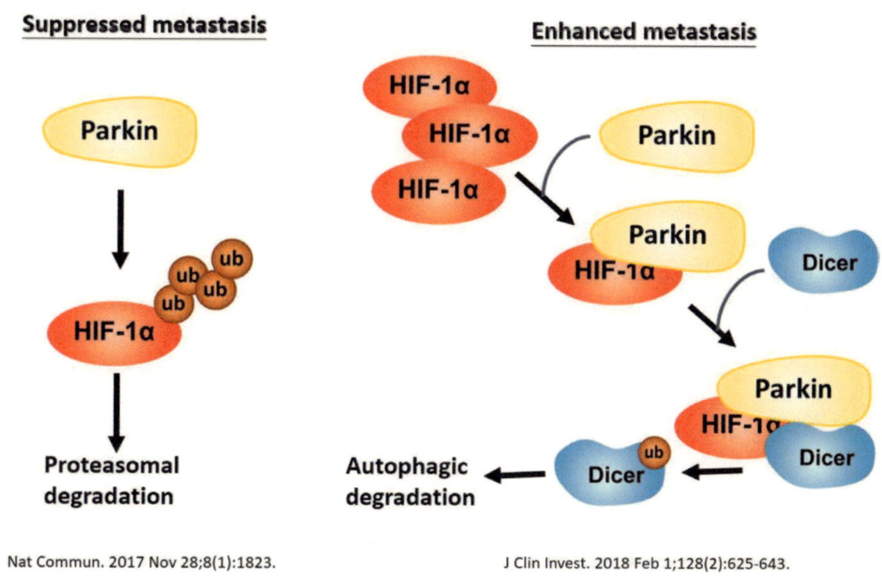

Figure 4.
Dual role of Parkin in cancer metastasis.

2.3.1 RAS/MAPK pathway

RAS oncogenes encode the highly-conserved RAS proteins as GTPases functioning in oncogenic transformation through the activation of MAPK pathway [61, 62]. Similar to p53, RAS mutations have been identified in human cancers, while stabilized RAS protein at post-translational level is also observed. The E3 ubiquitin-protein ligase, neural precursor cell-expressed developmentally downregulated protein 4 (NEDD4), is known to ubiquitinate RAS proteins for proteasomal degradation. NEDD4 targets KRAS, HRAS, NRAS for UPS, while its transcription is also activated by RAS signaling, which in turn, serves as a negative feedback to prevent the hyperactivation of RAS pathway [63]. More interestingly, this feedback mechanism is disrupted in cells expressing oncogenic RAS with activating mutation, exemplifying how an oncogenic protein (RAS) can escape from E3 (NEDD)-mediated degradation in cancer. Moreover, PTEN, a PI3K/Akt inhibitor, is another NEDD4 target, meaning that overexpression of NEDD4 facilitates PTEN degradation and activation of PI3K/Akt pathway. Thus, NEDD4 is supposedly to act as a tumor suppressor, but in cells expressing activating mutated RAS, NEDD4 no longer suppresses for RAS suppression and the concomitant PI3K/Akt activation corporately to amplify oncogenic signaling. In this case, NEDD4 functions as an oncogenic E3 [63]. However, a sustained RAS activation might be observed in lung cancer due to an elevated expression of DUB OTU domain-containing ubiquitin aldehyde-binding protein 1 (OTUB1), a deubiquitinase removing the ubiquitination of RAS and promoting the activation of RAS-mediated oncogenic downstream [64]. In addition to targeting RAS, several E3s are also identified to regulate downstream molecules of RAS. For example, ring finger protein 149 (RNF149) is an E3 targeting BRAF, a downstream kinase of RAS [65]. These machineries expend the complexity from reciprocal regulation in RAS/MAPK pathway.

2.3.2 PI3K/Akt pathway

PI3K/Akt pathway is induced by extracellular signaling such as activation of receptor tyrosine kinase (RTK) or G protein-coupled receptors (GPCRs). The regulatory subunit p85 and catalytic subunit p110 form heterodimer of PI3K [66]. In addition to p85, the p110 subunit also binds to Grb2 or insulin receptor substrate (IRS), and the competition from free p85 binding serves as an inhibitor for PI3K signaling [66, 67]. P85β is one of the variants of p85 subunits and is a target of the SCF-F-box and leucine-rich repeat protein 2 (FBXL2) complex [68]. SCF-FBXL2 enhances free p85β degradation through UPS and, consequently, disrupts the inhibitory effect of p85β pool on PI3K activation. Another layer of the regulation on this mechanism is controlled by dephosphorylation of p85β by PTPL1, which facilitates p85β degradation through enhanced interaction with FBXL2 [68]. The mechanistic target of rapamycin (mTOR) is a core component of PI3K/Akt pathway. The expression of mTOR is regulated by SCF-FBXW7 complex that triggers the ubiquitination and degradation of mTOR [69]. Loss of a single copy of FBXW7 in several breast cancer cell lines is observed to be incompatible with the loss of a single copy of PTEN, a negative regulator of mTOR, which further confirmed the significance of the stabilization of FBXW7-mediated mTOR in tumorigenesis. Therefore, loss of SCF-FBXW7 may motivate anabolic processes for tumor progression [69]. In addition to mTOR regulation, the F-box protein FBXW7 is a well-known tumor suppressor which recognizes and facilitates UPS of other oncogenic substrates including c-Myc, Notch1, and cyclin E [70–81]. The mechanism of cyclin E regulation will be discussed later in this chapter.

2.4 E3s as cell cycle regulators

Uncontrolled cell growth is one of the hallmarks of cancer [82]. Cell cycle progression is a fundamental process for cell survival and proliferation. Properly regulated cell cycle progression is required for the maintenance of genome stability, organismal development, tissue homeostasis that when deregulation coordinately leads to defect of cell growth control. Signals that control cell cycle entry, progression, and arrest are commonly malfunctioned in cancer, and the subsequent disruption of DNA replication, DNA repair, and chromosomal segregation often lead to genomic instability [83]. There is aberrant degradation caused by improper E3 activity in cancer. For example, abnormal expression of E3s regulates the degradation of cyclins and cyclin-dependent kinase (CDK) inhibitor proteins by UPS. The Anaphase Promoting Complex/Cyclosome (APC/C) and Skp1, Cullin1 F-box (SCF) E3 complexes that regulate cell cycle progression are the best-studied E3s, which further form different complexes with specific co-activators and provide the functional diversity of E3s to recognize different substrates at different phases for orchestrating cell cycle progression. In mitosis and G1 phase of cell cycle, APC/C is active to suppress mitotic CDKs [84]. In contrast, various SCF complexes formed by different protein partners contribute to multifaceted functions during the cell cycle progression. Here, we will discuss these two ubiquitin-protein ligases, and how they cooperatively regulate cell cycle progression.

APC/C is a well-studied E3 that recognizes the D-box sequence of substrate proteins and ubiquitinates them for proteasome degradation [85]. The recognition of substrates by APC/C is known to rely on a short linear motif called degron (derived from degradation motif) including KEN-box, D-box, and ABBA motif [86]. KEN-box is the prominent signal among APC/C degron and is contained in substrate proteins, such as CDC20 and securin. The substrates of APC/C cover numerous cell cycle regulators. Thus, APC/C contributes to the cell cycle regulation, especially during M phase to G1 phase. Cell division cycle 20 (CDC20) or CDC20-like protein 1 (CDH1) are two known activators for APC/C activation [87]. Interestingly, the two activators show opposite functions in cancer as oncogenic CDC20 or tumor-suppressive CDH1 interact with APC/C to exert their spatial and temporal functions during cell cycle [87]. It is widely observed that CDC20 is highly expressed in human malignancies and associates with poor prognosis of cancer patients [88–93]. Mechanistically, CDC20 recognizes securin [94], Cyclin A [95, 96], Cyclin B1 [97, 98], Nek2A [99], Mcl-1 [100], and p21 [101] as it substrates for ubiquitination. Cdc20 is primarily active in mid to late mitosis to promote ubiquitination and degradation of securin and cyclins to coordinately facilitate mitotic progression [87]. Thereafter, CDC20 degradation is triggered through ubquitination by APC/C-CDH1 or by itself in late M phase. As a result, the APC/C complex shifts from APC/C-CDC20 to APC/C-CDH1. APC/C-CDH1 is activated at late mitosis phase to degrade mitotic regulators, such as cyclins and kinases, and thereby promotes cells to exit from M phase and enter G1 phase to further prevent premature S phase entry [87]. Mutation or abnormal expression of CDH1 leads to genomic instability and premature S-phase entry [87, 102].

S-phase kinase-associated protein (Skp), cullin, and F-box domain containing proteins (F-box proteins) form SCF E3 ubiquitin ligase complex. Aurora kinase A and Cyclin E are substrates for SCF-FBXW7, thus inactivation of this complex causes defect in DNA repair system and sustained cell growth [103] suggesting the tumor-suppressive roles of SCF-FBXW7. As a key factor in SCF complex, dysregulation of F-box protein is frequently observed to affect SCF activity. SCF-Skp2 regulates cell cycle from G1/S to G2/M phase by targeting multiple CDK inhibitors for UPS degradation and consequently leads to enhanced cell cycle progression and tumorigenesis [104–111].

Interestingly, SCF-βTRCP complex also mediates the crosstalk between APC/C and SCF complex during cell cycle. Degradation of the APC/C inhibitor Emi1 during prometaphase and degradation of the Cdk1 inhibitor Wee1 during prophase are facilitated by SCF-βTRCP in cell cycle progression [112, 113]. Activation of SCF-Skp2 complex during G1 to S phase degrades cyclin-dependent kinase inhibitors (CKIs), thereby induces CDK activity. The CDK2-mediated phosphorylation and inactivation of CDH1 further stabilize Skp2 by phosphorylation. SKP2 is also a D-box-containing target of APC/C-CDH1 as an autoregulatory loop [114]. It is also noticed that the casein kinase I (CKI)-phosphorylated MDM2 is targeted and degraded by SCF-βTRCP complex and results in p53 stabilization and cell cycle arrest [115].

Parkin is another E3 participating in cell cycle regulation. Mutations and loss of copy number of *PARK2,* a gene encoding Parkin, are observed in cancer, which implies its tumor-suppressive role [116–121]. Loss of Parkin expression respectively results in the elevation of Cyclin D and Cyclin E owing to the suppression of FBXW7-containing Parkin-cullin-RING or F-box only protein 4 (FBXO4)-containing complexes [117]. In animal models, $Park2^{+/-}Apc^{Min/+}$ mice have higher rate of tumor formation than $Apc^{Min/+}$ mice, which may result from the accumulation of Cyclin E and uncontrolled cell growth when Parkin expression is lost [118, 120]. Similar associations between PARK2 mutations and Cyclin D, Cyclin E, and CDK4 are also observed in human cancers [117]. Therefore, Parkin may also regulate several cell cycle or mitotic regulators including CDC20, CDH1, Aurora kinase A, Aurora kinase B, NEK2, PLK1, Cyclin B1, and securin, suggesting its function in maintaining genomic stability and growth control to suppress tumor formation [119].

3. Conclusions

Malfunction of UPS machinery, especially the target selection factor E3, has been observed in cancer for a period of time. Abnormal expression, mutation, distribution of E3s, or even the degradation of themselves may affect the affinity or activity on substrate recognition and ubiquitination, and thus consequently regulate proteasomal degradation and cellular behaviors depending on the normal functions of dysregulated targets. Although we have focused on the selective proteolysis through UPS, E3-mediated ubiquitination is not the only way for proteasomal degradation and also, proteasomal degradation is not the only fate for ubiquitinated proteins. Oftentimes, these proteins undergo autophagic degradation, intracellular localization, functional inhibition, or activation. Moreover, the lysosomal and autolysosomal (autophagy-lysosomal) degradation, which are not described in detail in this chapter, are responsible for another side of selective proteolysis. In concert with the landscape of post-translational modification, the crosstalk and cooperation among these proteolysis systems enable our cells to maintain biological functions in control. Simply speaking, proteolysis serves as a dead end for protein, thus the selection of target substrates should be tightly controlled. This chapter introduces several pathways as examples of selective UPS. In addition, there are several clinical trials for drugs designed to target proteolysis. As we know more about the mechanisms, we are moving a step forward in developing strategies to fix the proteolytic chaos of cells.

Acknowledgements

We would like to thank Jie-Ning Li, Yu-Jhen Lyu, and Yuchieh Chien for their help on content editing.

Conflict of interest

The author declares that there is no conflict of interest.

Author details

Pai-Sheng Chen[1,2]

1 Institute of Basic Medical Sciences, College of Medicine, National Cheng Kung University, Tainan, Taiwan

2 Department of Medical Laboratory Science and Biotechnology, College of Medicine, National Cheng Kung University, Tainan, Taiwan

*Address all correspondence to: bio.benson@gmail.com

References

[1] Ciechanover A. Intracellular protein degradation: From a vague idea through the lysosome and the ubiquitin-proteasome system and onto human diseases and drug targeting. Cell Death and Differentiation. 2005;**12**(9):1178-1190

[2] Chhangani D, Joshi AP, Mishra A. E3 ubiquitin ligases in protein quality control mechanism. Molecular Neurobiology. 2012;**45**(3):571-585

[3] Eletr ZM, Wilkinson KD. Regulation of proteolysis by human deubiquitinating enzymes. Biochimica et Biophysica Acta-Molecular Cell Research. 2014;**1843**(1):114-128

[4] Mukhopadhyay D, Riezman H. Proteasome-independent functions of ubiquitin in endocytosis and signaling. Science. 2007;**315**(5809):201-205

[5] Liu PD, Gan WJ, Su SY, Hauenstein AV, Fu TM, Brasher B, et al. K63-linked polyubiquitin chains bind to DNA to facilitate DNA damage repair. Science Signaling. 2018;**11**(533)

[6] Jackson SP, Durocher D. Regulation of DNA damage responses by ubiquitin and SUMO. Molecular Cell. 2013;**49**(5):795-807

[7] Chen ZJJ. Ubiquitin signalling in the NF-kappa B pathway. Nature Cell Biology. 2005;**7**(8):758-765

[8] Sun LJ, Deng L, Ea CK, Xia ZP, Chen ZJJ. The TRAF6 ubiquitin ligase and TAK1 kinase mediate IKK activation by BCL10 and MALT1 in T lymphocytes. Molecular Cell. 2004;**14**(3):289-301

[9] Raasi S, Varadan R, Fushman D, Pickart CM. Diverse polyubiquitin interaction properties of ubiquitin-associated domains. Nature Structural & Molecular Biology. 2005;**12**(8):708-714

[10] Varadan R, Assfalg M, Raasi S, Pickart C, Fushman D. Structural determinants for selective recognition of a lys48-linked polyubiquitin chain by a UBA domain. Molecular Cell. 2005;**18**(6):687-698

[11] Hicke L, Schubert HL, Hill CP. Ubiquitin-binding domains. Nature Reviews Molecular Cell Biology. 2005;**6**(8):610-621

[12] Koegl M, Hoppe T, Schlenker S, Ulrich HD, Mayer TU, Jentsch S. A novel ubiquitination factor, E4, is involved in multiubiquitin chain assembly. Cell. 1999;**96**(5):635-644

[13] Crosas B, Hanna J, Kirkpatrick DS, Zhang DP, Tone Y, Hathaway NA, et al. Ubiquitin chains are remodeled at the proteasome by opposing ubiquitin ligase and deubiquitinating activities. Cell. 2006;**127**(7):1401-1413

[14] Kim HT, Kim KP, Lledias F, Kisselev AF, Scaglione KM, Skowyra D, et al. Certain pairs of ubiquitin-conjugating enzymes (E2s) and ubiquitin-protein ligases (E3s) synthesize nondegradable forked ubiquitin chains containing all possible isopeptide linkages. Journal of Biological Chemistry. 2007;**282**(24):17375-17386

[15] Johnson ES, Ma PCM, Ota IM, Varshavsky A. A Proteolytic pathway that recognizes ubiquitin as a degradation signal. Journal of Biological Chemistry. 1995;**270**(29):17442-17456

[16] Kirkpatrick DS, Hathaway NA, Hanna J, Elsasser S, Rush J, Finley D, et al. Quantitative analysis of in vitro ubiquitinated cyclin B1 reveals complex chain topology. Nature Cell Biology. 2006;**8**(7):700-U121

[17] Grice GL, Lobb IT, Weekes MP, Gygi SP, Antrobus R, Nathan JA. The proteasome distinguishes between

heterotypic and Homotypic Lysine-11-linked polyubiquitin chains. Cell Reports. 2015;**12**(4):545-553

[18] Dammer EB, Na CH, Xu P, Seyfried NT, Duong DM, Cheng DM, et al. Polyubiquitin linkage profiles in three models of proteolytic stress suggest the etiology of Alzheimer disease. Journal of Biological Chemistry. 2011;**286**(12):10457-10465

[19] Ohtake F, Tsuchiya H, Saeki Y, Tanaka K. K63 ubiquitylation triggers proteasomal degradation by seeding branched ubiquitin chains. Proceedings of the National Academy of Sciences of the United States of America. 2018;**115**(7):E1401-E1408

[20] Husnjak K, Dikic I. Ubiquitin-binding proteins: Decoders of ubiquitin-mediated cellular functions. Annual Review of Biochemistry. 2012;**81**:291-322

[21] Komander D, Rape M. The ubiquitin code. Annual Review of Biochemistry. 2012;**81**:203-229

[22] Shang F, Deng GJ, Liu Q, Guo WM, Haas AL, Crosas B, et al. Lys(6)-modified ubiquitin inhibits ubiquitin-dependent protein degradation. Journal of Biological Chemistry. 2005;**280**(21):20365-20374

[23] Bieging KT, Mello SS, Attardi LD. Unravelling mechanisms of p53-mediated tumour suppression. Nature Reviews Cancer. 2014;**14**(5):359-370

[24] Wade M, Li YC, Wahl GM. MDM2, MDMX and p53 in oncogenesis and cancer therapy. Nature Reviews Cancer. 2013;**13**(2):83-96

[25] Vousden KH, Lu X. Live or let die: The cell's response to p53. Nature Reviews Cancer. 2002;**2**(8):594-604

[26] Banin S, Moyal L, Shieh S, Taya Y, Anderson CW, Chessa L, et al. Enhanced phosphorylation of p53 by ATM in response to DNA damage. Science. 1998;**281**(5383):1674-1677

[27] Canman CE, Lim DS, Cimprich KA, Taya Y, Tamai K, Sakaguchi K, et al. Activation of the ATM kinase by ionizing radiation and phosphorylation of p53. Science. 1998;**281**(5383):1677-1679

[28] Shieh SY, Ikeda M, Taya Y, Prives C. DNA damage-induced phosphorylation of p53 alleviates inhibition by MDM2. Cell. 1997;**91**(3):325-334

[29] Maya R, Balass M, Kim ST, Shkedy D, Leal JF, Shifman O, et al. ATM-dependent phosphorylation of Mdm2 on serine 395: Role in p53 activation by DNA damage. Genes & Development. 2001;**15**(9):1067-1077

[30] Lahav G, Rosenfeld N, Sigal A, Geva-Zatorsky N, Levine AJ, Elowitz MB, et al. Dynamics of the p53-Mdm2 feedback loop in individual cells. Nature Genetics. 2004;**36**(2):147-150

[31] Zhao YJ, Aguilar A, Bernard D, Wang SM. Small-molecule inhibitors of the MDM2-p53 protein-protein interaction (MDM2 inhibitors) in clinical trials for cancer treatment. Journal of Medicinal Chemistry. 2015;**58**(3):1038-1052

[32] Li MY, Chen DL, Shiloh A, Luo JY, Nikolaev AY, Qin J, et al. Deubiquitination of p53 by HAUSP is an important pathway for p53 stabilization. Nature. 2002;**416**(6881):648-653

[33] Li MY, Brooks CL, Kon N, Gu W. A dynamic role of HAUSP in the p53-Mdm2 pathway. Molecular Cell. 2004;**13**(6):879-886

[34] Cummins JM, Rago C, Kohli M, Kinzler KW, Lengauer C, Vogelstein B. Disruption of HAUSP gene stabilizes p53. Nature. 2004;**428**(6982)

[35] Yuan J, Luo KT, Zhang LZ, Cheville JC, Lou ZK. USP10 regulates p53 localization and stability by deubiquitinating p53. Cell. 2010;**140**(3):384-396

[36] Dornan D, Wertz I, Shimizu H, Arnott D, Frantz GD, Dowd P, et al. The ubiquitin ligase COP1 is a critical negative regulator of p53. Nature. 2004;**429**(6987):86-92

[37] Brooks CL, Gu W. p53 ubiquitination: Mdm2 and beyond. Molecular Cell. 2006;**21**(3):307-315

[38] Leng RP, Lin YP, Ma WL, Wu H, Lemmers B, Chung S, et al. Pirh2, a p53-induced ubiquitin-protein ligase, promotes p53 degradation. Cell. 2003;**112**(6):779-791

[39] Sheng Y, Laister RC, Lemak A, Wu B, Tai E, Duan S, et al. Molecular basis of Pirh2-mediated p53 ubiquitylation. Nature Structural & Molecular Biology. 2008;**15**(12):1334-1342

[40] Wang ZH, Yang B, Dong LL, Peng BW, He XH, Liu WH. A novel oncoprotein Pirh2: Rising from the shadow of MDM2. Cancer Science. 2011;**102**(5):909-917

[41] Duan WR, Gao L, Druhan LJ, Zhu WG, Morrison C, Otterson GA, et al. Expression of Pirh2, a newly identified ubiquitin protein ligase, in lung cancer. Journal of the National Cancer Institute. 2004;**96**(22):1718-1721

[42] Logan IR, Gaughan L, McCracken SRC, Sapountzi V, Leung HY, Robson CN. Human PIRH2 enhances androgen receptor signaling through inhibition of histone deacetylase 1 and is overexpressed in prostate cancer. Molecular and Cellular Biology. 2006;**26**(17):6502-6510

[43] Semenza GL. Targeting HIF-1 for cancer therapy. Nature Reviews. Cancer. 2003;**3**(10):721-732

[44] Latif F, Tory K, Gnarra J, Yao M, Duh FM, Orcutt ML, et al. Identification of the von Hippel-Lindau disease tumor suppressor gene. Science. 1993;**260**(5112):1317-1320

[45] Kibel A, Iliopoulos O, DeCaprio JA, and Kaelin WG, Jr., Binding of the von Hippel-Lindau tumor suppressor protein to Elongin B and C. Science 1995;**269**(5229):1444-1446

[46] Pause A, Lee S, Worrell RA, Chen DY, Burgess WH, Linehan WM, et al. The von Hippel-Lindau tumor-suppressor gene product forms a stable complex with human CUL-2, a member of the Cdc53 family of proteins. Proceedings of the National Academy of Sciences of the United States of America. 1997;**94**(6):2156-2161

[47] Kamura T, Koepp DM, Conrad MN, Skowyra D, Moreland RJ, Iliopoulos O, et al. Rbx1, a component of the VHL tumor suppressor complex and SCF ubiquitin ligase. Science. 1999;**284**(5414):657-661

[48] Ohh M, Park CW, Ivan N, Hoffman MA, Kim TY, Huang LE, et al. Ubiquitination of hypoxia-inducible factor requires direct binding to the beta-domain of the von Hippel-Lindau protein. Nature Cell Biology. 2000;**2**(7):423-427

[49] Tanimoto K, Makino Y, Pereira T, Poellinger L. Mechanism of regulation of the hypoxia-inducible factor-1 alpha by the von Hippel-Lindau tumor suppressor protein. The EMBO Journal. 2000;**19**(16):4298-4309

[50] Triner D, Shah YM. Hypoxia-inducible factors: A central link between inflammation and cancer. Journal of Clinical Investigation. 2016;**126**(10):3689-3698

[51] Mahon PC, Hirota K, Semenza GL. FIH-1: A novel protein that interacts with HIF-1 alpha and VHL to mediate repression of HIF-1 transcriptional

activity. Genes & Development. 2001;**15**(20):2675-2686

[52] Zhang J, Zhang Q. VHL and hypoxia signaling: Beyond HIF in cancer. Biomedicine. 2018;**6**(1): pii: E35. DOI: 10.3390/biomedicines6010035.

[53] Kaelin WG Jr, The VHL. Tumor suppressor gene: Insights into oxygen sensing and cancer. Transactions of the American Clinical and Climatological Association. 2017;**128**:298-307

[54] Bento CF, Fernandes R, Ramalho J, Marques C, Shang F, Taylor A, et al. The chaperone-dependent ubiquitin ligase CHIP targets HIF-1 alpha for degradation in the presence of methylglyoxal. PLoS One. 2010;**5**(11):e15062. DOI: 10.1371/journal.pone.0015062

[55] Nakayama K, Frew IJ, Hagensen M, Skals M, Habelhah H, Bhoumik A, et al. Siah2 regulates stability of prolyl-hydroxylases, controls HIF1 alpha abundance, and modulates physiological responses to hypoxia. Cell. 2004;**117**(7):941-952

[56] Altun M, Zhao B, Velasco K, Liu HY, Hassink G, Paschke J, et al. Ubiquitin-specific protease 19 (USP19) regulates hypoxia-inducible factor 1 alpha (HIF-1 alpha) during hypoxia. Journal of Biological Chemistry. 2012;**287**(3):1962-1969

[57] Calzado MA, de la Vega L, Moller A, Bowtell DDL, Schmitz ML. An inducible autoregulatory loop between HIPK2 and Siah2 at the apex of the hypoxic response. Nature Cell Biology. 2009;**11**(1):85-U180

[58] Ma B, Chen Y, Chen L, Cheng HC, Mu CL, Li J, et al. Hypoxia regulates Hippo signalling through the SIAH2 ubiquitin E3 ligase. Nature Cell Biology. 2015;**17**(1):95-103

[59] Liu J, Zhang C, Zhao YH, Yue XT, Wu H, Huang S, et al. Parkin targets HIF-1 alpha for ubiquitination and degradation to inhibit breast tumor progression. Nature Communications. 2017;**8**

[60] Lai HH, Li JN, Wang MY, Huang HY, Croce CM, Sun HL, et al. HIF-1alpha promotes autophagic proteolysis of dicer and enhances tumor metastasis. The Journal of Clinical Investigation. 2018;**128**(2):625-643

[61] Simanshu DK, Nissley DV, McCormick F. RAS proteins and their regulators in human disease. Cell. 2017;**170**(1):17-33

[62] Pylayeva-Gupta Y, Grabocka E, Bar-Sagi D. RAS oncogenes: Weaving a tumorigenic web. Nature Reviews Cancer. 2011;**11**(11):761-774

[63] Zeng TL, Wang Q, Fu JY, Lin Q, Bi J, Ding WC, et al. Impeded Nedd4-1-mediated RAS degradation underlies Ras-driven tumorigenesis. Cell Reports. 2014;**7**(3):871-882

[64] Baietti MF, Simicek M, Abbasi Asbagh L, Radaelli E, Lievens S, Crowther J, et al. OTUB1 triggers lung cancer development by inhibiting RAS monoubiquitination. EMBO Molecular Medicine. 2016;**8**(3):288-303

[65] Hong SW, Jin DH, Shin JS, Moon JH, Na YS, Jung KA, et al. Ring finger protein 149 is an E3 ubiquitin ligase active on wild-type v-Raf murine sarcoma viral oncogene homolog B1 (BRAF). Journal of Biological Chemistry. 2012;**287**(28):24017-24025

[66] Hemmings BA, Restuccia DF. PI3K-PKB/Akt Pathway. Cold Spring Harbor Perspectives in Biology. 2012;**4**(9):a011189. DOI: 10.1101/cshperspect.a011189

[67] Luo J, Field SJ, Lee JY, Engelman JA, Cantley LC. The p85 regulatory subunit of phosphoinositide 3-kinase down-regulates IRS-1 signaling via

the formation of a sequestration complex. Journal of Cell Biology. 2005;**170**(3):455-464

[68] Kuchay S, Duan SS, Schenkein E, Peschiaroli A, Saraf A, Florens L, et al. FBXL2-and PTPL1-mediated degradation of p110-free p85 beta regulatory subunit controls the PI(3)K signalling cascade. Nature Cell Biology. 2013;**15**(5):472-480

[69] Mao JH, Kim IJ, Wu D, Climent J, Kang HC, DelRosario R, et al. FBXW7 targets mTOR for degradation and cooperates with PTEN in tumor suppression. Science. 2008;**321**(5895):1499-1502

[70] Yada M, Hatakeyama S, Kamura T, Nishiyama M, Tsunematsu R, Imaki H, et al. Phosphorylation-dependent degradation of c-Myc is mediated by the F-box protein Fbw7. EMBO Journal. 2004;**23**(10):2116-2125

[71] King B, Trimarchi T, Reavie L, Xu LY, Mullenders J, Ntziachristos P, et al. The ubiquitin ligase FBXW7 modulates leukemia-initiating cell activity by regulating MYC stability. Cell. 2013;**153**(7):1552-1566

[72] Reavie L, Buckley SM, Loizou E, Takeishi S, Aranda-Orgilles B, Ndiaye-Lobry D, et al. Regulation of c-Myc ubiquitination controls chronic myelogenous leukemia initiation and progression. Cancer Cell. 2013;**23**(3):362-375

[73] Ntziachristos P, Lim JS, Sage J, Aifantis I. From fly wings to targeted cancer therapies: A centennial for notch signaling. Cancer Cell. 2014;**25**(3):318-334

[74] Weng AP, Ferrando AA, Lee W, Morris JP, Silverman LB, Sanchez-Irizarry C, et al. Activating mutations of NOTCH1 in human T cell acute lymphoblastic leukemia. Science. 2004;**306**(5694):269-271

[75] Malyukova A, Dohda T, von der Lehr N, Akhondi S, Corcoran M, Heyman M, et al. The tumor suppressor gene hCDC4 is frequently mutated in human T-cell acute lymphoblastic leukemia with functional consequences for notch signaling. Cancer Research. 2007;**67**(12):5611-5616

[76] Strohmaier H, Spruck CH, Kaiser P, Won KA, Sangfelt O, Reed SI. Human F-box protein hCdc4 targets cyclin E for proteolysis and is mutated in a breast cancer cell line. Nature. 2001;**413**(6853):316-322

[77] Minella AC, Grim JE, Welcker M, Clurman BE. P53 and SCFFbw7 cooperatively restrain cyclin E-associated genome instability. Oncogene. 2007;**26**(48):6948-6953

[78] Rajagopalan H, Jallepalli PV, Rago C, Velculescu VE, Kinzler KW, Vogelstein B, et al. Inactivation of hCDC4 can cause chromosomal instability. Nature. 2004;**428**(6978):77-81

[79] Hubalek MM, Widschwendter A, Erdel M, Gschwendtner A, Fiegl HM, Muller HM, et al. Cyclin E dysregulation and chromosomal instability in endometrial cancer. Oncogene. 2004;**23**(23):4187-4192

[80] Koepp DM, Schaefer LK, Ye X, Keyomarsi K, Chu C, Harper JW, et al. Phosphorylation-dependent ubiquitination of cyclin E by the SCFFbw7 ubiquitin ligase. Science. 2001;**294**(5540):173-177

[81] Welcker M, Singer J, Loeb KR, Grim J, Bloecher A, Gurien-West M, et al. Multisite phosphorylation by Cdk2 and GSK3 controls cyclin E degradation. Molecular Cell. 2003;**12**(2):381-392

[82] Hanahan D, Weinberg RA. Hallmarks of cancer: The next generation. Cell. 2011;**144**(5):646-674

[83] Jeggo PA, Pearl LH, Carr AM. DNA repair, genome stability and cancer: A historical perspective. Nature Reviews Cancer. 2016;**16**(1):35-42

[84] Zachariae W, Nasmyth K. Whose end is destruction: Cell division and the anaphase-promoting complex. Genes & Development. 1999;**13**(16):2039-2058

[85] Morgan DO. The D box meets its match. Molecular Cell. 2013;**50**(5):609-610

[86] He J, Chao WCH, Zhang ZG, Yang J, Cronin N, Barford D. Insights into degron recognition by APC/C coactivators from the structure of an Acm1-Cdh1 complex. Molecular Cell. 2013;**50**(5):649-660

[87] Pines J. Cubism and the cell cycle: The many faces of the APC/C. Nature Reviews Molecular Cell Biology. 2011;**12**(7):427-438

[88] Jiang JH, Jedinak A, Sliva D. Ganodermanontriol (GDNT) exerts its effect on growth and invasiveness of breast cancer cells through the down-regulation of CDC20 and uPA. Biochemical and Biophysical Research Communications. 2011;**415**(2):325-329

[89] Rajkumar T, Sabitha K, Vijayalakshmi N, Shirley S, Bose MV, Gopal G, et al. Identification and validation of genes involved in cervical tumourigenesis. BMC Cancer. 2011;**11**:80. DOI: 10.1186/1471-2407-11-80

[90] Marucci G, Morandi L, Magrini E, Farnedi A, Franceschi E, Miglio R, et al. Gene expression profiling in glioblastoma and immunohistochemical evaluation of IGFBP-2 and CDC20. Virchows Archiv. 2008;**453**(6):599-609

[91] Ouellet V, Guyot MC, Le Page C, Filali-Mouhim A, Lussier C, Tonin PN, et al. Tissue array analysis of expression microarray candidates identifies markers associated with tumor grade and outcome in serous epithelial ovarian cancer. International Journal of Cancer. 2006;**119**(3):599-607

[92] Zaravinos A, Lambrou GI, Boulalas I, Delakas D, Spandidos DA. Identification of common differentially expressed genes in urinary bladder Cancer. PLoS One. 2011;**6**(4):e18135. DOI: 10.1371/journal.pone.0018135

[93] Kim JM, Sohn HY, Yoon SY, Oh JH, Yang JO, Kim JH, et al. Identification of gastric cancer-related genes using a cDNA microarray containing novel expressed sequence tags expressed in gastric cancer cells. Clinical Cancer Research. 2005;**11**(2):473-482

[94] Zur A, Brandeis M. Securin degradation is mediated by fzy and fzr, and is required for complete chromatid separation but not for cytokinesis. EMBO Journal. 2001;**20**(4):792-801

[95] Geley S, Kramer E, Gieffers C, Gannon J, Peters JM, Hunt T. Anaphase-promoting complex/cyclosome-dependent proteolysis of human cyclin a starts at the beginning of mitosis and is not subject to the spindle assembly checkpoint. Journal of Cell Biology. 2001;**153**(1):137-147

[96] Ohtoshi A, Maeda T, Higashi H, Ashizawa S, Hatakeyama M. Human p55(CDC)/Cdc20 associates with cyclin a and is phosphorylated by the cyclin A-Cdk2 complex. Biochemical and Biophysical Research Communications. 2000;**268**(2):530-534

[97] Lim HH, Goh PY, Surana U. Cdc20 is essential for the cyclosome-mediated proteolysis of both Pds1 and Clb2 during M phase in budding yeast. Current Biology. 1998;**8**(4):231-234

[98] Shirayama M, Toth A, Galova M, Nasmyth K. APC(Cdc20) promotes exit from mitosis by destroying the anaphase

inhibitor Pds1 and cyclin Clb5. Nature. 1999;**402**(6758):203-207

[99] Hames RS, Wattam SL, Yamano H, Bacchieri R, Fry AM. APC/C-mediated destruction of the centrosomal kinase Nek2A occurs in early mitosis and depends upon a cyclin A-type D-box. EMBO Journal. 2001;**20**(24):7117-7127

[100] Harley ME, Allan LA, Sanderson HS, Clarke PR. Phosphorylation of Mcl-1 by CDK1-cyclin B1 initiates its Cdc20-dependent destruction during mitotic arrest. EMBO Journal. 2010;**29**(14):2407-2420

[101] Amador V, Ge S, Santamaria PG, Guardavaccaro D, Pagano M. APC/C-Cdc20 controls the ubiquitin-mediated degradation of p21 in prometaphase. Molecular Cell. 2007;**27**(3):462-473

[102] Sudo T, Ota Y, Kotani S, Nakao M, Takami Y, Takeda S, et al. Activation of Cdh1-dependent APC is required for G(1) cell cycle arrest and DNA damage-induced G(2) checkpoint in vertebrate cells. EMBO Journal. 2001;**20**(22):6499-6508

[103] Grim JE, Knoblaugh SE, Guthrie KA, Hagar A, Swanger J, Hespelt J, et al. Fbw7 and p53 cooperatively suppress advanced and chromosomally unstable intestinal cancer. Molecular and Cellular Biology. 2012;**32**(11):2160-2167

[104] Carrano AC, Eytan E, Hershko A, Pagano M. SKP2 is required for ubiquitin-mediated degradation of the CDK inhibitor p27. Nature Cell Biology. 1999;**1**(4):193-199

[105] Carrano AC, Pagano M. Role of the F-box protein Skp2 in adhesion-dependent cell cycle progression. Journal of Cell Biology. 2001;**153**(7):1381-1389

[106] Latres E, Chiarle R, Schulman BA, Pavletich NP, Pellicer A, Inghirami C, et al. Role of the F-box protein Skp2 in lymphomagenesis. Proceedings of the National Academy of Sciences of the United States of America. 2001;**98**(5):2515-2520

[107] Loda M, Cukor B, Tam SW, Lavin P, Fiorentino M, Draetta GF, et al. Increased proteasome-dependent degradation of the cyclin-dependent kinase inhibitor p27 in aggressive colorectal carcinomas. Nature Medicine. 1997;**3**(2):231-234

[108] Shim EH, Johnson L, Noh HL, Kim YJ, Sun H, Zeiss C, et al. Expression of the F-box protein SKP2 induces hyperplasia, dysplasia, and low-grade carcinoma in the mouse prostate. Cancer Research. 2003;**63**(7):1583-1588

[109] Delogu S, Wang CM, Cigliano A, Utpatel K, Sini M, Longerich T, et al. SKP2 cooperates with N-Ras or Akt to induce liver tumor development in mice. Oncotarget. 2015;**6**(4):2222-2234

[110] Lin HK, Chen ZB, Wang GC, Nardella C, Lee SW, Chan CH, et al. Skp2 targeting suppresses tumorigenesis by Arf-p53-independent cellular senescence. Nature. 2010;**464**(7287):374-U66

[111] Zhao HL, Bauzon F, Fu H, Lu ZL, Cui JH, Nakayama K, et al. Skp2 deletion unmasks a p27 safeguard that blocks tumorigenesis in the absence of pRb and p53 tumor suppressors. Cancer Cell. 2013;**24**(5):645-659

[112] Vodermaier HC. APC/C and SCF: Controlling each other and the cell cycle. Current Biology. 2004;**14**(18):R787-R796

[113] Cappell SD, Mark KG, Garbett D, Pack LR, Rape M, Meyer T. EMI1 switches from being a substrate to an inhibitor of APC/C-CDH1 to start the cell cycle. Nature. 2018;**558**(7709):313-317

[114] Nakayama KI, Nakayama K. Ubiquitin ligases: Cell-cycle control and cancer. Nature Reviews Cancer. 2006;**6**(5):369-381

[115] Inuzuka H, Tseng A, Gao DM, Zhai B, Zhang Q, Shaik S, et al. Phosphorylation by casein kinase I promotes the turnover of the Mdm2 oncoprotein via the SCF beta-TRCP ubiquitin ligase. Cancer Cell. 2010;**18**(2):147-159

[116] Beroukhim R, Mermel CH, Porter D, Wei G, Raychaudhuri S, Donovan J, et al. The landscape of somatic copy-number alteration across human cancers. Nature. 2010;**463**(7283):899-905

[117] Gong YX, Zack TI, Morris LGT, Lin K, Hukkelhoven E, Raheja R, et al. Pan-cancer genetic analysis identifies PARK2 as a master regulator of G1/S cyclins. Nature Genetics. 2014;**46**(6):588-594

[118] Veeriah S, Taylor BS, Meng S, Fang F, Yilmaz E, Vivanco I, et al. Somatic mutations of the Parkinson's disease-associated gene PARK2 in glioblastoma and other human malignancies. Nature Genetics. 2010;**42**(1):77-U98

[119] Lee SB, Kim JJ, Nam HJ, Gao BW, Yin P, Qin B, et al. Parkin regulates mitosis and genomic stability through Cdc20/Cdh1. Molecular Cell. 2015;**60**(1):21-34

[120] Poulogiannis G, McIntyre RE, Dimitriadi M, Apps JR, Wilson CH, Ichimura K, et al. PARK2 deletions occur frequently in sporadic colorectal cancer and accelerate adenoma development in Apc mutant mice. Proceedings of the National Academy of Sciences of the United States of America. 2010;**107**(34):15145-15150

[121] Yeo CWS, Ng FSL, Chai C, Tan JMM, Koh GRH, Chong YK, et al. Parkin pathway activation mitigates glioma cell proliferation and predicts patient survival. Cancer Research. 2012;**72**(10):2543-2553

Chapter 6

Ubiquitin Carboxyl-Terminal Hydrolase L1 in Parkinson's Disease

Dang Thi Phuong Thao

Abstract

Ubiquitin plays the crucial roles to maintain the ubiquitin proteasome system (UPS) functions, which were suggested that involved in Parkinson's diease (PD). Ubiquitin C-terminal hydrolase L1 (UCHL1), which was detected in Lewy bodies of nerve cells in PD brains, plays an important role for maintaining ubiquitin pool in UPS. The first UCHL1 mutation (UCHL1I93M) was found in two siblings of a PD family. By contrast, UCHL1S18Y mutation was recognized to reduce the risk of developing PD by its specific antioxidant protective function. The studies of UCHL1 in mouse models showed that lack of UCHL1 resulted in motor ataxia, degeneration of axons, and instability of free ubiquitin level. Transgenic mice expressing UCHL1I93M mutant exhibited dopaminergic neuron (DA) degeneration in MPTP-treated conditions. In this chapter, we provide a summary on recent findings related to roles of UCH-L1 in PD. Knockdown dUCH, a homolog of human UCHL1, in fly dopaminergic neuron resulted as some Parkinson's disease—like phenotype such as: (1) the underdevelopment and/or degeneration of DA neurons; (2) the shortage of dopamine in the brain; and (3) the locomotor dysfunctions. Those finding indicated that dUCH (ortholog of human UCH-L1 in *Drosophila*) plays an important role in Parkinson's disease.

Keywords: UCH-L1, Parkinson's diseases, PD model

1. Introduction

Parkinson's disease (PD) was first described in 1817 by Dr. James Parkinson. PD is considered as the second most common neurodegenerative disease which impacts 1% of the population over 60 years old [1]. The basic symptoms of Parkinson's disease are difficulty walking, slow movement, stiff and trembling limbs, balance disorders, and facial paralysis. Symptoms appear gradually and are not marked; it is difficult to recognize and often may be confused with other diseases [2]. Causes are attributed to lack of dopamine, a chemical that plays an important role in nerve signal transmission, due to degeneration/loss of dopaminergic neurons. Besides, the presence of Lewy body was also reported as one of PD symptoms although it is not clear to be a cause or a result of PD [2–4]. The complex interaction between environmental and genetic factors is also thought to be a cause of PD. However, the interaction between these factors in the PD remains unclear [5]. Previous studies have shown that mitochondrial dysfunction, oxidative stress, altered protein proteolysis, and inflammation are responsible for PD pathogenesis [6–8]. In addition, the relation to PD of many genes and their variants such as α-synuclein, PINK-1, DJ-1, LRRK2, and UCH-L1 has been reported [9, 10].

Ubiquitin carboxyl-terminal hydrolase L1 (UCH-L1) is an abundant protein in neurons. The UCH-L1 polypeptide is 24,824 Da, contains 223 amino acids, and accounts for 1–2% of brain protein in humans [11]. In addition to the brain, UCH-L1 is also expressed strongly in the peripheral nervous system, including sensory and nervous system activity. UCH-L1 functions as an important enzyme in ubiquitin proteasome system. In a form of monomer, UCH-L1 hydrolyzes the peptide bond between two ubiquitin molecules [12]. In dimer form, it plays a function as a ligase [13]. However, the functions of UCH-L1 in living cell and tissue still remain unclear. UCH-L1 has been suggested to have its functions via the role of ubiquitin proteasome system by maintaining a pool of free ubiquitin molecules [14]. Dysfunction of UCH-L1 resulted in reduction of protein degradation, consequenced by the accumulation of ubiquitinated proteins which has been believed as the cause of cell degeneration [15–17]. UCH-L1, therefore, involves in many biological processes such as cell signaling, cell cycle, DNA repair, and other ubiquitination-dependent biological processes [14–16]. Consequently, UCH-L1 had been reported as close relevant to neurodegenerative diseases, diabetes, as well as cancer [14–16, 18].

2. UCH-L1 in Parkinson's disease

In PD, there are some evidences which reveal that ubiquitin carboxyl-terminal hydrolase L1 (UCH-L1 or PGP9.5) is associated with PD. First, an UCH-L1I93M mutant was identified in two siblings from a German family with autosomal dominant PD in 1998 [18]. After that, UCH-L1S18Y mutant was discovered by Lincoln et al. [19]. UCH-L1S18Y mutant, in some cases, has been believed to have the potential in decreasing the risk of developing PD [20, 21] by its specific antioxidant protective function [22]. Moreover, UCH-L1 is also localized in Lewy bodies [23]; inclusions were found in nerve cells of PD. Although UCH-L1 had shown to have close link to PD, roles of the protein in PD are still controversial. Previous studies showed that not all mutant carriers manifest the phenotype of PD or show the protective effect to PD. The homozygous mutation of UCH-L1 (UCH-L1E7A), which also shows the decrease in hydrolytic activity, was found in three siblings of a Turkish family with progressive visual loss due to optic atrophy but neither the patients homozygous for UCH-L1E7A nor their heterozygous parents or siblings exhibited PD features on neurological examination [24]. In addition, Healy et al. and Ragland et al. showed that UCH-L1S18Y does not exhibit any protective effects against PD [25, 26].

Recently, research on UCH-L1 cellular and animal models has revealed many important findings of UCH-L1 functions in PD. An in-frame deletion of UCH-L1 gene encoding a truncated UCH-L1 lacking catalytic residue [17] in gracile axonal dystrophy (gad) mouse exhibits some PD pathogenesis such as locomotor ataxia, tremor, and difficulty in moving, and these symptoms are progressively severe [27]. Analysis of transcriptomic, proteomic, and histochemical in the brain of gad mouse revealed some prominent genes and proteins, which contribute to PD pathogenesis [28–30].

In PD research, *Drosophila* melanogaster has served as a valuable model to get insight into important features of PD pathogenesis [31–33]. The *Drosophila* model of PD provided a useful tool for tracking the integrity of the whole dopaminergic neuron system, analyzing neurodegeneration with a large number of animals to study PD in the population level, high-throughput genetic and drug screening.

In our study, knockdown *Drosophila* homolog of human UCH-L1 (dUCH) in dopaminergic neuron system of the fly brain exerted a fly model of Parkinson's disease [34]. The fly model mimics all of the main PD-related symptoms including locomotor behaviors, dopamine production, DA neuron integrity, as well as the progression of DA neuron degeneration.

2.1 Loss function of UCH-L1 homolog in *Drosophila melanogaster* resulted in locomotor dysfunction, one of most important PD phenotypes

Parkinson's disease is the most common movement disorder which is normally featured by motor symptoms. These symptoms include tremor, rigidity, bradykinesia, and postural instability. In the early stage of *Drosophila* development, the effects of dUCH knockdown on the third instar larval wandering behavior were examined by crawling assay. Heterozygous dUCH knockdown larvae displayed a tremor-like behavior which was tracked as tight wavy line when moving horizontally on agar plates. Additionally, these larvae accomplished a shorter moving path (**Figure 1A**, right panel) comparing to driver controls (**Figure 1A**, left panel) in an identical interval of time. The mean velocity of knockdown larvae was reduced to 62% of the controls (**Figure 1B**). The reduction was statistically significantly different when comparing two means (Student's t test with Welch's correction, N= 10, p< 0.01). In the adult stage, dUCH knockdown resulted in decline locomotion of *Drosophila*. Both knockdown and control flies showed age-related decline in the climbing ability (**Figure 1C**). However, the climbing ability of knockdown flies dropped sharply at day 25 and led to the difference between the control and knockdown to statistically significance at this time

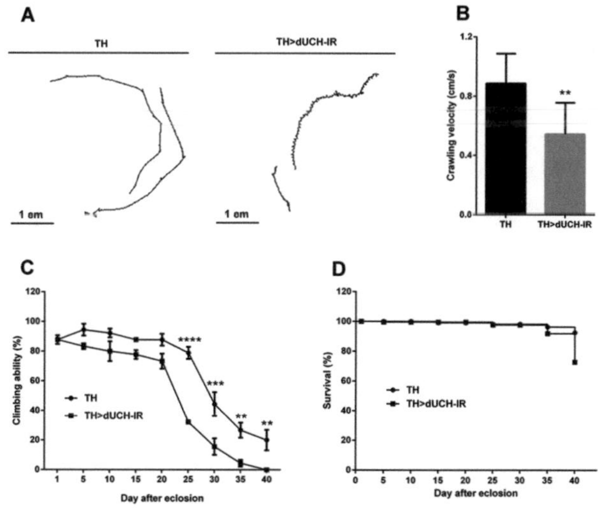

Figure 1.
*The dysfunction in locomotor behavior of dopaminergic neuron-specific dUCH knockdown flies. (A) Motion paths of control larvae (TH) and dUCH knockdown larvae (TH>dUCH-IR). Knockdown larvae exhibit shorter and disorder crawling paths (right panel) compared to control (left panel). (B) Crawling velocity of control (TH) and knockdown larvae (TH>dUCH-IR). Knockdown larvae show the reduction in crawling pace and parametric unpaired t test with Welch's correction: ** $p < 0.01$, error bars present SD. (C) The climbing ability of control (TH) and dUCH knockdown adult flies (TH>dUCH-IR). Knockdown flies start to exhibit the decline in the climbing ability at day 25 after eclosion (repeated measures two-way ANOVA with Bonferroni's post hoc test, **$p < 0.01$, ***$p < 0.001$, ****$p < 0.0001$; error bars present SEM). (D) Survival curve of control (TH) and dUCH knockdown (TH>dUCH-IR). Control and knockdown flies do not show the difference in survival (Kaplan-Meier method with log-rank test, $p > 0.05$. TH (+; +; TH-GAL4/+), and TH>dUCH-IR (+; +; TH-GAL4/UAS-dUCH-IR)).*

point (repeated measures two-way ANOVA with Bonferroni's post hoc test, $p < 0.0001$). The decline in climbing function of knockdown flies still sustained on day 30 onward (repeated measures two-way ANOVA with Bonferroni's post hoc test, $p < 0.001$ at day 30 and $p < 0.01$ at days 35 and 40) and was struck down to 0% at day 40 in which no fly can climb across 10 ml mark in 10 s (**Figure 1C**). Furthermore, the survival analysis was carried out to determine the toxicity of the reduction of dUCH. There is no significant difference in survival curve of control and knockdown flies (Kaplan-Meier method with log-rank test, $p > 0.05$) (**Figure 1D**). It illustrated that the knockdown of dUCH played no effect on *Drosophila* life span. The analysis also proved that there were no effects of death events in climbing analysis. Taken together, those data demonstrated that the reduction of dUCH specifically in dopaminergic neurons of *Drosophila melanogaster* leads to the disorder in crawling behavior and decline in locomotor ability but does not affect *Drosophila*'s life span.

2.2 Loss function of UCH-L1 homolog in *Drosophila melanogaster* exerted PD phenotype of dopaminergic neuron degeneration

Forno [35] and Thomas [2] have shown that locomotor dysfunction in PD patients may be caused by the degeneration of dopaminergic neurons (DA neurons) [2, 35]. These neurons play important roles in dopamine production for central nervous system and control multiple functions of the brain including voluntary movement. In *Drosophila*, the locomotor deficit was observed in many PD-related genes such as SNCA [36], LRRK2 [37], and PARKINR275W [37] ectopic expression followed by the degeneration of DA neurons. The study of Budnik and White showed that DA neurons assembled into some different clusters with the differences in projection and number of DA neurons [38]. In addition, DA neurons in PPL2 cluster in adult brain were demonstrated to originate from DL2a cluster [39]. It seems to be that the development of DA neurons not only occurs in embryonic to larval stages but also in larval to adult stages. In *Drosophila* model of PD, dopaminergic neurons in both larval and adult dUCH knockdown brains showed its degeneration. The DA neuronal system in the third instar larval brain lobe was classified into six clusters: DM1a, DM1b, DM2, DL1, DL2a, and DL2b (**Figure 2A**) [38, 39]. The pattern, shape, and number of DA neurons in most of clusters in dUCH knockdown and control are similar except on DL1 cluster (**Figure 2B**). In DL1 cluster, dUCH knockdown brain (TH>dUCH-IR) exhibited the reduction in numbers of DA neurons compared to driver control (TH). It indicates that the reduction of dUCH may cause the incomplete loss or underdevelopment of DA neurons in DL1 clusters of the third instar larval central brain.

On the other hand, in the adult *Drosophila*, when dUCH was specifically knocked down in dopaminergic neurons, the PPM2 dopaminergic cluster lost its neurons (**Figure 3A2, A2'** and **A3, A3'**). This loss occurred in 1-day-old flies, increased progressively by age, reached 50% on age 20, and affected all the individuals of 40-day-old flies (**Figure 3A4**). In PPM3 dopaminergic cluster, the neuron loss occurred in 1-day-old flies with highly prevalence proportion nearly 40% of the population (**Figure 3B2, B2', B3**). The number of PD disease—like flies—reached 50% as early as 10 days old and got nearly maximum prevalence on the age of 20 (**Figure 3B3**). In PPL2 dopaminergic cluster, the loss of a specific neuron was seen in knockdown brain TH>dUCH-IR (**Figure 3C2, C2'**) compared to TH-GAL4 control TH (**Figure 3C1, C1'**). DA neuron in PPL2 cluster was lost in highly prevalence proportion; 40% of knockdown flies exhibited DA loss at 1 day old. However, unlike PPM3 cluster the number of disease flies increased steadily through the age

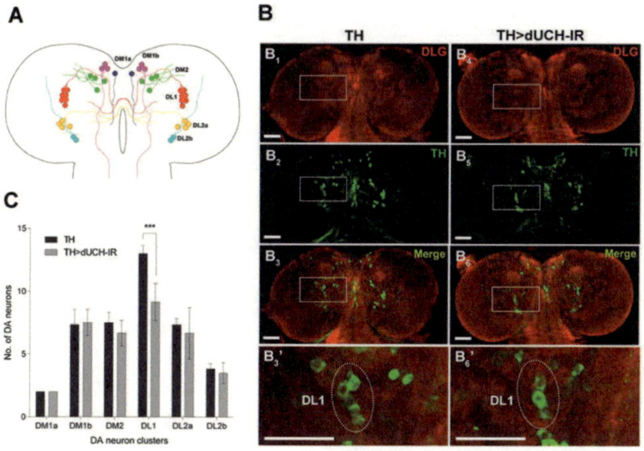

Figure 2.
*The abnormality in the number of DL1 dopaminergic neurons in dUCH knockdown larval brain. (A) A schematic representation of six DA neuron clusters DM1a, DM1b, DM2, DL1, DL2a, and DL2b and projection in Drosophila larval central brain were redrawn based on the study of Blanco et al. [39]. (B) Representative confocal images show that DA neuron clusters in the third instar larval central brain were stained with anti-TH (green). The whole brain was counterstained with anti-DLG (red). The heterozygous driver control TH-GAL4/+ (TH) on the left panel (B1, B2, B3, B3') and heterozygous dUCH knockdown TH-GAL4/ UAS-dUCH-IR (TH>dUCH-IR) on the right panel (B4, B5, B6, B6'). The boxed area in merge image (B3, B6) marks that DL1 cluster was magnified in (B'3, B'6), respectively. The number of DA neurons in DL1 clusters in dUCH knockdown brain was less than those in driver control (B'3, B'6). (C) Quantification of DA neurons in each cluster in driver control (black bars) and dUCH knockdown (gray bars). Only the difference in the number of DA neurons in DL1 clusters between dUCH knockdown and driver control was significantly different (parametric unpaired Student's t-test with Welch's correction, ***$p < 0.001$, $n = 6$). Scale bars, 50 μm. DA neuron, dopaminergic neuron; DM, dorsal medial; DL, dorsal lateral; TH, tyrosine hydroxylase; DLG, Drosophila discs large.*

and reached maximum prevalence on the age of 40 (**Figure 3C3**). The DA neuron susceptibility to dUCH reduction depended on the age and neuronal type of *Drosophila* adult brain.

2.3 Loss function of UCH-L1 homolog in *Drosophila melanogaster* resulted in dopamine shortage

The reduction of neurotransmitter, dopamine, was observed in PD patients' brain which has been thought to be a direct cause leading to PD symptoms. The production of dopamine mainly occurs in DA neurons according to catecholamine biosynthesis pathway (**Figure 4A**). In addition, studies on *Drosophila* model have demonstrated that some *Drosophila* life activities such as locomotor activity [33, 40], olfactory conditioning [41], sleep and arousal regulation [42–44], and memory and learning process [45, 46] involve dopamine (**Figure 4B**). In the *Drosophila* model of PD, the underdevelopment or degeneration of DA neurons was detected in the brain of dUCH knockdown flies. Therefore, the dopamine level in the brain of knockdown flies may be affected by these impairments. Quantification of dopamine in dUCH knockdown brain showed that dopamine was reduced at every time point of examination (1, 10, 15, 20, and 25 days after eclosion) in dUCH knockdown flies compared to driver control (**Figure 4C**) (ordinary two-way ANOVA with Tukey's multiple comparisons test, ****$p < 0.0001$). The statistical analysis indicated that there is no significant difference in the amount of dopamine at every time point in the driver control flies. However, there are significant differences in the dopamine level from days 1 to 10 and 20 to 25 in the knockdown flies (ordinary two-way ANOVA with Tukey's multiple comparisons test, ***$p < 0.001$, *$p < 0.05$). The data

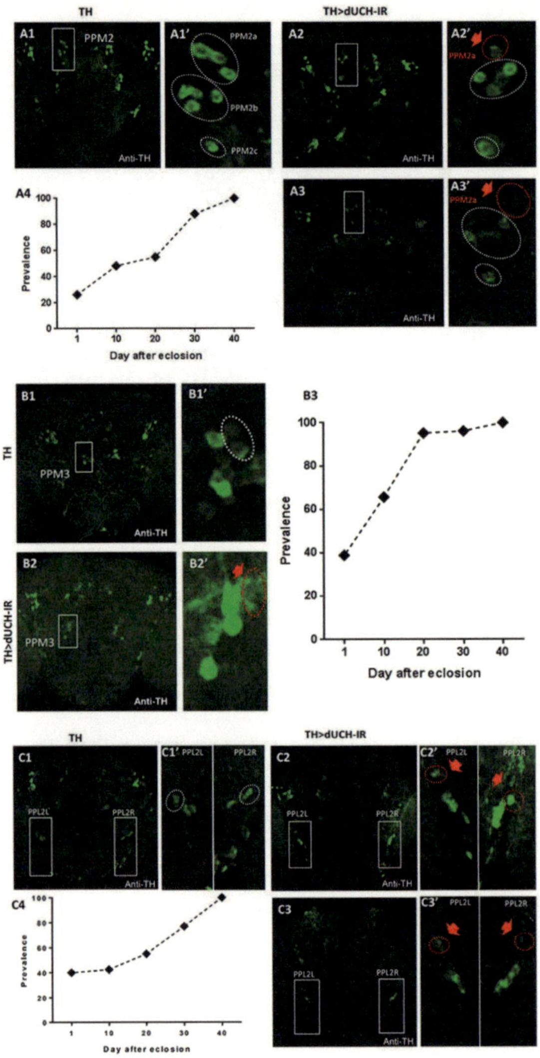

Figure 3.
The susceptibility of DA neurons in each cluster depends on age and neuronal type. Confocal images showed TH-positive neurons in PPM2 (A1–A3), PPM3 (B1–B2), and PPL2 (C1–C3) clusters in adult central brain. The prevalence proportion of dUCH knockdown flies on each cluster was described as a progressive graph, PPM2 (A4), PPM3 (B3), and PPL2 (C4). In PPM2 cluster, two kinds of partial loss of DA neurons (two to three neurons) were observed in heterogeneous dUCH knockdown flies TH>dUCH-IR (A2, A2'; A3, A3') compared to heterogeneous driver control TH-GAL4 (TH) (A1, A1'). The prevalence of PPM2 in dUCH knockdown flies increased with age; 20-day-old flies reached nearly 50% of population. In PPM3 cluster, the loss of two DA neurons was specifically seen in knockdown flies (B2, B2') compared to control (B1, B1'). However, the number of flies with this loss rose dramatically and reached 50% of population before 10 days old (B3). The loss also occurred partially in a specific DA neurons in PPL2 cluster in knockdown flies (C2, C2'; C3, C3') compared to TH-GAL4 control (C1, C1'). Loss of DA neurons in PPL2 cluster happened steadily through aging brain was described in C4; 50% of population suffered from PPL2 DA neuron loss around 20 days old (C4).

indicated that knockdown of dUCH leads to the reduction of dopamine beginning at the first day of eclosion and continuing on the following days. Interestingly, there are two significant periods (1 to 10 and 20 to 25 days) showed the reduction of dopamine in dUCH knockdown brain which may involve in the DA neuron integrity. The reduction of dopamine in dUCH knockdown flies suggested the connection between DA neuron impairment and locomotor deficit. These results can be modeled as the reduction of dUCH caused the impairment of DA neurons which leads to the reduction of dopamine followed by the dysfunction in locomotor behaviors (**Figure 4D**).

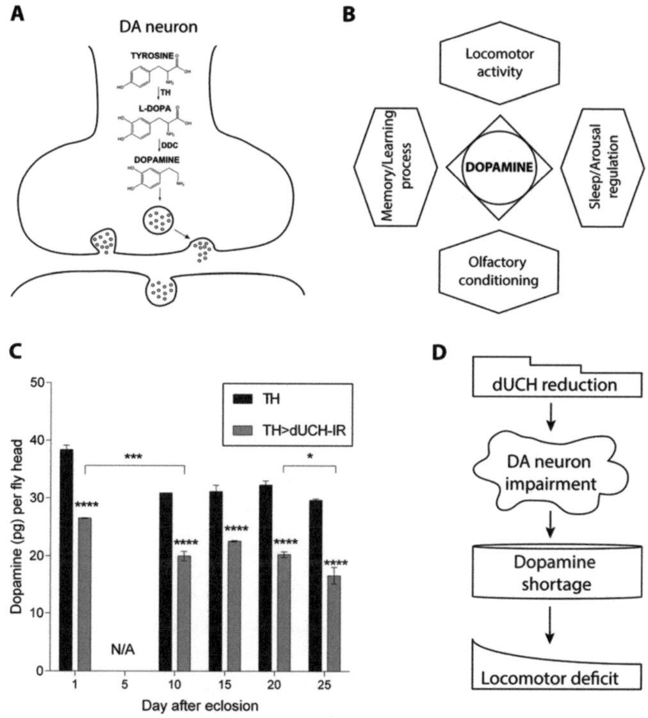

Figure 4.
*Dopamine shortage in adult dUCH knockdown brain. (A) The production of dopamine through catecholamine biosynthesis pathway in DA neuron. (B) The functions of dopamine in life activities of fruit fly [40, 41, 43–46]. (C) The quantity of dopamine per fly head in dUCH knockdown (TH>dUCH-IR) and driver control (TH). Knockdown flies show the reduction of dopamine in the brain in every time point compared to driver control (two-way ANOVA with Tukey's multiple comparisons test, ****$p < 0.0001$). Knockdown flies also show the reduction of the dopamine level in 1- to 10- and 20- to 25-day period (two-way ANOVA with Tukey's multiple comparisons test, ***$p < 0.001$, *$p < 0.05$). (D) The intermediate role of dopamine in the process of DA neuron impairment to locomotor deficit.*

3. Material and methods

3.1 Fly stocks

Fly stocks were maintained at 25°C on standard food containing 0.7% agar, 5% glucose, and 7% dry yeast. Wild-type strain Canton-S was obtained from the Bloomington Drosophila Stock Center (BDSC). RNAi lines carrying UAS-dUCH-IR fusion (GD#26468) for knockdown *Drosophila* ubiquitin carboxyl-terminal hydrolase (dUCH, CG4265) were received from the Vienna *Drosophila* Resource Center (VDRC). GAL4 drivers were used to perform the targeted knockdown of dUCH in dopaminergic neuron of *D. melanogaster*: TH-GAL4 (BDSC#8848).

3.2 Immunostaining

Larval and adult brains were dissected in cold phosphate-buffered saline (PBS) and fixed in 4% paraformaldehyde at 25°C for 15 min. After washing with 0.3% PBS-T (PBS containing 0.3% Triton-X100) twice, the samples were blocked in blocking solution (0.15% PBS-T containing 10% normal goat serum) at 25°C for 20 min. Samples were then incubated with the following primary antibodies diluted in blocking solution: rabbit anti-*Drosophila* ubiquitin carboxyl-terminal hydrolase (anti-dUCH; 1:500) at 4°C for 16 h or rabbit anti-tyrosine hydroxylase (anti-TH; 1:250; Millipore, AB152) at 4°C for 20 h. After washing with 0.3% PBS-T, samples were incubated with secondary antibodies conjugated with Alexa 488 or FITC (1,500; Invitrogen) at 25°C for 2 h and then washed and mounted in Vectashield Antifade Mounting Medium (Vector Laboratories, Japan). Finally, the samples were inspected by a confocal laser scanning microscope (Olympus Fluoview FV10i) or Olympus BX41 Microscope.

3.3 Crawling assay

Male larvae in the early third instar stage were collected randomly and washed with PBS to discard food traces. After that, larvae were transferred to agar plates containing 2% agar with a density of two to four larvae per plate. The movement of larvae was recorded by a digital camera for 60 s. The recorded videos were then converted into the AVI type by MOV to AVI converter (Pazera Jacek, Poland) and then analyzed by ImageJ (NIH, USA) with wrMTrck plugin (developed by Dr. Jesper Søndergaard Pedersen) to track larval movement and draw motion paths.

3.4 Climbing assay

Newly eclosed adult male flies were collected and transferred to conical tubes which have heights of 15 cm and diameters of 2 cm. After that, the tubes were tapped to collect the flies to the bottom, and the length of time to record the movement of flies was 30 s. The procedures were repeated five times and recorded by a digital camera. For all of the climbing experiments, the height which each fly climbed to was scored as follows: 0 (less than 2 cm), 1 (between 2 and 4 cm), 2 (between 4 and 6 cm), 3 (between 6 and 8 cm), 4 (between 8 and 10 cm), and 5 (more than 10 cm). The climbing assay was performed every 5 days until all flies lose their locomotor abilities.

3.5 Dopamine quantification

Dopamine quantification procedure was performed as described [45] with the following modifications. Thirty fly heads were homogenized in 600 μl homogenization buffer (0.1 M perchloric acid/3% trichloroacetic acid) on ice and sonicated 5 times for 30 s each and then placed on ice for 30 min. Debris were removed by centrifugation at 15,000 g for 15 min at 4°C. Fifty microliter of supernatant was utilized for HPLC analysis using Nanospace SI-2 (Shiseido, Japan) with running buffer containing 180 mM chloroacetic acid, 50 μM EDTA, 160 mM sodium hydroxide, and 8.5% acetonitrile. Sample was separated in CapCell Pak C18 UG120 column (Shiseido, Japan) at 0.5 ml/min flow rate. Dopamine was electrochemically detected by Electrochemical Detector 3005 (Shiseido, Japan). Dopamine (H8502, Sigma-Aldrich) was used to build the standard curve at 0.0025, 0.005, 0.01, 0.02, and 0.04 μM. The differences in the dopamine level of examined samples were statistically analyzed using ordinary two-way ANOVA with Tukey's multiple comparison test and graphed by GraphPad Prism 6.0 (GraphPad Software, USA).

4. Conclusion and perspective

Ubiquitin carboxyl-terminal hydrolase L1 (UCH-L1) is a protein that may play multiple roles in the cell through the effect on ubiquitin system. UCH-L1 had been found as a PD-related protein. However, the exact mechanism remains unclear. In the *Drosophila* model, specific knockdown dUCH in dopaminergic neuron exerted PD-lie phenotypes including locomotor dysfunctions, DA degeneration, and dopamine shortage. Interestingly, the degeneration of DA neurons in dUCH knockdown adult brain which occurred progressively and severely during the course of aging mimics the epidemiology of PD. These results provided one more evidence of the UCH-L1 in PD and suggest that the dUCH knockdown *Drosophila* is a promising model for studying both PD pathogenesis and epidemiology. The major advantages of the *Drosophila* model are a complex nervous system with DA neuron clusters and a conservation of the basic biological process and PD-related genes and can exhibit many PD features. The *Drosophila* model also can be utilized for either high-throughput screen.

Acknowledgements

I am grateful to Professor Yamaguchi Masamitsu and Professor Tran Linh Thuoc for their great supports to our research. I would also like to thank my students Tran Hoang Hiep, Dang Ngoc Anh Suong, and Huynh Man Anh for their great contribution in the research.

Author details

Dang Thi Phuong Thao
University of Science, Laboratory of Molecular Biotechnology, VNU-HCMC, Ho Chi Minh City, Vietnam

*Address all correspondence to: dtpthao@hcmus.edu.vn

References

[1] Polymeropoulos MH, Lavedan C, Leroy E, Ide SE, Dehejia A, Dutra A, et al. Mutation in the alpha-synuclein gene identified in families with Parkinson's disease. Science. 1997;**276**(5321):2045-2047

[2] Thomas B, Beal MF. Parkinson's disease. Human Molecular Genetics. 2007;**16**:R183-R194

[3] Seidel K, Schols L, Nuber S, Petrasch-Parwez E, Gierga K, Wszolek Z, et al. First appraisal of brain pathology owing to A30P mutant alpha-synuclein. Annals of Neurology. 2010;**67**(5):684-689

[4] Zimprich A, Biskup S, Leitner P, Lichtner P, Farrer M, Lincoln S, et al. Mutations in LRRK2 cause autosomal-dominant parkinsonism with pleomorphic pathology. Neuron. 2004;**44**(4):601-607

[5] Ross CA, Smith WW. Gene-environment interactions in Parkinson's disease. Parkinsonism & Related Disorders. 2007;**13**(Suppl 3):S309-S315

[6] Finkel T, Holbrook NJ. Oxidants, oxidative stress and the biology of ageing. Nature. 2000;**408**(6809):239-247

[7] Park J, Kim SY, Cha GH, Lee SB, Kim S, Chung J. Drosophila DJ-1 mutants show oxidative stress-sensitive locomotive dysfunction. Gene. 2005;**361**:133-139

[8] Vincent A, Briggs L, Chatwin GF, Emery E, Tomlins R, Oswald M, et al. Parkin-induced defects in neurophysiology and locomotion are generated by metabolic dysfunction and not oxidative stress. Human Molecular Genetics. 2012;**21**(8):1760-1769

[9] Goldwurm S, Zini M, Mariani L, Tesei S, Miceli R, Sironi F, et al. Evaluation of LRRK2 G2019S penetrance: Relevance for genetic counseling in Parkinson disease. Neurology. 2007;**68**(14):1141-1143

[10] Rana HQ, Balwani M, Bier L, Alcalay RN. Age-specific Parkinson disease risk in GBA mutation carriers: Information for genetic counseling. Genetics in Medicine. 2013;**15**(2):146-149

[11] Wilkinson KD, Lee KM, Deshpande S, Duerksen-Hughes P, Boss JM, et al. The neuron-specific protein PGP 9.5 is a ubiquitin carboxyl-terminal hydrolase. Science. 1989;**246**:670-673

[12] Larsen CN, Krantz BA, Wilkinson KD. Substrate specificity of deubiquitinating enzymes: Ubiquitin C-terminal hydrolases. Biochemistry. 1998;**37**(10):3358-3368

[13] Liu Y, Fallon L, Lashuel HA, Liu Z, Lansbury PT Jr. The UCH-L1 gene encodes two opposing enzymatic activities that affect alpha-synuclein degradation and Parkinson's disease susceptibility. Cell. 2002;**111**(2):209-218

[14] Hurst-Kennedy J, Chin LS, Li L. Ubiquitin C-terminal hydrolase l1 in tumorigenesis. Biochemistry Research International. 2012;**123706**

[15] Tan YY, Zhou HY, Wang ZQ, Chen SD. Endoplasmic reticulum stress contributes to the cell death induced by UCH-L1 inhibitor. Molecular and Cellular Biochemistry. 2008;**318**(1-2):109-115

[16] Bifsha P, Landry K, Ashmarina L, Durand S, Seyrantepe V, Trudel S, et al. Altered gene expression in cells from patients with lysosomal storage disorders suggests impairment of the ubiquitin pathway. Cell Death and Differentiation. 2007;**14**(3):511-523

[17] Saigoh K, Wang YL, Suh JG, Yamanishi T, Sakai Y, Kiyosawa H,

et al. Intragenic deletion in the gene encoding ubiquitin carboxy-terminal hydrolase in gad mice. Nature Genetics. 1999;**23**(1):47-51

[18] Leroy E, Boyer R, Auburger G, Leube B, Ulm G, Mezey E, et al. The ubiquitin pathway in Parkinson's disease. Nature. 1998;**395**(6701):451-452

[19] Harhangi BS, Farrer MJ, Lincoln S, Bonifati V, Meco G, De Michele G, et al. The Ile93Met mutation in the ubiquitin carboxy-terminal-hydrolase-L1 gene is not observed in European cases with familial Parkinson's disease. Neuroscience Letters. 1999;**270**(1):1-4

[20] Maraganore DM, Lesnick TG, Elbaz A, Chartier-Harlin MC, Gasser T, Kruger R, et al. UCHL1 is a Parkinson's disease susceptibility gene. Annals of Neurology. 2004;**55**(4):512-521

[21] Carmine Belin A, Westerlund M, Bergman O, Nissbrandt H, Lind C, Sydow O, et al. S18Y in ubiquitin carboxy-terminal hydrolase L1 (UCH-L1) associated with decreased risk of Parkinson's disease in Sweden. Parkinsonism & Related Disorders. 2007;**13**(5):295-298

[22] Kyratzi E, Pavlaki M, Stefanis L. The S18Y polymorphic variant of UCH-L1 confers an antioxidant function to neuronal cells. Human Molecular Genetics. 2008;**17**(14):2160-2171

[23] Lowe J, McDermott H, Landon M, Mayer RJ, Wilkinson KD. Ubiquitin carboxyl-terminal hydrolase (PGP 9.5) is selectively present in ubiquitinated inclusion bodies characteristic of human neurodegenerative diseases. The Journal of Pathology. 1990;**161**(2):153-160

[24] Bilguvar K, Tyagi NK, Ozkara C, Tuysuz B, Bakircioglu M, Choi S, et al. Recessive loss of function of the neuronal ubiquitin hydrolase UCHL1 leads to early-onset progressive neurodegeneration. Proceedings of the National Academy of Sciences of the United States of America. 2013;**110**(9):3489-3494

[25] Healy DG, Abou-Sleiman PM, Casas JP, Ahmadi KR, Lynch T, Gandhi S, et al. UCHL-1 is not a Parkinson's disease susceptibility gene. Annals of Neurology. 2006;**59**(4):627-633

[26] Ragland M, Hutter C, Zabetian C, Edwards K. Association between the ubiquitin carboxyl-terminal esterase L1 gene (UCHL1) S18Y variant and Parkinson's Disease: A HuGE review and meta-analysis. American Journal of Epidemiology. 2009;**170**:1344-1357

[27] Yamazaki K, Wakasugi N, Tomita T, Kikuchi T, Mukoyama M, et al. Gracile axonal dystrophy (GAD), a new neurological mutant in the mouse. Proceedings of the Society for Experimental Biology and Medicine. 1988;**187**:209-215

[28] Bonin M, Poths S, Osaka H, Wang YL, Wada K, Riess O. Microarray expression analysis of gad mice implicates involvement of Parkinson's disease associated UCH-L1 in multiple metabolic pathways. Brain Research. Molecular Brain Research. 2004;**126**(1):88-97

[29] Castegna A, Thongboonkerd V, Klein J, Lynn BC, Wang YL, Osaka H, et al. Proteomic analysis of brain proteins in the gracile axonal dystrophy (gad) mouse, a syndrome that emanates from dysfunctional ubiquitin carboxyl-terminal hydrolase L-1, reveals oxidation of key proteins. Journal of Neurochemistry. 2004;**88**(6):1540-1546

[30] Goto A, Wang YL, Kabuta T, Setsuie R, Osaka H, Sawa A, et al. Proteomic and histochemical analysis of proteins involved in the dying-back-type of axonal degeneration in the gracile axonal dystrophy (gad) mouse. Neurochemistry International. 2009;**54**(5-6):330-338

[31] Haywood AF, Staveley BE. Parkin counteracts symptoms in a Drosophila model of Parkinson's disease. BMC Neuroscience. 2004;**5**:14

[32] Liu Z, Wang X, Yu Y, Li X, Wang T, Jiang H, et al. A Drosophila model for LRRK2-linked parkinsonism. Proceedings of the National Academy of Sciences of the United States of America. 2008;**105**(7):2693-2698

[33] Riemensperger T, Issa AR, Pech U, Coulom H, Nguyen MV, Cassar M, et al. A single dopamine pathway underlies progressive locomotor deficits in a Drosophila model of Parkinson disease. Cell Reports. 2013;**5**(4):952-960

[34] Tran HH et al. Drosophila ubiquitin C-terminal hydrolase knockdown model of Parkinson's disease. Scientific Reports. 2018;**8**(1):4468

[35] Forno LS. Neuropathology of Parkinson's disease. Journal of Neuropathology and Experimental Neurology. 1996;**55**(3):259-272

[36] Feany MB, Bender WW. A Drosophila model of Parkinson's disease. Nature. 2000;**404**(6776):394-398

[37] Wang C, Lu R, Ouyang X, Ho MW, Chia W, Yu F, et al. Drosophila overexpressing parkin R275W mutant exhibits dopaminergic neuron degeneration and mitochondrial abnormalities. The Journal of Neuroscience. 2007;**27**(32):8563-8570

[38] Budnik V, White K. Catecholamine-containing neurons in *Drosophila melanogaster*: Distribution and development. The Journal of Comparative Neurology. 1988;**268**(3):400-413

[39] Blanco J, Pandey R, Wasser M, Udolph G. Orthodenticle is necessary for survival of a cluster of clonally related dopaminergic neurons in the Drosophila larval and adult brain. Neural Development. 2011;**6**:34

[40] Riemensperger T, Isabel G, Coulom H, Neuser K, Seugnet L, Kume K, et al. Behavioral consequences of dopamine deficiency in the Drosophila central nervous system. Proceedings of the National Academy of Sciences of the United States of America. 2011;**108**(2):834-839

[41] Selcho M, Pauls D, Han KA, Stocker RF, Thum AS. The role of dopamine in Drosophila larval classical olfactory conditioning. PLoS One. 2009;**4**(6):e5897

[42] Andretic R, van Swinderen B, Greenspan RJ. Dopaminergic modulation of arousal in Drosophila. Current Biology. 2005;**15**(13):1165-1175

[43] Kume K, Kume S, Park SK, Hirsh J, Jackson FR. Dopamine is a regulator of arousal in the fruit fly. The Journal of Neuroscience. 2005;**25**(32):7377-7384

[44] Ueno T, Tomita J, Tanimoto H, Endo K, Ito K, Kume S, et al. Identification of a dopamine pathway that regulates sleep and arousal in Drosophila. Nature Neuroscience. 2012;**15**(11):1516-1523

[45] Aso Y, Siwanowicz I, Bracker L, Ito K, Kitamoto T, Tanimoto H. Specific dopaminergic neurons for the formation of labile aversive memory. Current Biology. 2010;**20**(16):1445-1451

[46] Liu C, Placais PY, Yamagata N, Pfeiffer BD, Aso Y, Friedrich AB, et al. A subset of dopamine neurons signals reward for odour memory in Drosophila. Nature. 2012;**488**(7412):512-516

Section 2

Enzymes and Mechanisms within the UPS

Chapter 7

New Insights into the Mechanisms Underlying NEDD8 Structural and Functional Specificities

Elena Santonico

Abstract

Ubiquitin (Ub) and ubiquitin-like (Ubl) proteins are small polypeptides that are conjugated to substrates affecting their activity and stability. Cells encode "receptors" containing Ub-/Ubl-binding domains that interpret and translate each modification into appropriate cellular responses. Among the different Ubls, NEDD8, which is the ubiquitin's closest relative, retains many of the structural determinants that enable ubiquitin the ability to target proteins to degradation. Nevertheless, the direct involvement of NEDD8 conjugation to proteasome recruitment has been proved only in a few cases. To date, well-defined major NEDD8 substrates are primarily members of the cullin family, and cullin neddylation does not appear to mark these proteins for degradation. Various studies have demonstrated that selectivity between ubiquitin and NEDD8 is guaranteed by small but substantial differences. Nevertheless, several issues still need to be addressed, mainly concerning which interaction surfaces mediate NEDD8 function and what domains recognize them. Recently, two novel domains identified in KHNYN and N4BP1 proteins have shed new light on this research area. Here, I discuss some recent reports that contributed to shed light on the mechanisms underlining the discrimination between ubiquitin and NEDD8. Understanding the details of these molecular mechanisms represents a prominent facet for the identification of new therapeutic targets.

Keywords: NEDD8, ubiquitination, neddylation, ubiquitin-binding domains, KHNYN, N4BP1

1. Introduction

Protein diversity in living organisms is the result of several mechanisms acting at different steps of gene expression. Alternative splicing determines the production of a variety of proteins from a single pre-mRNA, and different promoters and termination sites increase the protein diversity during gene transcription. An additional level of complexity is achieved through posttranslational modifications (PTMs). More than 90,000 individual PTMs have been detected using biochemical and biophysical methods [1]. These modifications extend proteome diversity by inducing structural changes, such as the covalent binding of functional groups (phosphate, acetyl, methyl, lipids, and others) or the cleavage and the selective degradation of regulatory subunits. Such modifications in turn play a central role in regulating protein function, as they finely tune intermolecular interactions that modulate almost all biological processes. Thus, it is not surprising that almost 5% of the human genome

IntechOpen

encodes enzymes in charge of catalyzing reactions leading to PTMs. Different from the majority of the biological processes that participate in increasing protein diversification, a key feature of PTMs is the reversibility. Indeed, with only a few exceptions such as proteolysis, deamidation [2], and the recently reported eliminylation catalyzed by phosphothreonine lyases [3], which are irreversible, PTMs are typically regulated by a set of enzymes that coordinate the temporary addition and removal of protein modifications, thus ensuring a finely tuned control of the process.

Typically, signal processing requires a third component that reads the PTM and transmits the signal to the downstream effectors of the signaling pathway. Since different signal transducers can recognize the same type of posttranslational modification, specificity is generally ensured by a context-specific recognition, meaning that the transducer interacts with the target molecule by recognizing a posttranslationally modified sequence motif that is unique for that target.

The specificity and cross talk underlining the signaling mechanisms have been illustrated by Wendell Lim and Tony Pawson, who coined the paradigm of the "three-part toolkit" applied to tyrosine phosphorylation [4]. Kinases "write" the modification mark, SH2 domain-containing proteins "read" it, and finally phosphatases "erase" it. Such a simple model, implying for each modification writer, eraser, and reader modules, can be easily extended to most modification systems. As might be expected, these three module toolkits are not isolated systems, and similar modifications have different degrees of cross talk. For instance, in the case of Ser/Thr or Tyr phosphorylation, writers and erasers display some promiscuity with a large number of kinases and phosphatases being able to phosphorylate or dephosphorylate both Ser/Thr and Tyr with low selectivity. The reader modules, on the other hand, are rather selective with SH2/PTB and 14-3-3/FHA being able to bind only pTyr- or pSer-/Thr-containing peptides, respectively.

The balance between specificity and promiscuity in modifications toolkits is particularly relevant in the ubiquitination process, where up to 16 different ubiquitin-like peptides can be covalently linked to a variety of proteins to support very diverse cellular events [5].

Ubiquitination is an ATP-dependent process that involves the action of at least three enzymes: an ubiquitin-activating enzyme, also known as E1 enzyme, which catalyzes the first step in the ubiquitination reaction; an ubiquitin-conjugating enzyme, dubbed E2, which performs the second step in the ubiquitination reaction; and the E3 ligase involved in the ligation step, consisting in the transfer of ubiquitin from the E2 to an internal lysine residue, thus generating an isopeptide bond. If the "three-part toolkit" paradigm is applied to the ubiquitination process, E3 ligases (writers), deubiquitinating enzymes (erasers), and ubiquitin-binding domains (readers), respectively, create, transmit, and cancel the ubiquitin signal. Moreover, the mechanisms underlying the conjugation, recognition, and removal of different Ubls have been modified by evolution into distinct systems with their associated, not interchangeable, modification toolkits, with the consequence that each Ubl-associated path has evolutionary gained a specific biological function.

The case of ubiquitin and neural precursor cell expressed developmentally downregulated protein 8 (NEDD8) is particularly intriguing. NEDD8 is the ubiquitin's closest relative (58% sequence identity and 80% sequence similarity) and can be conjugated to target substrates in a process, called neddylation, which is similar to ubiquitination, but relies on its own enzymatic kit and targets a specific and limited set of substrates. The clear discrimination between conjugation and deconjugation of ubiquitin or NEDD8 is the result of a few subtle differences between these two molecules, primarily aimed at ensuring that the substrates targeted for neddylation are not modified by the addition of ubiquitin. In this way, the neddylation pathway remains insulated, only affecting defined biological processes.

Strikingly, alongside a clear function for neddylation, the ubiquitin readers are in general rather promiscuous and bind to either modifications, though not necessarily with similar affinity. Moreover, neddylation modulates ubiquitination efficiency. Indeed, NEDD8 conjugation to cullins is an essential step in the stimulation of the ubiquitin ligase activity of the Cullin RING-ligase (CRL) complexes, clearly indicating a cross talk between the two PTMs.

Here the current comprehension of the mechanisms underlining the discrimination between ubiquitin and NEDD8 is recapitulated, together with the emerging understanding of NEDD8 recognition domains.

2. The neddylation pathway

The *NEDD8* gene encodes a small protein of 81 amino acids, which is 58% identical and 80% homologous to ubiquitin. Like ubiquitin, the maturation of the precursor protein requires a proteolytic step that allows the exposure of the Gly_{75}-Gly_{76} residues, which are critical for the conjugation process (**Figure 1A**). This reaction is catalyzed by two enzymes: ubiquitin C-terminal hydrolase isozyme 3 (UCHL3), which can also process the ubiquitin precursor [6], and NEDD8-specific protease 1 (NEDP1), also called deneddylase 1 (DEN1) or SUMO-1-/sentrin-/SMT3-specific peptidase

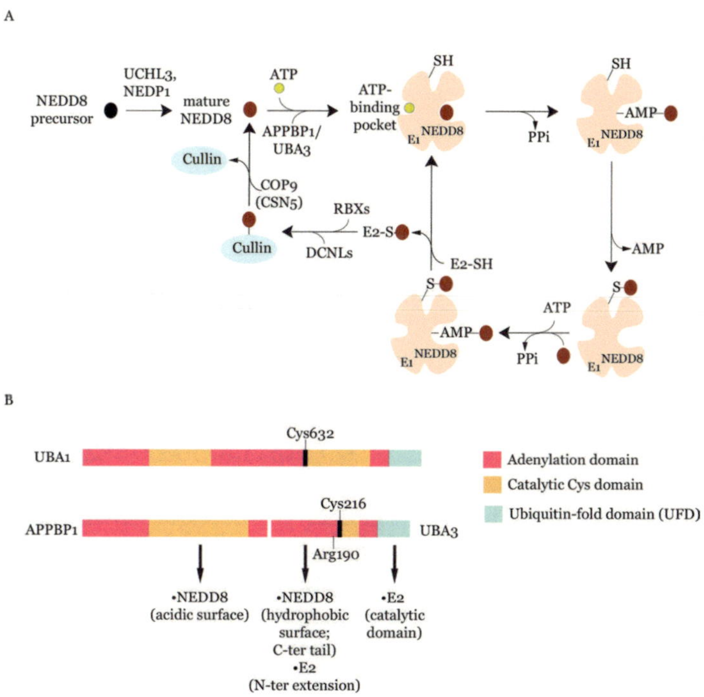

Figure 1.
(A) Overview of NEDD8 activation, conjugation, and deneddylation. The NEDD8 precursor is cleaved at the carboxyl terminus to allow the exposure of the Gly75-Gly76 residues. Free mature NEDD8 is adenylated by APPBP1/UBA3 in an ATP-dependent process; the APPBP1-UBA3-bound NEDD8 is transferred from AMP to the active-site Cys with the release of AMP; the ATP-dependent adenylation of a second free NEDD8 by the Cys-NEDD8-loaded APPBP1-UBA3 is followed by the transfer of the activated NEDD8 from APPBP1/UBA3 to the catalytic Cys of the E2 enzyme. The E2-bond NEDD8 is transferred to a Lys residue at the carboxyl-terminal ends of cullin proteins by RBXs RING-ligases in the context of the CRL complexes; this step requires the activity of the DCNL proteins. Deneddylation of cullins by CSN5 subunit of COP9 complex releases free NEDD8 for another cycle of conjugation/deconjugation. (B) Organization of the APPBP1/UBA3 heterodimer and comparison of the domain structures of NEDD8's E1 with the E1 for ubiquitin. Direct interactions between the heterodimer APPBP1-UBA3 and NEDD8, ATP, and the E2 enzymes are shown.

8 (SENP8). NEDP1 shows a remarkable specificity for the NEDD8 precursor as it cannot cleave ubiquitin nor the small ubiquitin-like modifier (SUMO) precursors [7]. Once converted into the mature form, NEDD8 is activated by the NEDD8-activating E1 enzyme, a heterodimer composed of amyloid-β precursor protein-binding protein 1 (APPBP1, also called NAE1) and the ubiquitin-activating enzyme 3 (UBA3), the two subunits corresponding, respectively, to the N-terminal and C-terminal halves of a typical single-chain E1 [8–10]. Again, like all the E1 enzymes, three different domains can be identified in NAE1: an adenylation domain, a catalytic cysteine-containing domain harboring the Cys residue involved in the E1-NEDD8 thioester linkage, and a domain structurally resembling ubiquitin (called ubiquitin-fold domain, UFD) that binds the E2 [11]. The first step of the reaction requires the ATP-dependent adenylation of the NEDD8 C-terminus, mediated by the UBA3 adenylation domain. The Ubl is then transferred to the catalytic cysteine in UBA3 with the formation of a thioester intermediate and the release of AMP. At this step, the Cys~NEDD8-loaded E1 molecule goes through a second cycle of adenylation of a free NEDD8, which is followed by transfer of the first activated NEDD8 from the active cysteine of the E1 to the catalytic cysteine of the E2. Being NAE1 the only E1 enzyme for NEDD8, it has received attention as a promising target in new cancer therapeutic strategies, and several drugs have been developed in order to inhibit its activity. Among them, MLN4924, an analogue of adenosine 5′-monophosphate, binds to the ATP-binding site in UBA3 and forms an irreversible MLN4924-UBA3 adduct that inhibits NAE1, thus causing the disruption of protein turnover and cellular apoptosis by deregulation of DNA synthesis [12]. MLN4924 has shown significant anticancer efficacy in preclinical studies, and it has been advanced into several phase I clinical trials for certain solid tumors and hematological malignancies [12].

Structural studies of the heterodimer APPBP1-UBA3 in complex with NEDD8 showed that the interface between the two partners requires three contact sites: a hydrophobic surface in NEDD8 that contacts the adenylation domain portion of UBA3, an electrostatic interaction between NEDD8 and a charged surface on the E1-specific catalytic cysteine domain portion of APPBP1, and, finally, the partial stiffening of NEDD8's C-terminal tail that extends away from the globular domain and sits in a channel on the surface of UBA3 (**Figure 1B**). As shown below, each of these interactions participates in determining the specificity of NEDD8 for its E1 enzyme [9]. A second trans-thiolation reaction involves the transfer of NEDD8 from the active-site Cys of APPBP1-UBA3 onto the active-site Cys of the NEDD8 E2 enzyme, which in metazoans can be UBC12 (also known as UBE2M) or UBE2F. The E1-E2 interface involves two surfaces. The first one is shared with other E1-E2 pairs and mediates the interaction between the catalytic core domain of the E2 and the UBA3's UFD [13]. The second surface, which is unique for the NEDD8's E1-E2 pair, consists in an amino terminus extension in the NEDD8 E2s that interacts with a binding groove in the adenylation domain of UBA3. The combination of these two binding sites explains why the NEDD8 pathway remains substantially isolated from the ubiquitin enzymatic cascade despite sharing a common structural scaffold.

Once the E2 enzyme is charged, NEDD8 is conjugated to different substrates through the catalytic activity of several E3 ligases, resulting in the formation of an isopeptide bond linking the terminal carboxyl group of NEDD8 with the ε-NH2 group of a lysine residue of the substrate [14–16]. Interestingly, all the NEDD8 E3 enzymes that have been characterized so far also function as E3 ligases for ubiquitin. Among them, the best studied are the RING domain subunits RING-box proteins 1 and 2 (RBX1 and 2), which are components of the CRL complexes. RBX ligases transfer NEDD8 onto the cullins, a protein family acting as molecular scaffold of CRL complexes [17–19]. Cullin neddylation increases CRL ubiquitination activity via conformational changes that optimize ubiquitin transfer to the target proteins [20]. Considering that cullins can potentially assemble hundreds of CRL complexes by utilizing different substrate receptors and switching from one receptor to another thanks to a finely tuned mechanism, CRLs are responsible for targeting a myriad of cellular substrates to degradation. Differently from RBX1, which recruits both NEDD8 E2s, RBX2 shows a clear preference for UBE2F [21]. Interestingly, biochemical and structural studies have demonstrated that binding of E1 and E3 to NEDD8 E2 UBC12 are mutually exclusive, so that the unidirectionality of the E1-E2-E3 conjugation cascade requires a switch based on the different binding affinities of NEDD8 toward charged and uncharged enzymes [13, 22–24].

RBX1 and 2 require auxiliary E3 factors to direct NEDD8 toward the correct lysine residue. These factors are encoded by proteins of the defective in cullin neddylation 1 (DCN1) family [25]. Human genome encodes for five DCN1-like proteins termed DCNL1–DCNL5. These DCNLs have distinct N-terminal domains but share a conserved C-terminal potentiating neddylation (PONY) domain that directly binds to cullins. Moreover, all DCNLs interact strongly with the CRL regulator CAND1, which promotes the release of old and the association of new substrate adaptors to cullin–RING core complexes [20, 26, 27]. These complexes including CAND1/deneddylated cullin/DCNL act as "reserve" CRLs that are ready to be rapidly activated [28]. Finally, it has been recently shown that additional factors can be required to ensure the efficient neddylation of specific cullins. TFB3, an essential subunit of the TFIIH complex, which mediates transcription initiation and is also involved in nucleotide excision repair (NER) [29], is needed for efficient neddylation of CUL3 and for neddylation and ubiquitylation of Rtt101, two yeast cullins regulating DNA-associated processes [30].

The COP9 signalosome (CSN) complex performs cullins deneddylation. Of the eight subunits that make up the complex, CSN5 is the one carrying the deneddylating activity. Furthermore, the CSN complex stably binds deneddylated CRLs and sterically inhibits RBX1-mediated E2 activation [31, 32]. The experimental evidences indicate that CSN5 is kept in an auto-inhibited state within the CSN complex and that its activation requires the interaction with neddylated CRLs [33]. It has been also observed that, despite being the major cullin deneddylase, CSN5 does not deconjugate hyper-neddylated cullins [33]. On the contrary, the deneddylating enzyme NEDP1 is responsible for the processing of non-cullin substrates [34], but it can also convert the hyper-neddylated cullin in the mononeddylated form [33]. Intriguingly, in *Arabidopsis*, NEDP1 has been shown to be involved in recovering NEDD8 moieties from autoneddylated NAE1 subunits in order to maintain the NEDD8 pathway activity, thus suggesting a more complex involvement of the deneddylase in NEDD8 homeostasis [34, 35]. Finally, NEDD8 peptides and fusion proteins can be also cleaved by USP21 [36], ataxin-3 [37], the parasite hydrolase PfUCH54 [38], UCH-L3 [6], and its orthologue in *S. cerevisiae* Yuh1 [39], all exhibiting a dual specificity for ubiquitin and NEDD8 precursors.

3. Structure of the ubiquitin-like protein NEDD8

Similar to ubiquitin and other Ubls, the structure of NEDD8 consists of a globular core called the β-grasp fold, comprising five-stranded mixed β-sheet and an α-helix, and a flexible C-terminal tail that projects away from the body of the globular domain and terminates with the Gly-Gly motif (Gly_{75}-Gly_{76}) [40]. The surface distribution of charges is asymmetric, as in the ubiquitin structure, with a predominantly acidic face and a hydrophobic surface characterized by the Ile44-centered patch, including residues Leu8, Ile44, His68, and Val70. The Ile44 patch, which mediates the recognition of ubiquitin by the majority of the ubiquitin-binding domains, is perfectly conserved in NEDD8.

Only three amino acids differ between the ubiquitin orthologs in *H. sapiens* and *S. cerevisiae*. Similarly, a high conservation is also evident for NEDD8, with approximately 80% identity between the human and yeast orthologs. The sequence alignments of ubiquitin and NEDD8 from yeast to human, performed with protein visualization (ProViz) [41], are shown in **Figure 2A**. Below the alignments, colored boxes that mark identical residues, conservative substitutions, or non-conserved positions represent the sequence conservation degree of ubiquitin and NEDD8 in eukaryotes. In **Figure 2B**, residues that are divergent between the two molecules and not necessarily conserved during evolution within each sequence are highlighted [40]. As shown, positions 4, 12, 14, 31, 63, 64, and 72 share the common feature of being highly conserved in ubiquitin and NEDD8, while differing between the two molecules. Among them, Lys63 in ubiquitin is a substrate for the conjugation of polyubiquitin chains that are involved in several biological processes, primarily "proteasome-independent," such as inflammatory signal transduction, DNA repair, and endocytosis [42, 43] (**Figure 2B**). Interestingly, unlike most of the lysine residues, which are conserved between ubiquitin and NEDD8, Lys63 is always a Gly in the Ubl, thus indicating that this position has acquired a completely different role in the Ubl. On the contrary, positions 22, 25, 28, 51, and 53, despite being divergent between the two molecules, show a certain degree of variability, to a different extent from one position to the other. For example, position 22 is almost always a Thr in ubiquitin, with the only exception of the Ser22 in *C. intestinalis*, representing a conservative substitution. In NEDD8, a Lys occupies position 22, with the single

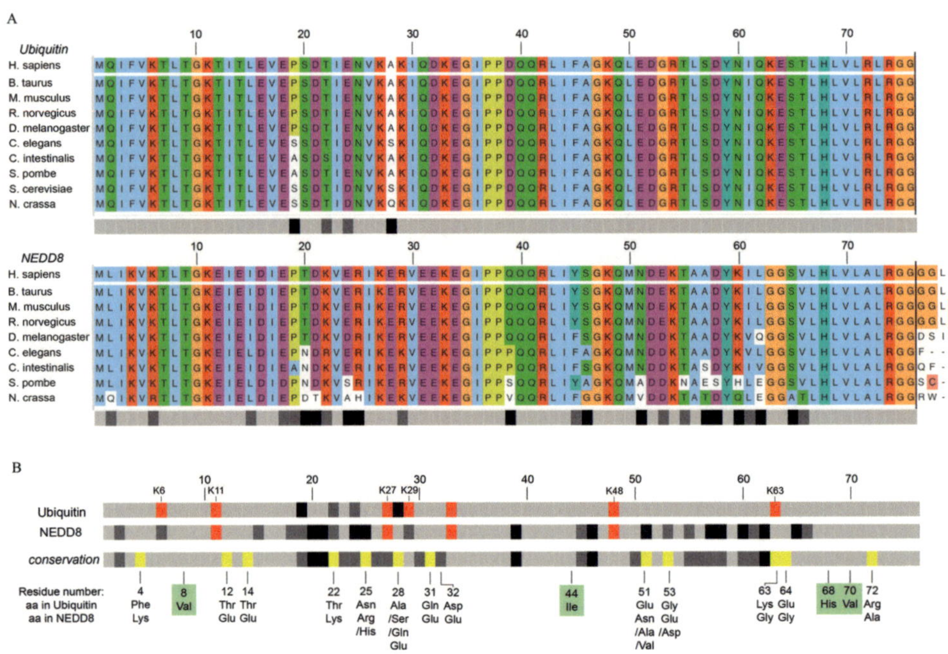

Figure 2.
(A) Amino acid sequence alignment of ubiquitin and NEDD8 orthologs performed with the protein visualization (ProViz). The vertical black line indicates the C-termini of the mature proteins. Below the alignment conserved and non-conserved positions are shown as boxes marked with different colors. Invariant positions are shown in light gray; positions having similar chemical properties are marked in dark gray; divergent positions are shown in black. (B) Conservation between ubiquitin and NEDD8 sequences in human. The same colored boxes described in (A) are here represented with the addition of red boxes corresponding to the Lys residues. The conservation is shown below. Invariant positions are shown in light gray; positions having similar chemical properties in dark gray; conserved/divergent positions in yellow, while divergent positions are shown in black. The amino acids occupying the positions that are described in the text are indicated (aa in ubiquitin and aa in NEDD8). Residues included in the hydrophobic patch are in shown in green boxes.

exception of *S. cerevisiae* that has an Arg at that position. The negatively charged Glu in NEDD8 invariably occupies position 28, while in ubiquitin the small nonpolar Ala and Ser or Gln, which are both polar but uncharged, occupy this position. Position 51 is always a Glu in ubiquitin, a negatively charged residue; conversely, in NEDD8 there is almost always an Asn, which is a polar residue, or alternatively an Ala or a Val, which are both nonpolar aliphatic amino acids, suggesting that these positions allow a greater variability compared to the same position in ubiquitin. Finally, position 52 is always a Gly in ubiquitin, a residue that acts as a flexibility source, while a negatively charged residue (Glu or Asp) occupies this position in NEDD8.

The surface distribution of these residues on ubiquitin and NEDD8 is shown in **Figure 3**, together with the hydrophobic patch (shown in green). As previously observed, when mapped on the ubiquitin and NEDD8 surfaces, the conserved/divergent residues are organized in clusters along each side of the molecule [40]. In one cluster, positions 22, 25, 28, and 31 form a line that is characterized by positive (Arg, Lys) and negative (Glu, Glu) charges in NEDD8. In ubiquitin, this region is mainly uncharged, with three polar residues (Thr, Asn, Gln) and the nonpolar aliphatic Ala.

Positions 51 and 53 appear to be in continuity with the residue in position 22 and arranged like the short arm of an "L," with the cluster constituting the longer arm. This "short arm" is charged/polar in NEDD8 (Glu, Asn) and nonpolar/charged in ubiquitin (Gly, Glu).

The second cluster is positioned at the opposite side of the Ubl like a stripe of aligned residues and includes positions 4, 12, 14, 63, and 64. The chemical nature of this pattern, looking from the top down in **Figure 3**, is nonpolar (Gly, Gly) and charged (Lys, Glu, Glu) in NEDD8 and charged (Lys, Glu), aromatic (Phe), and polar (Thr, Thr) in ubiquitin. Therefore, with the exception of Arg72 which is surrounded by residues that are identical between ubiquitin and NEDD8, the remaining divergent positions are structurally assembled in patterns characterizing different surface regions on NEDD8 and ubiquitin and having chemical features that are distinctive for each molecule. The features of these two groups of residues clearly suggest that they could mediate interactions with different NEDD8-specific partners. Accordingly, together with Ala72, most of these positions have been demonstrated to be involved in the discrimination between ubiquitin and NEDD8 by enzymes that are involved in the conjugation and deconjugation reactions, as described below.

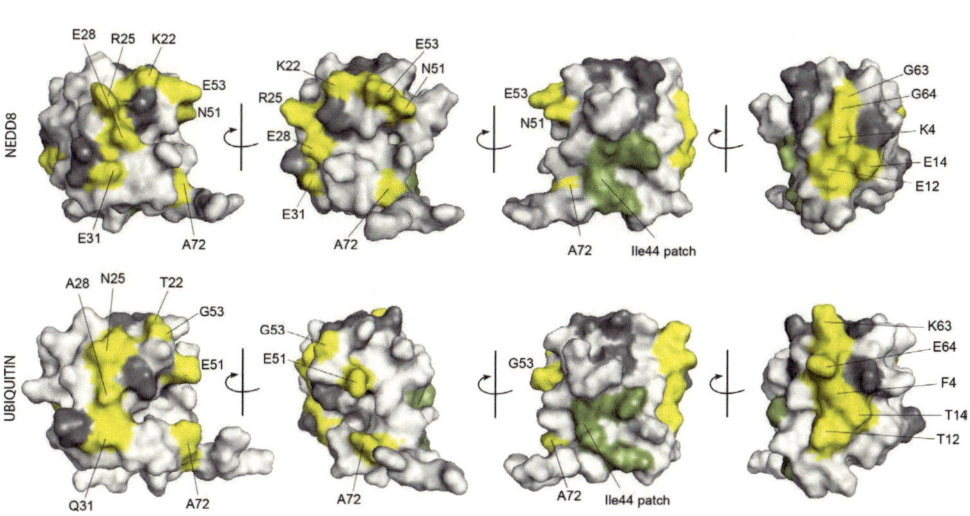

Figure 3.
Ribbon representation of NEDD8 (upper panels) and ubiquitin (lower panels). The conserved/divergent surfaces and the main amino acid differences are shown.

4. Molecular mechanisms underlining the discrimination between ubiquitin and NEDD8 by neddylating and deneddylating enzymes

The current knowledge regarding the mechanisms allowing the discrimination between ubiquitin and NEDD8 by neddylating and deneddylating enzymes is summarized in **Figure 2B**. The first key difference concerns position 72, which is the unique divergent residue in ubiquitin and NEDD8 C-terminal tails. As previously described, an Ala in NEDD8 and an Arg in ubiquitin occupy this position. Comparison of the crystal structures of NEDD8 alone [40] and within the complex with NEDP1 [44] or the E1-activating enzyme [45] reveals that, unlike both enzymes, NEDD8 does not undergo large conformational changes upon binding. The main exception is given by significant conformational changes observed in the NEDD8 C-terminal tail. Particularly, in the complex of NEDD8 with UBA3, the last three residues in the flexible tail, which are disordered in free NEDD8, adopt an extended more rigid conformation as it docks into the binding pocket in UBA3. This rearrangement allows the direct interaction of Leu71 and Ala72 with UBA3's residues Leu206 and Tyr207 in the so-called crossover loop, which is essential for the binding preference shown by UBA3 toward NEDD8 [40, 45]. Since all the remaining residues in NEDD8 that contact UBA3 are conserved in ubiquitin, the discrimination between NEDD8 and ubiquitin, at the level of the first step in the neddylation process, is entirely dependent on a single amino acid difference that is sufficient to ensure that ubiquitin is not mistaken for NEDD8. Accordingly, the substitution Ala72Arg can disrupt the specificity of the NEDD8 E1-activating enzyme [45]. On the other hand, it has been ascertained that NEDD8 is qualitatively competent in the interaction with the E1 ubiquitin enzyme. Indeed, the presence of an alanine in position 72 does not cause any repulsion with the binding groove on the E1 enzyme, but the kinetic of the reaction is significantly slower, thus making the NEDD8 thioesterification reaction by the ubiquitin E1 enzyme a kinetically disadvantaged process [40, 46, 47]. This intrinsic difference between process that is clearly prevented and another that is only disadvantaged indicates that NEDD8 performs functions that are strictly specific and not interchangeable. On the other hand, there are biological conditions in which, to a certain extent, NEDD8 can perform the functions typically absolved with ubiquitin. For example, under diverse stress conditions, the ubiquitin E1 enzyme UBE1 can activate NEDD8 [48]. This would suggest that the differences between the two molecules must be sufficient to guarantee their engagement in diverse functional contexts. At the same time, a high degree of similarity ensures a very tight cross talk when specific biological conditions take over.

Once transferred to the E2 enzymes UBC12 or UBE2F, the E3-ligase activity of RBX1/2 promotes cullin neddylation that switches the target preference and activates substrate polyubiquitination. A recent report described the crystal structure representing the RBX1-UBC12~NEDD8-CUL1-DCN1 intermediate and showed the mechanism of NEDD8 ligation to CUL1 by Rbx1 [49]. This study demonstrated that NEDD8 directs the juxtaposition of the UBC12~NEDD8 active site and the CUL1 acceptor site. This specificity is due to the side chain differences in positions 31 and 32, respectively, Glu-Glu in NEDD8, and Gln-Asp in ubiquitin [49]. As expected, at least one of these residues—specifically Glu31 in NEDD8 and Gln31 in ubiquitin—is divergent between ubiquitin and the Ubl. Both residues contact RBX1's Trp35, which is a key position in the NEDD8-binding site. Accordingly, mutation of Trp35 to Ala or to the aromatic residues Tyr and Phe only slightly decreases NEDD8 ligation to a cullin, whereas the Trp35Asp substitution abrogates the interaction with NEDD8 due to the repulsion with NEDD8's Glu31 and 32 [49]. On the other hand, the subtle differences in the side chain length between Glu and Asp in position 32 may be a determinant in the recognition by RBX1. Indeed, while the aliphatic portion of

NEDD8's Glu32 makes hydrophobic contacts with RBX1's Trp35, the latter would repel the ubiquitin's shorter Asp32. Notably, the swapping of residues 31 and 32 in ubiquitin with the corresponding amino acids in NEDD8 is sufficient to promote the ubiquitination of the C-terminal end of CUL1 by the E2 enzyme UBCH5, thus highlighting the crucial role of this pair in the discrimination between NEDD8 and ubiquitin by the CRL component RBX1.

CRL complexes recruit the E2 enzymes UBCH5 and CDC34, together with specific adaptor components, to the ubiquitin-conjugation machinery, and they carry out the substrate ubiquitination [50]. Given its dual capability to conjugate both NEDD8 and ubiquitin, depending on the specific E2 enzyme that is recruited (UBC12, UBE2F, UBCH5, and CDC34), Rbx1/2 constitutes a clear example of a multifunctional RING-ligase. This dual property is shared with other E3 ligases such as Mdm2, c-Cbl, IAPs, and RNF111, indicating that ubiquitination and neddylation of substrates are two closely connected processes.

As previously discussed, a dual specificity is also common among NEDD8 proteases, with few enzymes showing a clear selectivity for NEDD8 (CSN and NEDP1) and the majority catalyzing the C-terminal cleavage of both molecules (such as USP21, PfUCH54, Ataxin-3, UCH- L1, and UCH-L3 [51]). To date, no activity against neddylated cullins has yet been reported for the hydrolases of the second group [52]. Moreover, in addition to its role in cullin deneddylation, NEDP1 appears to control the deneddylation of many non-cullin proteins such as p53, Mdm2, Tap73, BCA3, and E2F1 [51, 53–56].

The NEDD8/NEDP1 structure has been solved by several groups [44, 55, 57]. In these complexes, Ala72 is oriented away from the NEDP1 surface [44, 55] with the majority of the contacts with NEDP1 involving the main chain of the C-terminal tail (Ala72-Gly76) and two amino acid stretches in the NEDP1 enzyme, including residues Trp26-Asp29 and Ala99-Thr101. Like in the case of UBA3 that has been previously described, NEDP1 undergoes a dramatic conformational change upon NEDD8 binding [55]. Conversely, the structure of NEDD8 is a little changed from the native protein [40]. Mutational studies have shown that two key residues, Ala72 and Asn51 (respectively, Arg and Glu in ubiquitin), determine the clear preference of this protease for NEDD8. Interestingly, while the A72R mutant is cleaved more slowly by NEDP1 than the wild-type NEDD8, cleavage is similarly abrogated in the presence of the single-mutation N51E or the double mutant N51E, A72R. Moreover, the mutation R72A in ubiquitin is not sufficient to redirect NEDP1 specificity, while the single-substitution E51N promotes recognition and cleavage of the mutant by NEDP1. Accordingly, the higher catalytic activity of NEDP1 is observed in the presence of the double mutant E51N, R72A. Therefore, at least in this case, the conserved/divergent position 51 seems to be the primary discrimination site for the recognition by NEDP1.

Additional information can be obtained from the work of Yung-Cheng Shin and collaborators. By using a biochemical approach, the authors demonstrated that the discrimination between ubiquitin and NEDD8 by the Ub-specific peptidase USP2 depends on the recognition of the conserved/divergent pattern including residues Phe4, Thr12, and Thr14 together with the C-terminal Arg72. Based on their model, USP2 firstly binds the N-terminus of ubiquitin allowing a stable interaction, which is followed by the recognition of the C-terminus of ubiquitin that ensures the substrate specificity. The relevance of both binding sites in the discrimination process is demonstrated by the observation that the NEDD8 Thr12/Thr14/Arg72 and the NEDD8 Phe4/Thr12/Thr14/Arg72 mutants are both accessible for hydrolysis by USP2, while the ubiquitin Lys4/Glu12/Glu14/Ala72 mutant completely prevents it [58].

Concluding, available data move toward demonstrating that divergences in the NEDD8 sequence, which have been fixed during the evolution of the Ubl, primarily concern the acquisition of binding sites that mediate the recognition by enzymes that selectively recognize NEDD8. These binding properties are not at all associated with loss of ubiquitin recognition but rather with the acquisition of a potential dual-recognition mode, which could be possibly modulated by the specific asset of binding partners.

5. Old and new NEDD8 substrates

In recent years, it has become increasingly evident that CRLs are the main targets of neddylation, but not the only ones (**Tables 1** and **2**). Indeed, nowadays a broad range of proteins besides cullins are known to be modified by NEDD8, and several non-RBX-family NEDD8 E3 ligases have been characterized, including Mdm2/HDM2, HUWE1, RNF111, c-Cbl, IAP1, and Parkin [59–68]. How neddylation controls protein function is still not fully understood, as well as how these dual specificity ligases distinguish signals promoting neddylation or ubiquitination of the same substrate. Nevertheless, the evidence that many of the reported non-cullin targets include key cell cycle regulators, tumor suppressors, signaling receptors, components of the apoptotic machinery, ribosomal proteins, and histones highlights the potential role for NEDD8 in controlling diverse cancer-related processes and the urgency of reaching a deeper understanding of these regulative mechanisms. Moreover, the auto-neddylation of several ubiquitin E3 ligases have revealed a more complex level of regulation of these enzymes, demonstrating that our knowledge about the spectrum of processes that are cooperatively regulated by ubiquitin and Ubls is still largely incomplete. A detailed description of the novel NEDD8 targets has been already carried out by others [23, 30, 69, 70]. Here, we recapitulate the current knowledge regarding the effects of E3-ligase neddylation and the involvement of NEDD8 in the cellular stress response.

5.1 General overview

Similar to other posttranslational modifications, neddylation causes the structural modification of target proteins, thus affecting the enzymatic activity, the interaction with binding partners, and/or the subcellular localization. In addition to the well-known role in the activity of CRL complexes, NEDD8 has been shown to be also involved in the regulation of several transcription factors, by modulating their intracellular distribution and/or transcriptional activity. By comparing the effects that are associated to the conjugation of NEDD8 to non-cullin targets, the most frequent consequence of neddylation is a switch in protein stability, with neddylated targets usually being stabilized by the conjugation of the Ubl. For example, neddylation of p53 prevents the nuclear translocation and inhibits p53 transcriptional activity [59]. This effect is also common to TAp73β, BCA3, the ribosomal proteins L11 and L14, HBx, HuR, TGFβRII, and Pink1 55 kDa fragment. Conversely, ITCH neddylates its substrate JunB and promotes its ubiquitin-dependent degradation, thus attenuating its transcriptional activity [71]. Differently from cullins that are neddylated on a single conserved lysine residue, all these novel substrates have been shown to be ubiquitinated as well as neddylated at multiple residues, which are in most cases overlapping. The only known exception is given by BRAP2 in which the neddylation site Lys432 is within an amino acid sequence that resembles the consensus neddylation sequence conserved in all cullin family proteins [72]. Consequently, while ubiquitination of these residues promotes the proteasomal degradation of target proteins, the addition of NEDD8 moieties to the same residues prevents substrate degradation. In line with this effect, the insulation of the substrate from the degradative pathway is frequently associated with the relocalization in a compartment in which the

E3-ligase	E3-ligase function	Target	Substrate function	Effect of neddylation	Ref.
MDM2	E3 ubiquitin ligase localized in the nucleus. Targets tumor suppressor proteins for proteasomal degradation	MDM2		Neddylation promotes MDM2 protein stabilization and is reverted by NEDP1 activity. Cysteine C462, which is required for the E3 ubiquitin ligase activity of Mdm2, is also required for Mdm2-dependent neddylation.	[51, 59]
		p53	Short-lived tumor suppressor protein having transcriptional activity that responds to diverse cellular stresses	Neddylation of p53 inhibits its transcriptional activity and causes nuclear localization	[59, 153, 154]
		TAp73β	Member of the p53 family involved in cellular responses to stress and development	Neddylation of TAp73 promotes cytoplasmic localization and inhibits its transcriptional activity	[53]
		VHL	Component of the E3-ligase complex including elongin B, elongin C, and cullin-2. It is involved in the ubiquitination and degradation of HIF1α	Neddylation of VHL promotes binding to fibronectin and prevents the incorporation of VHL within a CRL2 complex	[74, 75]
		L11	Ribosomal protein component of the 60S subunit. It is located in the cytoplasm	Neddylation protects L11 from degradation by ensuring its nucleolar localization. Neddylation is reverted by NEDP1 activity	[115, 155, 156]
		L14	Ribosomal protein component of the 60S subunit. It is located in the cytoplasm	Neddylated RPL14 localizes in the nucleolus. Neddylation is reverted by NEDP1	[115, 155]
		MDMX	Structural homolog of MDM2 lacking ligase activity. In complex with MDM2 binds the transcriptional activation domain of p53 and inhibits its activity. It inhibits MDM2 degradation		[84]
HDM2	Human homolog of MDM2	HBx	Transcriptional activator that modulates the expression of HBV and inflammatory genes	Neddylation enhances HBx stability by inhibiting its ubiquitination and promotes chromatin localization. Neddylation of HBx is reverted by NEDP1 activity	[60]
		HuR	RNA-binding protein that selectively binds AU-rich elements (ARE) and stabilizes ARE-containing mRNAs.	Neddylation promotes HuR nuclear localization and protection from degradation. Deneddylation by NEDP1 reduces the nuclear localization of HuR	[61]

E3-ligase	E3-ligase function	Target	Substrate function	Effect of neddylation	Ref.
FBXO11	Subunit of the ubiquitin-protein ligase SKP1-cullin-F-box (SCF)	p53		Neddylation of p53 inhibits its transcriptional activity	[86]
c-Cbl	RING finger E3 ligase acting as negative regulator of signal transduction. It interacts with Y-phosphorylated substrates and targets them for proteasome degradation	EGFR	Transmembrane glycoprotein with kinase activity that binds to epidermal growth factor. Binding of the protein to a ligand induces receptor dimerization and tyrosine autophosphorylation and leads to cell proliferation	NEDD8 conjugation enhances ligand-induced ubiquitination of the EGF receptor and clathrin-mediated endocytosis for lysosomal degradation.	[64]
		TGFβRII	Kinase that forms a heterodimeric complex with TGFβRI. Once activated by TGF-β, the complex phosphorylates target proteins, which enter the nucleus and regulate the transcription of several genes	Neddylation protects the receptor from ubiquitination by promoting the clathrin-mediated endocytosis of TβRII into EEA1-positive early endosomes. Neddylation is reverted by NEDP1	[118]
RBX1/2 (alias RNF7)	RING finger protein playing a key role in CRL complexes	CRL1, CRL2, CRL3, CRL4, CRL5	Cullins are scaffold proteins of CRL complexes that control the stability of proteins with diverse functions	Neddylation enhances the activity of CRLs and subsequent ubiquitination and degradation of the regulated substrates	[11, 21]
XIAP (alias IAP3)	RING-ligase is a negative regulator of apoptosis pathway	XIAP			[65, 157]
		Caspase-7	Cysteine-aspartic acid protease acting in the execution phase of cell apoptosis. Casp7 is activated upon cell death stimuli	Neddylation suppresses caspase activity, while deneddylation is required to execute apoptosis	[65, 157]
DIAP	DIAP is the Drosophila homolog of hIAP1	DIAP		Auto-neddylation of DIAP does not seem to affect its E3-ligase activity	[65]
		Drice	It is involved in the activation cascade of caspases in *Drosophila melanogaster*	Neddylation inhibits its proteolytic activity. The inhibitory effect is reverted by NEDP1	[65]
CRL2	E3 ubiquitin-protein ligase consisting of elongin C, elongin B, RBX1, cullin 2, and an E2 ubiquitin-conjugating enzyme	HIF1a	HIF-1 functions as a master regulator of cellular and systemic homeostatic response to hypoxia by activating transcription of many genes, including those involved in energy metabolism, angiogenesis, apoptosis, and other genes whose protein products increase oxygen delivery or facilitate metabolic adaptation to hypoxia	The effect of neddylation on HIF-1 stability is contradictory. It has been suggested that the neddylation process is required for the ubiquitination and subsequent degradation of HIF-1	[79]

E3-ligase	E3-ligase function	Target	Substrate function	Effect of neddylation	Ref.
SMURF1	E3 ligase that is specific for receptor-regulated SMAD proteins in the bone morphogenetic protein (BMP) pathway	SMURF1		The auto-neddylation of SMURF1 stimulates the activation of its ubiquitin E3 ligase and the degradation of substrates. Neddylation of SMURF1 is reverted by NEDP1 activity	[66]
RSP5	Homolog of SMURF1 in *S. cerevisiae*	RSP5		Neddylation of Rsp5 is important for the Rsp5-mediated ubiquitylation of Vps9	[66]
RNF111	Nuclear RING E3 ligase promoting the ubiquitination and proteosomal degradation of inhibitor SMADs in the TGFβ/NODAL signaling pathway	Histone H4	Basic nuclear protein that is a component of the nucleosome structure	DNA damage causes an increase in RNF111-mediated H4 neddylation and the recognition of the NEDD8 chains by RNF168, which is required for the subsequent recruitment of the latter to the sites of DNA damage	[63]
RNF168	RING finger E3 ligase involved in DNA double-strand break (DSB) repair	Histone H2A	Basic nuclear protein that is a component of the nucleosome structure	DNA damage causes a decrease in H2A-neddylation that allows its ubiquitination and the redistribution of RNF168 from H2A to H4	[63, 145]
TRIM40	Member of the tripartite motif (TRIM) protein family that plays a role as a negative regulator against inflammation and carcinogenesis	IKKγ	IKKγ (alias NEMO) is the regulatory subunit of the inhibitor of kappaB kinase (IKK) complex, which activates NFκB resulting in activation of genes involved in inflammation, immunity, cell survival, and other pathways	TRIM40-mediated neddylation of IKKγ inhibits the transcriptional activity of NFκB. Neddylation has been proposed to oppose the activation of IKKγ mediated by conjugation of K63-linked and linear polyubiquitin chains	[143]
FANCA	Member of the FANC complementation group, a nuclear protein complex required for the monoubiquination and relocalization of FANCD2 to nuclear foci in response to DNA damage	CXCR5	Cytokine receptor that is expressed in mature B cells and Burkitt's lymphoma. Binds to B-lymphocyte chemoattractant and is involved in B-cell migration	Neddylation of CXCR5 receptor promotes membrane localization and is required for cell motility and migration of B lymphocytes to the germinal centers in response to its ligand, CXCL13	[158]

Table 1.
Neddylated substrates and E3 ligases responsible for the modification.

proteasomal degradation is precluded. NEDD8-conjugated proteins predominately reside in the nucleus, thus supporting the idea that nuclear compartmentalization is a feature frequently associated with the neddylation process. Examples are the Mdm2-mediated neddylation of p53, which causes its nuclear localization [59], while p53 degradation is mainly cytoplasmic. Neddylation of the ribosomal proteins L11

E2F-1	Transcription factors playing a crucial role in the control of cell cycle and tumor suppressor proteins. E2F binds to retinoblastoma protein pRB in a cell cycle-dependent manner. It can mediate both cell proliferation and p53-dependent/p53-independent apoptosis	Neddylation negatively regulates E2F-1 activity. The effect is reverted by the action of the NEDP1 enzyme or upon DNA damage. Neddylation sites are also targets of methylation	[54, 78]
BRAP2	Cytoplasmic RING E3 ligase which may regulate nuclear targeting by retaining proteins with a nuclear localization signal in the cytoplasm. It binds to the nuclear localization signal of BRCA1 and other proteins	Unknown	[72]
AICD	Fragment produced by γ-secretases; it binds to the transcriptional coactivator Fe65 and forms a complex in conjunction with Tip60 to activate the transcription of a target genes	Neddylation of AICD inhibits its interaction with FE65 and Tip60 transcriptional co-regulator, thus resulting in the impairment of AICD–Fe65–Tip60 complex formation	[159]
Parkin	Component of a multiprotein E3 ubiquitin ligase complex that mediates the targeting of substrate proteins for proteasomal degradation	Neddylation promotes ubiquitin E3-ligase activity toward synphilin-1 and increases Parkin auto-ubiquitination	[67, 68]
Pink1	Serine/threonine protein kinase that localizes to mitochondria. It is thought to protect cells from stress-induced mitochondrial dysfunction	PINK1 neddylation increases the stability of the PINK1 55 kDa fragment, a processed form of PINK1 that is found in the cytoplasm and forms a complex with Parkin	[67]
BCA3 (alias AKIP1)	Nuclear protein that interacts with protein kinase A catalytic subunit and regulates the effect of the cAMP-dependent protein kinase signaling pathway on the NF-kappa-B activation cascade	BCA3 neddylation suppresses NFκB transcription via its ability to associate with nuclear p65. Moreover, neddylated BCA3 associates with cyclin D1 promoter	[56]

Table 2.
Neddylated substrates for which the E3 ligase is unknown.

and L14 ensures their localization in the nucleolar compartment, protecting L11 from degradation occurring in the nucleus. Relocalization following neddylation is also observed for TAp73β that, unlike p53, accumulates in the cytoplasm [53]. Moreover, the RNA-binding protein HuR relocalizes in the nucleus following Mdm2-mediated neddylation, and it is protected from degradation. Pink1 55 kDa fragment is stabilized following neddylation and relocalizes from mitochondria to the cytoplasm, where it forms a complex with Parkin [67]. If NEDD8 conjugation promotes the relocalization of the target protein, thus it is plausible to assume that the differential intracellular distribution would be dependent on the recognition of the NEDD8 modification by shuttling proteins (spatial sequestration). Interestingly, the majority of the transcription factors that are neddylated by the same E3 ligases responsible for their ubiquitination show reduced transcriptional activity and intracellular relocalization. A molecular switch affecting the enzymatic activity (conformational inactivation) could explain this effect. However, neddylation may also control the differential incorporation in complexes that, similarly to the case of shuttling proteins, drive the transcriptional activity mediated by the neddylated targets toward different promoters (selective recruitment), thus inhibiting the main downstream effects and promoting alternative pathways that must be activated in specific cellular conditions. The observation that transcription factors such as p53 are also involved in cytoplasmic processes like endocytic degradative pathways of membrane receptors and actin remodeling [73] supports the notion that differential posttranslational modifications could be key events underlying triggering of alternative pathways. As an example, the Hdm2-mediated neddylation of hepatitis B virus X protein (HBx), a transcriptional activator that modulates the expression of hepatitis B virus (HBV) genes and inflammatory genes, enhances HBx stability and promotes chromatin localization, which in turn favors HBx-dependent transcriptional regulation, cell proliferation, and HBV-driven tumor growth [60].

Another example of switch mediated by target neddylation is given by the Von Hippel-Lindau (VHL) tumor suppressor [59, 74]. VHL is a component of the VHL tumor suppressor-containing E3 ubiquitin ligase complex (ECV), a class of CRL2 complexes that controls the stability of the transcription factor hypoxia-inducible factor 1-alpha (HIF1α) upon hypoxic conditions [75]. The function of VHL is to recruit the substrate HIF1α by directly interacting with the α-subunits of the transcription factor, thus promoting its degradation by the CRL2 complex. Interestingly, VHL can also interact with fibronectin and promote extracellular matrix assembly [76]. The switch between these two functions (negative regulator of HIF1α and positive regulator of fibronectin) is regulated by the neddylation of VHL. Indeed, NEDD8 conjugation precludes ECV complex formation by steric hindrance, causing the exclusion of CUL2 from the complex and the stabilization of HIF1α. At the same time, neddylation of VHL confers the ability to interact with fibronectin and to promote extracellular matrix assembly [75].

Neddylation of several membrane receptors has been recently described. The EGFR and TGFβRII have been shown to be neddylated by c-Cbl, a RING E3 ligase containing an N-terminal phosphotyrosine-binding domain, which allows it to interact with numerous tyrosine-phosphorylated substrates, targeting them for proteasomal degradation. Following NEDD8 modification, two opposite effects have been described: while EGFR degradation is promoted following ligand stimulation, neddylation of TGFβRII reduces the receptor degradation rate by shifting the receptor internalization from caveolae to clathrin-coated pits. In both cases, neddylation of these receptors results in the inhibition of proliferation and cell cycle arrest. Therefore, despite promoting different outcomes, the resulting effects of neddylation converge toward the same signaling output. Shc, an adaptor that transduces signals from several receptor tyrosine kinases and cytokine receptors, has been recently identified as a NEDD8 target. Neddylation of Shc promotes downstream signaling by favoring the formation of a ZAP70-Shc-Grb2 complex that is required for downstream Erk activation [77].

It must be pointed out that several reports analyzed the neddylation of substrates using an experimental approach that is based on the transient overexpression of tagged NEDD8, which is known to cause nonphysiological effects due to "atypical" neddylation of substrates [47]. This is particularly relevant in those papers in which the E3 ligase responsible for target neddylation has not been identified [78–80]. Nevertheless, the parallel use of complementary approaches, such as knockdown or inhibition of APP-BP1 activity, the use of a dominant-negative mutant of Ubc12, and NEDP1-mediated deneddylation, supports the main statement and encourages in the direction of deepening the unexpected functions of neddylation and how ubiquitin and Ubl conjugations are combined in modulating biological processes.

5.2 The Mdm2/p53 pathway

In unstressed conditions, Mdm2/Hdm2, which heterodimerize with MdmX, mainly controls the levels of the transcription factor p53. Mono-ubiquitination of p53, promoted by Mdm2 low levels, is sufficient for nuclear export, whereas high levels of Mdm2 activity induce p53's polyubiquitination and proteasome-dependent degradation [81]. Several evidences suggest that neddylation, promoted by Mdm2, regulates p53 signaling at different levels. Under DNA damage conditions, the phosphorylation of Mdm2 and MdmX by c-Abl promotes the association of the heterodimer, with Mdm2 acting as a ubiquitinating enzyme that destabilizes MdmX resulting in p53 stabilization [82–84]. On the contrary, in growing conditions under c-Src activity, Mdm2 acts as neddylating enzyme that neddylates p53, Mdm2 itself, and MdmX [84, 85]. Neddylation of p53 prevents the nuclear translocation and inhibits p53 transcriptional activity; in contrast the neddylation-resistant mutants retain the transcriptional activity [59].

Interestingly, also the F-box protein FBX011, a component of the CRL1 complex, promotes p53 neddylation by inhibiting p53 transactivation function. Nevertheless, differently from Mdm2, it does not ubiquitinate p53 nor affect its stability [86]. Deneddylation of p53 is promoted by two different mechanisms: the activity of NEDP1 [51] and the interaction of p53 with NUB1, a NEDD8-interacting protein that acts as negative regulator of NEDD8 and neddylated proteins [87]. NUB1 associates with neddylated p53, decreases p53 neddylation, and stimulates p53 mono-ubiquitination, resulting in p53 nuclear export. Recently, it has been found that NUB1 is a target of Mdm2-mediated ubiquitination. Instead of representing a degradative signal, Mdm2 conjugates a di-ubiquitin signal on NUB1 that is thought to be necessary for exposing the NEDD8-binding site following the intramolecular recognition between the UBA domain of NUB1 and the di-ubiquitin signal [87].

5.3 Dual E3 enzymes: insights into the regulation of HECT and RBR ligases

Based on their structural organization and conjugation mechanism, E3 enzymes are commonly grouped into three classes. Really interesting new gene (RING) E3s act as allosteric activators of E2 that promotes the transfer of ubiquitin from the E2 to the target protein. Homologous to E6AP C-terminus (HECT)- and RING-between-RING (RBR)-type ligases, although being structurally distinct, require an intermediate step where the ubiquitin is first transferred from the E2 to an active-site cysteine residue on the E3 ligase to be then conjugated to the target protein [88, 89]. The NEDD4 subfamily is the best-characterized family of the HECT E3s. It contains an N-terminal C2 domain that binds Ca2+ and phospholipids followed by two up to four WW domains, responsible for the substrate recognition [90, 91]. Auxiliary or inhibitory factors, respectively, facilitate or interfere with the substrate recruitment [91–93]. A distinct level of regulation instead affects the HECT domain and depends on intramolecular interactions between the HECT and the C2 domains that keep the E3 ligase in an auto-inhibited conformation. Typically, signals that trigger the addition of phosphorylations along the amino acid chains promote relief of the inhibited state and substrate recruitment [94]. In the only case of SMURF1, homo-dimerization inhibits the catalytic domain by promoting its association with the C2 domain of the partner in the dimer; the relief is achieved by the interaction with adaptor proteins including CKIP, CDH1, or CCM2 [95–97]. Among the PTMs that are known to regulate HECT E3 ligases, auto-ubiquitination, together with the reverse reaction promoted by deubiquitinating enzymes, defines the rate of HECT downregulation mediated by the proteasomal degradation [98]. Interestingly, a non-covalent interaction of the HECT domain with ubiquitin has also been reported, at least for Rsp5 and NEDD4 [99, 100]. The ubiquitin-binding site, called "exosite," is located in the N-terminal lobe of the HECT domain of these E3 ligases and contacts the hydrophobic patch residues of ubiquitin. Disruption of the interaction, obtained by mutagenesis of the HECT-/ubiquitin-binding interface, resulted in defects in the substrate polyubiquitination, leading to the interpretation that the N-lobe ubiquitin-binding site promotes the elongation of the ubiquitin chain by orienting the distal end of the ubiquitin chain in the optimal position for the next conjugation event.

In addition to ubiquitination, also neddylation regulates E3-ligase activity. Indeed, it has been shown that several E3 ligases have neddylating activity and are regulated by auto-neddylation. Typically, neddylation promotes both the activity toward substrates and the increase in auto-ubiquitination, leading to ubiquitination and subsequent degradation of the ligase. The first ubiquitin E3 ligases identified as subjected to such regulative mechanism were Mdm2, Parkin, and XIAP, all belonging to the RING-ligase family. Auto-neddylation has been then demonstrated to regulate also SMURF2, SMURF1, and its homolog in *S. cerevisiae* RSP5 [66, 101]. Even though the details of the activation mechanism have not been yet clarified, it is interesting to observe the active site responsible for SMURF1 auto-neddylation maps in the HECT N-lobe (Cys426), thus being different from the ubiquitination catalytic cysteine typically located in the C-lobe (Cys716). This spatial separation underlies the potential autonomy of the NEDD8-regulated mechanism. Conversely, the homolog RSP5 uses the same catalytic site Cys777 both for neddylation and ubiquitination [66]. A similar mechanism could also involve other members of the HECT family, such as ITCH, NEDL1, and NEDL2, but not in NEDD4.1 and NEDD4.2 [66]. Moreover, in addition to promoting its auto-neddylation, ITCH neddylates its substrate JunB and attenuates its transcriptional activity by promoting JunB ubiquitination-dependent degradation. This catalytic mechanism requires the same Cys residue (Cys830) that is involved in the ubiquitination reaction [71]. Whether the remaining HECT-type ubiquitin ligases can also catalyze the neddylation of protein substrate remains to be experimentally verified.

Interestingly, together with the capability to conjugate NEDD8, a Ubl non-covalent-binding site has been identified in SMURF2, and it has been shown to be conserved also in SMURF1 [102] (see below). The evidence that, following mutations that disrupt the non-covalent interaction with NEDD8, SMURF neddylation is reduced and the E3-ligase stabilized due to reduced self-ubiquitylation highlights the importance of this interaction in regulating SMURF enzymatic activity. Accordingly, NEDD8 binding to SMURF plays important roles in the regulation of cell migration and in the BMP and TGFβ signaling pathways, both biological processes in which SMURFs are known to play a key role [102].

5.4 Neddylation in the nucleolar stress response

In eukaryotes, ribosomes are preassembled in pre-ribosomal particles in a subnuclear compartment called nucleolus and then transferred to the cytoplasm where the assembly of the two subunits takes place. The limiting factor of ribosome assembly is the rate of rRNA transcription in the nucleus; therefore the formation and accumulation of unusable components are minimized through the rapid ubiquitination and proteasome-dependent degradation of free ribosomal proteins [103]. Since ribosome biogenesis is a complex and highly resource-consuming process, several control mechanisms are in charge of monitoring any insult that causes activation of stress signaling, thus leading to rapid repression of protein synthesis and ribosome biogenesis in response to stress, as well as rapid derepression in response to improved conditions. The nucleolus, long known primarily for its role in ribosome biosynthesis, has assumed a function in recent years as a sensor for those cellular stresses that lead to the impairment of ribosome biogenesis homeostasis. This function has led to the coining of the term "nucleolar stress" or "ribosomal stress," highlighting that perturbations in ribosome biogenesis are accompanied by morphological changes, functional defects, and eventually disruption of nucleolus. The nucleolus is in continuity with the surrounding nucleoplasm, so that any soluble molecule is potentially free to traffic between the nucleolus and the nucleoplasm. Accordingly, several proteins enriched in the nucleolus are frequently shuttled between the nucleolus and the nucleoplasm. Among them, nucleophosmin (NPM1), an abundant nucleolar protein, is massively moved to the nucleoplasm and cytoplasm following various cellular insults, so that its redistribution upon stress has been set as indicator of nucleolar stress [104].

Nucleolar stress is induced by various stressors such as heat shock, chemotherapeutic agents, UV, starvation, etc., resulting in the activation of both p53-dependent and p53-independent signaling pathways and cell cycle arrest. The p53-dependent response relies on the disruption of the interaction between p53 and Mdm2/Hdm2, which results in p53 stabilization and activation. For example, NPM1 undergoes S-glutathionylation upon nucleolar oxidative stress, translocates to the nucleoplasm, and disrupts the p53-Hdm2 interaction [105]. Interestingly, over the last years, several studies have shown that most of the p53-dependent but also p53-independent stress responses are dependent on the activity of free ribosomal proteins (RPs) that have been shown to perform multiple extra-ribosomal functions, including regulation of apoptosis, cell cycle arrest, cell proliferation, and DNA damage repair [106, 107]. In response to nucleolar stress, the RPs L5, L11, and L23 bind to Mdm2 and block Mdm2-mediated p53 ubiquitination and degradation, leading to p53 stabilization and activation [108–111]. Similarly, L5, L11, and L23 have been shown to bind and suppress c-Myc and E2F-1, two transcription factors that, respectively, control ribosome biogenesis and promote cell proliferation, thus leading to p53-independent inhibition of cell proliferation [112–114]. Interestingly, also the switch between the ribosome biogenesis pathway and the incorporation of free ribosomal proteins within transcription factor complexes is regulated by neddylation. Indeed, it has been shown that Mdm2 promotes the neddylation of L11, which accumulates in the nucleolus, thus protecting it by the ubiquitin-dependent degradation in the nucleoplasm. Under nucleolar stress, a decrease in L11 neddylation promotes its nucleoplasmic relocalization, where it binds and inhibits Mdm2. This not only provides a trigger for p53 activation but also makes L11 susceptible to ubiquitin-mediated degradation. Analogously to L11, the ribosomal protein L14 binds to Hdm2 and inhibits Hdm2-mediated p53 polyubiquitination and degradation, thus increasing p53 stability and activity. Upon neddylation, L14 associates with hCINAP, which in turn recruits NEDP1 on L14. Deneddylation by NEDP1 promotes the release of L14 from the complex and the relocalization of L14 to the nucleoplasm [115].

Recently, a novel function of NEDD8 in proteotoxic stress has been identified [62]. Exposure of cells to heat shock, proteasome inhibitors, or oxidative stress increases neddylation through an enzymatic chain that requires the ubiquitin E1-activating enzyme UBA1 instead of the NEDD8-activating E1 enzyme. It also involves the formation of hybrid NEDD8/ubiquitin conjugates [47, 48]. The authors demonstrated that NEDD8 promotes the transient and reversible nuclear protein aggregation during proteotoxic stress and colocalizes with ubiquitin in the nuclear aggregates but not in the cytoplasmic aggresomes. These nuclear aggregates, in which RNA transport and ribosomal proteins account for almost half of the total components, require the activation of NEDD8 by UBA1, as demonstrated by the observation that a drastic decrease of protein aggregation follows treatment of cells with the specific UBA1 inhibitor MLN7243 but not the NAE1 inhibitor MLN4924 [62]. Interestingly, even though they are also enriched in RPs, the NEDD8/ubiquitin nuclear aggregates are distinct from the stress-induced nucleolar structures. A proteomic study identified the HECT E3-ligase HUWE1 as a component of nuclear aggregates, and immunofluorescence and biochemical assays strongly suggested that HUWE1 is the major E3 ligase that specifically promotes neddylation during proteotoxic stress [62]. Accordingly, depletion of HUWE1 does not affect the global ubiquitination or the amount of ubiquitin staining in the nuclear aggregates while causing a dramatic decrease in neddylation, thus compromising the NEDD8-mediated cellular response to proteotoxic stress. Moreover, HUWE1 has been also reported to be a neddylation substrate in a proteomic analysis [116]. Finally, given that ubiquitin-NEDD8 mixed chains are resistant to proteasome degradation, the stress-induced aggregation of substrates modified with both molecules could be intended as a way to sequester a certain group of proteins from the ubiquitin-mediated degradative route, in order to promptly restore their functions once the external conditions return to be favorable to cellular growth and proliferation.

Concluding, neddylation is clearly emerging as an important regulator of several pathways, governing proliferation, differentiation, and survival. However, despite the importance of this posttranslation modification in cell biology, the mechanism through which neddylation controls protein properties is still a matter of debate. Neddylation influences the stability of the target protein, but it also determines its intracellular distribution by promoting the interaction with shuttling proteins. Most of all, neddylation seems to be involved in cellular stress signaling, by providing a regulatory mechanism that acts like a "switch," allowing the activation of alternative pathways dictating the key steps of the cellular stress response. This switch is promoted by the disruption of protein complexes acting in unstressed conditions and the assembly of new complexes in which seemingly secondary functions of key components become decisive for the correct signaling output.

6. NEDD8 chains and mixed ubiquitin-NEDD8 chains

Traditionally, neddylation consisted in the addition of a single NEDD8 molecule to a target protein. Nevertheless, a significant amount of experimental evidences gradually emphasized that the neddylation signal, as also happened for ubiquitin, was more detailed and complex than initially thought.

The first argument is the evidence that several lysine residues in NEDD8 can be used for chain extension [116] and that, at least in vitro, poly-NEDD8 chains can be synthesized [117]. Moreover, evidence that NEDD8 chains can be generated in vivo has been recently recovered, even though the biological significance of poly-neddylation remains unknown [116]. The second point that supports the existence of poly-neddylation is the evidence that high-molecular-weight NEDD8 conjugates have been observed through affinity purification and mass spectrometry [54, 67, 68, 118]. Again, whether these conjugates are the result of poly-neddylation or correspond to mono-neddylation events remains to be elucidated.

Finally, clear experimental evidence demonstrated that, upon proteotoxic stress, proteins are simultaneously modified by NEDD8 and ubiquitin mixed chains [48]. One possible explanation is that NEDD8 acts as a ubiquitin substitute, but, being less easily ubiquitinated, it caps the ubiquitin chain and prevents the excessive extension, a cellular response that could be a key event to avoid a further escalation of the proteotoxic stress, especially in those conditions in which the ubiquitin pool is depleted. Another possibility is based on the evidence that several ubiquitin-binding domains have been shown to interact poorly with NEDD8 [119] (see below). Based on this assumption, protein domains that recognize ubiquitin chains are disfavored when dealing with mixed NEDD8-ubiquitin chains. Consequently, the conjugation of these signals would trigger the exclusion of binding partners that transduce the ubiquitin signal under unstressed conditions, for example, favoring the interaction with protein domains that specifically recognize these mixed chains. This interpretation is consistent with the model that sees neddylation as a switch in the outcome of unstressed/stresses responses. Moreover, several proteomic studies have reported phosphorylation, acetylation, and succinylation sites on NEDD8 [120–122]. One could speculate that mixed chains are assembled and further modified in order to "create" an entirely new signal, which goes well beyond the simple random mix between ubiquitin and NEDD8. In order for poly-neddylation and mixed NEDD8-ubiquitin signals to be effective, however, the existence of a system that unambiguously discriminates between different combinations of two strictly similar molecules should be assumed. The identification of protein domains capable of recognizing unusual combinations of these signals and the need to clarify to what extent known NEDD8-binding proteins are capable to do it are research topics required for a deeper comprehension of the cross talk between these two posttranslational modifications.

7. Ubiquitin-binding domains: a general overview

The first ubiquitin receptors to be identified were intrinsic components of the proteasome that directly bind polyubiquitin chains driving the recruitment and the degradation of the substrates [123–125]. Subsequently, polyubiquitin-binding domains have been also found in mobile shuttling factors that direct polyubiquitinated proteins to the proteasome, and, in a relatively short space of time, it has been clear that ubiquitin functions spread in modern eukaryotes to play key roles in several cellular processes. Many of these functions are mediated by the association with ubiquitin-binding domains (UBDs), which usually bind to ubiquitin only weakly [126]. Nevertheless, several mechanisms work together in order to increase both affinity and avidity of the recognition. For example, specific binding can be greatly enhanced by polymerization of the ubiquitin signal, in the form of homologous, mixed, and branched ubiquitin chains. In this way, multiple surfaces can be generated on a target protein, thus increasing the number of different UBDs that can simultaneously recruit the substrate. Moreover, the arrangement in tandem of more UBDs along the same amino acid chain increases the contact sites between the substrate and the ubiquitin receptor.

To date, at least 20 structurally distinct UBD classes have been characterized that vary considerably in size (from ~30 to 150 residues) and tertiary structure [126]. They interact with ubiquitin monomers or chains, usually recognizing diverse ubiquitin signals. UBDs that bind to ubiquitin chains are often selective for a specific type of chain linkage. This selectivity may arise from the recognition of a unique orientation of the chain and distinct surfaces on the ubiquitin moieties or via direct interaction with the linker region connecting the two ubiquitins. Ubiquitin/UBD interactions have been thoroughly reviewed in recent years [126–128]. The canonical ubiquitin/UBD complex requires the recognition of the Ile44-centered hydrophobic patch in ubiquitin, including residues Ile44, Val8, His68, and Val70. Residues peripheral to the hydrophobic patch are also important for UBD binding. Indeed, distinct structural motifs have been shown to recognize the hydrophobic patch in a variety of orientations that can be attributed to specific contacts between each UBD and residues surrounding the hydrophobic patch on ubiquitin. Accordingly, mutations outside the hydrophobic patch that disrupt one ubiquitin/UBD pair do not necessarily affect the recognition by another domain [129]. In some cases, however, Ile44 does not constitute the "center" of the interaction site and contributes only marginally to the binding affinity of an interaction [130]. Moreover, although the hydrophobic patch plays a role in many interactions with UBDs, other regions can solve this function, among them the Ile36 patch (Leu8-Ile36-Leu71-Leu73), the C-terminal end (Gly75-Gly76), and the surface including residues of and around the flexible loop between β1 and β2 and Lys6 and Lys11. The Ile36-centered hydrophobic patch is the contact site for a relatively small number of binding partners.

Among them, contacts between the E3 enzyme and the Ile36 patch of the ubiquitin moiety conjugated to the E2 have been shown to favor the ubiquitin transfer [131]. Interestingly, due to the close proximity of different binding sites in ubiquitin, multiple contact sites can increase the interaction affinity for a given partner [132].

Like ubiquitin, the NEDD8 backbone is relatively rigid, although slight structural changes occur during binding. On the contrary, the C-terminus of both proteins is highly flexible in solution and adopts a more rigid conformation only following the interaction with the enzymes responsible for the conjugation process [44, 45]. Finally, while the main contact sites (Ile44-patch, Ile36 patch, and β1-β2 loop) are well conserved in NEDD8, polar and charged surfaces show a greater differentiation, and they are expected to influence specific binding properties [40].

8. NEDD8 interacting partners

Small ubiquitin-like modifier (SUMO), which is the Ubl that is more similar to ubiquitin after NEDD8, shows only 17% of identity and, similarly to NEDD8, is involved in many cellular processes where it exerts non-proteolytic functions. The conjugation of SUMO leads to consequences that are similar to NEDD8 conjugation. For example, sumoylation of transcription factors and histones is generally associated with decreased gene expression and intracellular relocalization. Moreover, SUMO and ubiquitin can compete for the same target lysines [133]. Both a sumoylation consensus motif and a non-covalent SUMO-interaction/SUMO-binding motif have been identified in selected SUMO enzymes, targets, and downstream effectors [134]. Despite a certain degree of cross talk between ubiquitin and these Ubls in several biological processes, it seems that NEDD8 and SUMO have been subjected to a very different evolutionary pressure, which have led to a clear pattern for the SUMO recognition by a specific set of binding partners, while maintaining a strong overlap in the recognition of NEDD8 versus ubiquitin. This "ambiguity" reflects the biological need to maintain a high degree of identity between ubiquitin and NEDD8, with few well-defined differences. But it also leaves open the question of whether neddylation is always functionally distinct from ubiquitination.

Despite the limited understanding of the NEDD8 interaction network, recent advances support the assumption that, similarly to ubiquitin, specific NEDD8-binding proteins recognize neddylated substrates acting as downstream effectors (**Table 3**). As shown, in most cases these proteins recognize non-cullin NEDD8-conjugated substrates or NEDD8-containing chains. In some cases they interact with NEDD8 only when conjugated to cullins, and, in a few cases, they also recognize free NEDD8.

Besides the enzymes of the neddylation pathway that has been previously discussed, the first NEDD8-binding protein to be identified was the negative regulator of the NEDD8 pathway NUB1 and its splicing variant NUB1L [135–138]. Following overexpression, NUB1 and NUB1L, which have both a nuclear localization signal, mainly localize in the nucleus [135]. Tanaka and collaborators characterized the NEDD8-binding sites in the two splicing variants: these signals localize at the C-terminal end of the amino acid chain and respond to the consensus sequence A(X4)L(X10)L(X3)L. Moreover, Leu to Ala substitutions abolish the interaction of NUB1 and NUB1L with NEDD8 in vitro as well as the capability of NUB1 to down-regulate NEDD8 conjugates. Another NEDD8 interactor, identified a short time later, was UBXN7 that belongs to the UBA-UBX family of substrate adaptors [139]. The UBA domain recruits proteasome substrates by interacting with the ubiquitin chains conjugated on the target protein, while the UBX takes contact with the p97

Interactor	Description	NEDD8-binding interface	Domain-binding interface	Ref.
NUB1/ NUB1L	Adaptor protein negatively regulates Nedd8 and neddylated substrates	Ile44 patch (predicted)	Consensus sequence A(X4) L(X10)L(X3)L	[135–138]
UBXD7	Involved in the degradation of misfolded or damaged proteins	Ile44 patch	UIM	[139, 140]
RNF168	RING E3 ligase involved in DNA double-strand break repair	NEDD8 chains	MIU2	[145]
TRIM40	Negative regulator of inflammation and carcinogenesis	Unknown	Unknown	[143]
DNMT3b	DNA methyltransferase that functions in de novo methylation	Unknown	Residues 532–583 of mouse DNMT3b	[142]
HGS	Adaptor involved in the endosomal sorting of membrane receptors	Ile44 patch (predicted)	UIM	[64]
TRIAD1	RBR ligase	Unknown	UBA-like domain	[147]
HHARI	RBR ligase	Unknown	UBA-like domain	[147]
hHR23a, UBQLN1 and Ddi1	Non-proteasomal ubiquitin receptors and shuttles	Ile44 patch	UBA2 (RAD23), UBA (UBQLN1), UBA (Ddi1)	[141]
BRAP2 (or RNF52)	Cytoplasmic protein, which may regulate nuclear targeting BRCA1 and other proteins with a nuclear localization signal, in the cytoplasm	Unknown	Multiple domains in BRAP2 show NEDD8-binding properties	[72]
AHR	Transcription factor localized in the cytoplasm that moves to the nucleus upon ligand binding and stimulates transcription of target genes involved in the response to planar aromatic hydrocarbons. Activated AhR is exported from the nucleus for degradation in the cytosol by the ubiquitin/proteasome pathway	Unknown	Unknown	[144]
SMURF1	HECT E3-ligase belonging to the NEDD4 family	Ile-patch	Consensus sequence L(X7) R(X5)F(X) ALQ.	[102]
SMURF2	HECT E3-ligase belonging to the NEDD4 family	Ile-patch	Consensus sequence L(X7) R(X5)F(X) ALQ.	[102]
NEDL1	HECT E3-ligase belonging to the NEDD4 family	Unknown	Unknown	[102]
NEDL2	HECT E3-ligase belonging to the NEDD4 family	Unknown	Unknown	[102]
Rpt6	Regulatory particle base subunit of the proteasome	Unknown	Unknown	[116]

Interactor	Description	NEDD8-binding interface	Domain-binding interface	Ref.
SMC1	Cohesins are required for nuclear division; they prevent premature separation of sister chromatids	Unknown	Unknown	[116]
DNA-PK	DNA-dependent protein kinase belonging to the PI3/PI4 kinase family, required for DNA double-strand break repair and recombination	Unknown	Unknown	[116]
KHNYN	Unknown function	Negatively charged surface centered on residues 31-EEKE-34	CUBAN domain	[119]

Table 3.
NEDD8 interacting proteins.

ATPase of the proteasome. UBXN7 also contains a UIM motif, which interacts with neddylated CUL2 and sequesters it, thus negatively regulating the ubiquitin ligase activity of the CRL complex. Accordingly, UBXN7 overexpression causes the accumulation of HIF1α in its non-ubiquitinated form, suggesting the involvement of the UIM-NEDD8 interaction in the processivity of the CRL ubiquitin ligase [139]. Subsequently, den Besten and colleagues reported that the replacement of NEDD8 with ubiquitin on CUL2 does not affect the interaction with UBXN7, indicating that this recognition is rather context dependent [140].

The non-proteasomal ubiquitin receptors hHR23a, UBQL1, and Ddi1 have been shown to interact non-covalently with NEDD8 [141]. These shuttle proteins deliver ubiquitinated cargoes from the cytoplasm and the nucleus to the proteasome. Their structures contain an N-terminal ubiquitin-like (UBL) domain that binds to the proteasome and a C-terminal UBA domain that binds to ubiquitin chains. It has been suggested that, under stress conditions in which heterologous chains are formed by the enzymes of the ubiquitin pathway in order to compensate the ubiquitin depletion, these chains would be recognized by the shuttles and processed by the proteasome. Since mixed chains are shorter than the polyubiquitin chains, due to the less efficiency in NEDD8 chain extension, the formation of the heterologous chains would counteract the further depletion of free ubiquitin within cells.

A heterogeneous group of proteins that non-covalently bind NEDD8 substrates includes the DNA methyltransferase DNMT3b [142], the RING-ligase TRIM40 [143], the HECT ligases NEDL1 and NEDL2 [102], the DNA-dependent protein kinase DNA-PK [116], cohesin SMC1 [116], the transcription factor AHR [144], the cytoplasmic retention factor for nuclear proteins BRAP2 [72], and a regulatory particle of the proteasome Rpt6 [116]. All of them have been identified by using yeast two-hybrid screenings or the immunoprecipitation of overexpressed tagged NEDD8. Moreover, with the exception of DNMT3b and BRAP2 for which an attempt, although not conclusive, to identify the NEDD8-binding region has been carried out, neither the identification of the NEDD8 recognition surface nor its function has been well established. This lack of knowledge does not allow the identification of a putative NEDD8 recognition module. Nevertheless, these works represent documented examples linking NEDD8 to the recruitment of effectors in processes like proteasome degradation, DNA methylation, transcription, and DNA damage repair.

A second class of NEDD8-interacting proteins comprises proteins in which a typical UBD has been shown to mediate NEDD8 binding with similar efficiency, thus leaving open the issue of whether NEDD8 and ubiquitin are effectively distinguished. These proteins exert downstream signaling effects by binding to NEDD8 chains and/or neddylated proteins. The motif interacting with ubiquitin (MIU) 2 domain of RNF168 binds NEDD8 chains and mediates the recruitment of the E3 ligase on sites of DNA damage, thus allowing the RNF168-mediated polyubiquitination of γ-H2AX [145]. The UIM motif of hepatocyte growth factor-regulated Tyr kinase substrate (HGS) promotes the recruitment of the activated EGF receptor, which is ubiquitinated and neddylated on several lysine targets in the cytoplasmic tails [64].

In the recent years, the HECT E3-ligases SMURF1 and SMURF2 have been demonstrated to require the direct interaction with free NEDD8 in order to promote their ubiquitin ligase activity toward both substrates and the enzyme itself. Interestingly, each ligase contains two NEDD8-binding sites, located, respectively, in the N-lobe and C-lobe, that are not embedded in a known UBD. Both sites conform to the consensus sequence L(X7)R(X5)F(X)ALQ [102]. Importantly, the NEDD8-binding sites in both ligases are clearly distinct from the ubiquitin-binding sites (the "exosites") that map in the N-lobe and recognize the Ile44 patch of ubiquitin [146]. This spatial separation supports the notion that, despite being closely associated and functionally interconnected, the non-covalent interactions with ubiquitin and NEDD8 play roles that are rather complementary more than simply overlapping.

Finally, it has been recently shown that two RING-between-RING E3 ligases, TRIAD1 and HHARI, engage a UBA-like domain to recruit different NEDD8-conjugated cullins [147]. Despite interacting with different CRL complexes, the outcome of both interactions is the same: the release of the auto-inhibited state and the subsequent stimulation of their E3-ligase activity, which also results in their own proteasomal degradation. Concluding, these observations reveal a novel role of NEDD8 in recruiting additional E3 ligases to existing CRL E3 ubiquitin ligase complexes.

9. CUBAN: a novel domain showing a preference for NEDD8

Recently, the characterization of several Ubl-binding domains has been performed by using the phage display approach [119]. By panning a human brain phage-displayed cDNA library, a number of ubiquitin- and NEDD8-binding domains were identified. In particular, panning with NEDD8 identified five putative binding partners, including the UBA domains of the ubiquitin shuttles RAD23 and UBQLN1. Both proteins have been previously shown to interact with NEDD8 ([141] and **Table 3**), and they have been suggested to be involved in the recognition of mixed ubiquitin-NEDD8 chains by the proteasome under stress conditions. Moreover, while the binding affinity for ubiquitin of UBA domain of RAD23 is sensibly higher than NEDD8, the corresponding domain in UBQLN1 shows a similar affinity for both molecules.

Even though the approach did not provide evidence for a protein domain strictly specific for NEDD8 (since all NEDD8-binding proteins are also ubiquitin interactors), it allowed the identification of a novel domain in the KHNYN (KH domain- and NYN domain-containing) protein, which has unique features compared to the previously characterized binding domains. The first peculiarity of this domain is to show a clear preference for NEDD8. Indeed, it interacts with monomeric NEDD8 with a binding affinity that has been evaluated to be around 24 ± 2 □M, a measure significantly higher than that detected for ubiquitin and that is comparable to the highest affinity for ubiquitin observed in some CUE and UBA domains [126]. Secondly, in addition to showing a clear preference for NEDD8 over monomeric

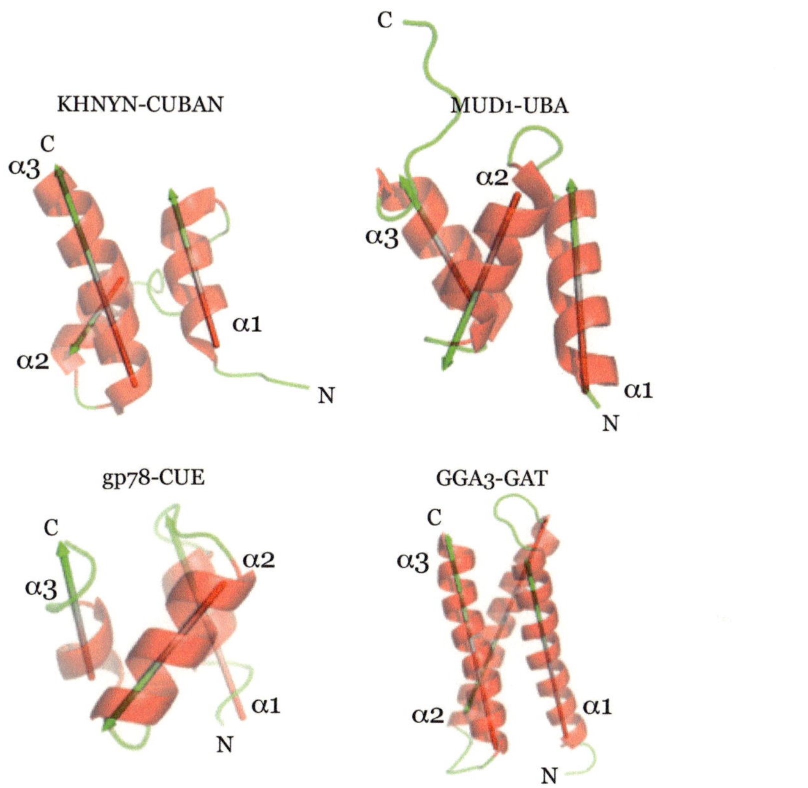

Figure 4.
Ribbon representation of CUBAN (KHNYN), UBA (MUD1), CUE (gp78), and GAT (GGA3) domains. Arrows indicate the relative spatial distributions of the three helices (⟨1, ⟨2, and ⟨3). N- and C-terminal ends are indicated.

ubiquitin, this domain—both isolated and in the context of the full-length protein—promotes the interaction with neddylated cullins, the reason for which it has been dubbed CUBAN, for cullin-binding domain associating with NEDD8. The investigation by NMR spectroscopy showed structural elements that are common in other UBDs belonging to the most populated category represented by a three-bundle helix, such as CUE and UBA. Nevertheless, unique features of CUBAN domain reveal the presence of a novel three-alpha-helix bundle domain, which is characterized by an unusually extended loop1. Moreover, by comparing the folding of the CUBAN domain with other known ubiquitin-binding domains, it can be observed that the central helix in CUE and UBA domains is located in front of the plane formed by helices 1 and 3, in CUBAN as well as in the GAT domain projected on the opposite side (**Figure 4**).

By studying the protein complex between NEDD8 and CUBAN, a third relevant feature emerged: CUBAN, although requiring the presence of an integral hydrophobic patch in NEDD8, binds to residues in the second β-strand (Ile13 and Glu14) and in the C-terminal end of helix α-1 (31-Glu-Glu-Lys-Glu-34) of NEDD8, indicating that the interaction is more electrostatic than hydrophobic. Accordingly, positively charged residues (His651, Arg652, Arg659, and Arg664) mapping in turn1, helix α2, and turn2 characterize the binding interface in CUBAN. Interestingly, the molecular details of this interaction are reminiscent of the electrostatic interaction between the acidic residues Glu31 and Glu32 in NEDD8 and the linker of RBX1 [148], suggesting that distinct interactors require the recognition of the same binding surface. A fourth consideration regarding CUBAN is that this domain also binds di-ubiquitins and this interaction disrupts the association with free NEDD8.

It must be pointed out that ubiquitin chains follow the principle of the higher avidity of binding, which is mediated through multivalent interactions that largely increase binding affinity [149]. Since these contact sites are in close proximity, it is not surprising that the di-ubiquitin competes out monomeric NEDD8, at least when analyzing the isolated domain. Nevertheless, it cannot be excluded that once NEDD8 is conjugated to cullins, KHNYN could interact with the CRL complex in a way that hampers the access of ubiquitin chains to the CUBAN domain. More important, more than being specific, the CUBAN domain demonstrates a discriminating capability, since it has gained structural features that clearly make it capable, differently from other ubiquitin-binding domains, to select and distinguish NEDD8 from its closer relative. This feature, which is the result of the evolution of two distinct binding sites, suggests that the interaction with NEDD8 and ubiquitin must be finely tuned, similarly to what has been shown for RBX1, which shares with CUBAN both the dual specificity toward ubiquitin and NEDD8 and the recognition of an electrostatic surface on the Ubl molecule.

Such discriminating capability is also revealed by the evidence that the A72R mutation in the C-terminal tail of NEDD8 partially affects binding of CUBAN to NEDD8, suggesting that the acquisition of ubiquitin features interferes with the ability of this domain to bind its specific target in KHNYN. On the contrary, the same mutation strongly favors the interaction with those UBDs showing an in vitro weak binding toward NEDD8, a consequence of the evolutionary conserved features mediating the ubiquitin recognition by the majority of UBDs. Therefore, among the residues that are divergent in NEDD8 and ubiquitin, Ala72 is responsible for the weak binding of NEDD8 to the UBDs that can potentially recognize both posttranslational modifications.

The NEDD8-binding region identified in KHNYN maps to the carboxyl-terminal end (aa 627–678). Interestingly, the full-length KHNYN is also consistently ubiquitinated in cells, and such covalent modification is abrogated in the absence of the CUBAN domain. The protein localizes both in the nucleus and the cytoplasm and contains a putative nuclear localization signal, and it has been found to be associated to membranes [150]. Interestingly, an evolutionary related protein called NEDD4-binding protein 1 (N4BP1) has been identified in the same panning experiment and shares with KHNYN the presence of a domain interacting with ubiquitin at the very carboxyl-terminal end. Despite the high level of identity (about 40%), the divergences are sufficient to determine a marked difference in their binding preferences, as highlighted by the evidence that only KHNYN can recognize monomeric and conjugated NEDD8 [119]. Both proteins contain an N-terminal evolutionary conserved KH domain (K homology), which is present in a wide variety of nucleic acid-binding proteins where it can function in RNA recognition, followed by the NYN domain (N4BP1, YacP nucleases), with predicted ribonuclease activity [151, 152]. The comprehension of the regulatory mechanisms that associate RNA recognition and degradation with the recruitment of activated cullin RING-ligase complexes will help in clarifying the biological functions and the cellular processes that are regulated by this protein.

10. Concluding remarks

Our understanding of the complexity within the NEDD8 pathway and how it cross-reacts with ubiquitin is rapidly growing. Numerous signaling pathways, such as apoptosis, DNA damage, and nucleolar stress signaling, are clearly emerging as biological contexts in which neddylation plays a key role. Nevertheless, an important question that needs to be further investigated is whether neddylation is always functionally distinct from ubiquitination. Current knowledge indicates that the two posttranslation modifications must be closely interconnected, as also testified by the evidence that NEDD8 has evolutionary gained few, but key, features that allow it to be distinguished from ubiquitin while maintaining a considerable degree of identity that is much higher than other Ubls. Moreover, the identification of novel NEDD8-binding motifs that are always close to ubiquitin-binding sites, for example, in SMURF1/2, RBX1 and KHNYN, would suggest that these interactions act cooperatively rather than in a mutually exclusive manner. A complete understanding of how these interactions are modulated and the mechanisms controlling the NEDD8 pathway will help in identifying new potential NEDD8 targets and inhibitors which can pave the way to clinical developments in the treatment of several diseases.

Acknowledgements

I would like to thank Prof. Maurizio Paci and Prof. Gianni Cesareni for revising the manuscript and Ridvan Nepravishta for helping in the figure preparation.

Conflict of interest

I have no conflict of interest to declare.

Author details

Elena Santonico
Department of Biology, University of Rome Tor Vergata, Rome, Italy

*Address all correspondence to: elena.santonico@uniroma2.it

References

[1] Khoury GA, Baliban RC, Floudas CA. Proteome-wide post-translational modification statistics: Frequency analysis and curation of the swiss-prot database. Scientific Reports. 2011;**1**:4

[2] Hao P, Sze SK. Proteomic analysis of protein deamidation. Current Protocols in Protein Science. 2014;**78**:24.5.1-24.5.14

[3] Brennan DF, Barford D. Eliminylation: A post-translational modification catalyzed by phosphothreonine lyases. Trends in Biochemical Sciences. 2009;**34**:108-114

[4] Lim WA, Pawson T. Phosphotyrosine signaling: Evolving a new cellular communication system. Cell. 2010;**142**:661-667

[5] van der Veen AG, Ploegh HL. Ubiquitin-like proteins. Annual Review of Biochemistry. 2012;**81**:323-357

[6] Wada H, Kito K. Cleavage of the C-terminus of NEDD8 by UCH-L3. Biochemical and Biophysical Research Communications. 1998;**251**:1-5

[7] Mendoza HM, Shen L-N, Botting C, Lewis A, Chen J, Ink B, et al. NEDP1, a highly conserved cysteine protease that deNEDDylates Cullins. The Journal of Biological Chemistry. 2003;**278**:25637-25643

[8] Walden H, Podgorski M, Schulman BA. Insights into the ubiquitin transfer cascade from the structure of the activating enzyme for NEDD8. Letters to Nature. 2003;**422**:1-5

[9] Schulman BA, Harper JW. Ubiquitin-like protein activation by E1 enzymes: The apex for downstream signalling pathways. Nature Reviews. Molecular Cell Biology. 2009;**10**:319-331

[10] Leyser HM, Lincoln CA, Timpte C, Lammer D, Turner J, Estelle M. Arabidopsis auxin-resistance gene AXR1 encodes a protein related to ubiquitin-activating enzyme E1. Nature. 1993;**364**:161-164

[11] Huang DT, Hunt HW, Zhuang M, Ohi MD, Holton JM, Schulman BA. Basis for a ubiquitin-like protein thioester switch toggling E1–E2 affinity. Nature. 2007;**445**:394-398

[12] Soucy TA, Smith PG, Milhollen MA, Berger AJ, Gavin JM, Adhikari S, et al. An inhibitor of NEDD8-activating enzyme as a new approach to treat cancer. Nature. 2009;**458**:732-736

[13] Huang DT, Miller DW, Mathew R, Cassell R, Holton JM, Roussel MF, et al. A unique E1-E2 interaction required for optimal conjugation of the ubiquitin-like protein NEDD8. Nature Structural & Molecular Biology. 2004;**11**:927-935

[14] Kumar S, Yoshida Y, Noda M. Cloning of a cDNA which encodes a novel ubiquitin-like protein. Biochemical and Biophysical Research Communications. 1993;**195**:393-399

[15] Kamitani T, Kito K, Nguyen HP, Yeh ETH. Characterization of NEDD8, a Developmentally Down-regulated Ubiquitin-like Protein. The Journal of Biological Chemistry. 1997;**272**:28557-28562

[16] Liakopoulos D, Doenges G, Matuschewski K, Jentsch S. A novel protein modification pathway related to the ubiquitin system. The EMBO Journal. 1998;**17**:2208-2214

[17] Kawakami T, Chiba T, Suzuki T, Iwai K, Yamanaka K, Minato N, et al. NEDD8 recruits E2-ubiquitin to SCF E3 ligase. The EMBO Journal. 2001;**20**:4003-4012

[18] Pan Z-Q, Kentsis A, Dias DC, Yamoah K, Wu K. Nedd8 on cullin: Building an expressway to protein destruction. Oncogene. 2004;**23**:1985-1997

[19] Lydeard JR, Schulman BA, Harper JW. Building and remodelling Cullin–RING E3 ubiquitin ligases. Nature Publishing Group. 2013;**14**:1050-1061

[20] Duda DM, Borg LA, Scott DC, Hunt HW, Hammel M, Schulman BA. Structural insights into NEDD8 activation of cullin-RING ligases: Conformational control of conjugation. Cell. 2008;**134**:995-1006

[21] Huang DT, Ayrault O, Hunt HW, Taherbhoy AM, Duda DM, Scott DC, et al. E2-RING expansion of the NEDD8 cascade confers specificity to cullin modification. Molecular Cell. 2009;**33**:483-495

[22] Huang DT, Paydar A, Zhuang M, Waddell MB, Holton JM, Schulman BA. Structural basis for recruitment of Ubc12 by an E2 binding domain in NEDD8's E1. Molecular Cell. 2005;**17**:341-350

[23] Enchev RI, Schulman BA, Peter M. Protein neddylation: Beyond cullin-RING ligases. Nature Publishing Group. 2015;**16**:30-44

[24] Eletr ZM, Huang DT, Duda DM, Schulman BA, Kuhlman B. E2 conjugating enzymes must disengage from their E1 enzymes before E3-dependent ubiquitin and ubiquitin-like transfer. Nature Structural & Molecular Biology. 2005;**12**:933-934

[25] Kurz T, Ozlü N, Rudolf F, O'Rourke SM, Luke B, Hofmann K, et al. The conserved protein DCN-1/Dcn1p is required for cullin neddylation in *C. elegans* and *S. cerevisiae*. Nature. 2005;**435**:1257-1261

[26] Liu J, Furukawa M, Matsumoto T, Xiong Y. NEDD8 modification of CUL1 dissociates p120(CAND1), an inhibitor of CUL1-SKP1 binding and SCF ligases. Molecular Cell. 2002;**10**:1511-1518

[27] Zheng J, Yang X, Harrell JM, Ryzhikov S, Shim EH, Lykke-Andersen K, et al. CAND1 binds to unneddylated CUL1 and regulates the formation of SCF ubiquitin E3 ligase complex. Molecular Cell. 2002;**10**:1519-1526

[28] Keuss MJ, Thomas Y, Mcarthur R, Wood NT, Knebel A, Kurz T. Characterization of the mammalian family of DCN-type NEDD8 E3 ligases. Journal of Cell Science. 2016;**129**:1441-1454

[29] Mydlikova Z, Gursky J, Pirsel M. Transcription factor IIH—the protein complex with multiple functions. Neoplasma. 2010;**57**:287-290

[30] Rabut G, Peter M. Function and regulation of protein neddylation. 'Protein Modifications: Beyond the Usual Suspects' review series. EMBO Reports. 2008;**9**:969-976

[31] Enchev RI, Scott DC, da Fonseca PCA, Schreiber A, Monda JK, Schulman BA, et al. Structural basis for a reciprocal regulation between SCF and CSN. Cell Reports. 2012;**2**:616-627

[32] Lingaraju GM, Bunker RD, Cavadini S, Hess D, Hassiepen U, Renatus M, et al. Crystal structure of the human COP9 signalosome. Nature. 2014;**512**:161-165

[33] Wu K, Yamoah K, Dolios G, Gan-Erdene T, Tan P, Chen A, et al. DEN1 is a dual function protease capable of processing the C terminus of Nedd8 and deconjugating hyper-neddylated CUL1. The Journal of Biological Chemistry. 2003;**278**:28882-28891

[34] Mergner J, Kuster B, Schwechheimer C. DENEDDYLASE1 protein counters automodification of neddylating enzymes to maintain NEDD8 protein homeostasis in *Arabidopsis*. Journal of Biological Chemistry. 3 Mar 2017;**292**(9):3854-3865. DOI: 10.1074/jbc.M116.767103

[35] Mergner J, Schwechheimer C. The NEDD8 modification pathway in plants. Frontiers in Plant Science. 2014;**5**:103

[36] Gong L, Kamitani T, Millas S, Yeh ETH. Identification of a novel isopeptidase

with dual specificity for ubiquitin- and NEDD8-conjugated proteins. The Journal of Biological Chemistry. 2000;**275**:14212-14216

[37] Ferro A, Carvalho AL, Teixeira-Castro A, Almeida C, Tomé RJ, Cortes L, et al. NEDD8: A new ataxin-3 interactor. Biochimica et Biophysica Acta. 2007;**1773**:1619-1627

[38] Artavanis-Tsakonas K, Misaghi S, Comeaux CA, Catic A, Spooner E, Duraisingh MT, et al. Identification by functional proteomics of a deubiquitinating/deNeddylating enzyme in *Plasmodium falciparum*. Molecular Microbiology. 2006;**61**:1187-1195

[39] Johnston SC, Riddle SM, Cohen RE, Hill CP. Structural basis for the specificity of ubiquitin C-terminal hydrolases. The EMBO Journal. 1999;**18**:3877-3887

[40] Whitby FG, Xia G, Pickart CM, Hill CP. Crystal structure of the human ubiquitin-like protein NEDD8 and interactions with ubiquitin pathway enzymes. The Journal of Biological Chemistry. 1998;**273**:34983-34991

[41] Jehl P, Manguy J, Shields DC, Higgins DG, Davey NE. ProViz-a web-based visualization tool to investigate the functional and evolutionary features of protein sequences. Nucleic Acids Research. 2016;**44**:W11-W15

[42] Komander D, Rape M. The ubiquitin code. Annual Review of Biochemistry. 2012;**81**:203-229

[43] Pickart CM. Mechanisms underlying ubiquitination. Annual Review of Biochemistry. 2001;**70**:503-533

[44] Reverter D, Wu K, Erdene TG, Pan Z-Q, Wilkinson KD, Lima CD. Structure of a complex between Nedd8 and the Ulp/Senp protease family member Den1. Journal of Molecular Biology. 2005;**345**:141-151

[45] Walden H, Huang DT, Schulman BA. The structure of the APPBP1-UBA3-NEDD8-ATP complex reveals the basis for selective ubiquitin-like protein activation by an E1. Molecular Cell. 2003;**12**:1-11

[46] Souphron J, Waddell MB, Paydar A, Tokgöz-Gromley Z, Roussel MF, Schulman BA. Structural dissection of a gating mechanism preventing misactivation of ubiquitin by NEDD8's E1. Biochemistry. 2008;**47**:8961-8969

[47] Hjerpe R, Thomas Y, Chen J, Zemla A, Curran S, Shpiro N, et al. Changes in the ratio of free NEDD8 to ubiquitin triggers NEDDylation by ubiquitin enzymes. The Biochemical Journal. 2012;**441**:927-936

[48] Leidecker O, Matic I, Mahata B, Pion E, Xirodimas DP. The ubiquitin E1 enzyme Ube1 mediates NEDD8 activation under diverse stress conditions. Cell Cycle. 2012;**11**:1142-1150

[49] Scott DC, Sviderskiy VO, Monda JK, Lydeard JR, Cho SE, Harper JW, et al. Structure of a RING E3 trapped in action reveals ligation mechanism for the ubiquitin-like protein NEDD8. Cell. 2014;**157**:1671-1684

[50] Petroski MD, Deshaies RJ. Function and regulation of cullin-RING ubiquitin ligases. Nature Reviews. Molecular Cell Biology. 2005;**6**:9-20

[51] Watson IR, Li BK, Roche O, Blanch A, Ohh M, Irwin MS. Chemotherapy induces NEDP1-mediated destabilization of MDM2. Oncogene. 2010;**29**:297-304

[52] Linghu B, Callis J, Goebl MG. Rub1p processing by Yuh1p is required for wild-type levels of Rub1p conjugation to Cdc53p. Eukaryotic Cell. 2002;**1**:491-494

[53] Watson IR, Blanch A, Lin DCC, Ohh M, Irwin MS. Mdm2-mediated NEDD8 modification of TAp73 regulates its transactivation function. The Journal of Biological Chemistry. 2006;**281**:34096-34103

[54] Aoki I, Higuchi M, Gotoh Y. NEDDylation controls the target specificity of E2F1 and apoptosis induction. Oncogene. 2012;**32**:1-11

[55] Shen L-N, Liu H, Dong C, Naismith JH, Hay RT. Structural basis of NEDD8 ubiquitin discrimination by the deNEDDylating enzyme NEDP1. The EMBO Journal. 2005;**24**:1341-1351

[56] Gao F, Cheng J, Shi T, Yeh ETH. Neddylation of a breast cancer-associated protein recruits a class III histone deacetylase that represses NFκB-dependent transcription. Nature Cell Biology. 2006;**8**:1171-1177

[57] Shin Y-C, Tang S-J, Chen J-H, Liao P-H, Chang S-C. The molecular determinants of NEDD8 specific recognition by human SENP8. PLoS One. 2011;**6**:e27742

[58] Shin Y-C, Chen J-H, Chang S-C. The molecular determinants for distinguishing between ubiquitin and NEDD8 by USP2. Scientific Reports. 2017;**7**:159

[59] Xirodimas DP, Saville MK, Bourdon J-C, Hay RT, Lane DP. Mdm2-mediated NEDD8 conjugation of p53 inhibits its transcriptional activity. Cell. 2004;**118**:83-97

[60] Liu N, Zhang J, Yang X, Jiao T, Zhao X, Li W, et al. HDM2 promotes NEDDylation of hepatitis B virus HBx to enhance its stability and function. Journal of Virology. 2017;**91**:1221

[61] Embade N, Fernández-Ramos D, Varela-Rey M, Beraza N, Sini M, Gutiérrez de Juan V, et al. Murine double minute 2 regulates Hu antigen R stability in human liver and colon cancer through NEDDylation. Hepatology. 2012;**55**:1237-1248

[62] Maghames CM, Lobato-Gil S, Perrin A, Trauchessec H, Rodriguez MS, Urbach S, et al. NEDDylation promotes nuclear protein aggregation and protects the Ubiquitin Proteasome System upon proteotoxic stress. Nature Communications. 2018;**9**:a004374-a004317

[63] Ma T, Chen Y, Zhang F, Yang C-Y, Wang S, Yu X. RNF111-dependent neddylation activates DNA damage-induced ubiquitination. Molecular Cell. 2013;**49**:897-907

[64] Oved S, Mosesson Y, Zwang Y, Santonico SK, Marmor MD, Kochupurakkal BS, et al. Conjugation to Nedd8 instigates ubiquitylation and down-regulation of activated receptor tyrosine kinases. The Journal of Biological Chemistry. 2006;**281**:21640-21651

[65] Broemer M, Tenev T, Rigbolt KTG, Hempel S, Blagoev B, Silke J, et al. Systematic in vivo RNAi analysis identifies IAPs as NEDD8-E3 ligases. Molecular Cell. 2010;**40**:810-822

[66] Xie P, Zhang M, He S, Lu K, Chen Y, Xing G, et al. The covalent modifier Nedd8 is critical for the activation of Smurf1 ubiquitin ligase in tumorigenesis. Nature Communications. 2014;**5**:1-16

[67] Choo YS, Vogler G, Wang D, Kalvakuri S, Iliuk A, Tao WA, et al. Regulation of parkin and PINK1 by neddylation. Human Molecular Genetics. 2012;**21**:2514-2523

[68] Um JW, Han KA, Im E, Oh Y, Lee K, Chung KC. Neddylation positively regulates the ubiquitin E3 ligase activity of parkin. Journal of Neuroscience Research. 2012;**90**:1030-1042

[69] Xirodimas DP. Novel substrates and functions for the ubiquitin-like molecule NEDD8. Biochemical Society Transactions. 2008;**36**:802-806

[70] Kandala S. Neddylation and deneddylation in cardiac biology. Review article. American Journal of Cardiovascular Disease. 2014;**4**:1-19

[71] Li H, Zhu H, Liu Y, He F, Xie P, Zhang L. Itch promotes the neddylation

of JunB and regulates JunB-dependent transcription. Cellular Signalling. 2016;**28**:1-10

[72] Takashima O, Tsuruta F, Kigoshi Y, Nakamura S, Kim J, Katoh MC, et al. Brap2 regulates temporal control of NF-κB localization mediated by inflammatory response. PLoS One. 2013;**8**:e58911-e58910

[73] Panni S, Salvioli S, Santonico E, Langone F, Storino F, Altilia S, et al. The adapter protein CD2AP binds to p53 protein in the cytoplasm and can discriminate its polymorphic variants P72R. Journal of Biochemistry. 2015;**157**:101-111

[74] Stickle NH, Chung J, Klco JM, Hill RP, Kaelin WG, Ohh M. pVHL modification by NEDD8 is required for fibronectin matrix assembly and suppression of tumor development. Molecular and Cellular Biology. 2004;**24**:3251-3261

[75] Russell RC, Ohh M. NEDD8 acts as a 'molecular switch' defining the functional selectivity of VHL. EMBO Reports. 2008;**9**:486-491

[76] Roberts AM, Ohh M. Beyond the hypoxia-inducible factor-centric tumour suppressor model of von Hippel-Lindau. Current Opinion in Oncology. 2008;**20**:83-89

[77] Jin H-S, Liao L, Park Y, Liu Y-C. Neddylation pathway regulates T-cell function by targeting an adaptor protein Shc and a protein kinase Erk signaling. Proceedings of the National Academy of Sciences of the United States of America. 2013;**110**:624-629

[78] Loftus SJ, Liu G, Carr SM, Munro S, La Thangue NB. NEDDylation regulates E2F-1-dependent transcription. EMBO Reports. 2012;**13**:811-818

[79] Ryu J-H, Li S-H, Park H-S, Park J-W, Lee B, Chun Y-S. Hypoxia-inducible factor α subunit stabilization by NEDD8 conjugation is reactive oxygen species-dependent. The Journal of Biological Chemistry. 2011;**286**:6963-6970

[80] Tan CY, Hagen T. Destabilization of CDC6 upon DNA damage is dependent on neddylation but independent of Cullin E3 ligases. The International Journal of Biochemistry & Cell Biology. 2013;**45**:1489-1498

[81] Li M, Brooks CL, Wu-Baer F, Chen D, Baer R, Gu W. Mono-versus polyubiquitination: Differential control of p53 fate by Mdm2. Science. 2003;**302**:1972-1975

[82] Waning DL, Lehman JA, Batuello CN, Mayo LD. c-Abl phosphorylation of Mdm2 facilitates Mdm2-Mdmx complex formation. The Journal of Biological Chemistry. 2011;**286**:216-222

[83] Maya R, Balass M, Kim ST, Shkedy D, Leal JF, Shifman O, et al. ATM-dependent phosphorylation of Mdm2 on serine 395: Role in p53 activation by DNA damage. Genes & Development. 2001;**15**:1067-1077

[84] Hauck PM, Wolf ER, Olivos DJ, McAtarsney CP, Mayo LD. The fate of murine double minute X (MdmX) is dictated by distinct signaling pathways through murine double minute 2 (Mdm2). Oncotarget. 2017;**8**:104455-104466

[85] Batuello CN, Hauck PM, Gendron JM, Lehman JA, Mayo LD. Src phosphorylation converts Mdm2 from a ubiquitinating to a neddylating E3 ligase. Proceedings of the National Academy of Sciences of the United States of America. 2015;**112**:1749-1754

[86] Abida WM, Nikolaev A, Zhao W, Zhang W, Gu W. FBXO11 promotes the Neddylation of p53 and inhibits its transcriptional activity. The Journal of Biological Chemistry. 2007;**282**:1797-1804

[87] Bonacci T, Audebert S, Camoin L, Baudelet E, Iovanna J-L, Soubeyran P. Regulation of NUB1 activity through non-proteolytic Mdm2-mediated ubiquitination. PLoS One. 2017;**12**:e0169988

[88] Sluimer J, Distel B. Regulating the human HECT E3 ligases. Cellular and Molecular Life Sciences. 2018;**75**:3121-3141

[89] Metzger MB, Pruneda JN, Klevit RE, Weissman AM. RING-type E3 ligases: Master manipulators of E2 ubiquitin-conjugating enzymes and ubiquitination. Biochimica et Biophysica Acta (BBA)—Molecular Cell Research. 2013;**1843**:1-14

[90] Dunn R, Klos DA, Adler AS, Hicke L. The C2 domain of the Rsp5 ubiquitin ligase binds membrane phosphoinositides and directs ubiquitination of endosomal cargo. The Journal of Cell Biology. 2004;**165**:135-144

[91] Rotin D, Kumar S. Physiological functions of the HECT family of ubiquitin ligases. Nature Reviews. Molecular Cell Biology. 2009;**10**:398-409

[92] Scheffner M, Staub O. HECT E3s and human disease. BMC Biochemistry. 2007;**8**:S6-S14

[93] Kee Y, Huibregtse JM. Regulation of catalytic activities of HECT ubiquitin ligases. Biochemical and Biophysical Research Communications. 2007;**354**:329-333

[94] Lorenz S. Structural mechanisms of HECT-type ubiquitin ligases. Biological Chemistry. 2018;**399**:127-145

[95] Wan L, Zou W, Gao D, Inuzuka H, Fukushima H, Berg AH, et al. Cdh1 regulates osteoblast function through an APC/C-independent modulation of Smurf1. Molecular Cell. 2011;**44**:721-733

[96] Lu K, Yin X, Weng T, Xi S, Li L, Xing G, et al. Targeting WW domains linker of HECT-type ubiquitin ligase Smurf1 for activation by CKIP-1. Nature Cell Biology. 2008;**10**:994-1002

[97] Crose LES, Hilder TL, Sciaky N, Johnson GL. Cerebral cavernous malformation 2 protein promotes smad ubiquitin regulatory factor 1-mediated RhoA degradation in endothelial cells. The Journal of Biological Chemistry. 2009;**284**:13301-13305

[98] de Bie P, Ciechanover A. Ubiquitination of E3 ligases: self-regulation of the ubiquitin system via proteolytic and non-proteolytic mechanisms. Cell Death and Differentiation. 2011;**18**:1393-1402

[99] French ME, Kretzmann BR, Hicke L. Regulation of the RSP5 ubiquitin ligase by an intrinsic ubiquitin-binding site. The Journal of Biological Chemistry. 2009;**284**:12071-12079

[100] Kim HC, Steffen AM, Oldham ML, Chen J, Huibregtse JM. Structure and function of a HECT domain ubiquitin-binding site. EMBO Reports. 2011;**12**:334-341

[101] Shu J, Liu C, Wei R, Xie P, He S, Zhang L. Nedd8 targets ubiquitin ligase Smurf2 for neddylation and promote its degradation. Biochemical and Biophysical Research Communications. 2016;**474**(1):30

[102] He S, Cao Y, Xie P, Dong G, Zhang L. The Nedd8 non-covalent binding region in the Smurf HECT domain is critical to its ubiquitin ligase function. Scientific Reports. 2017;**7**:41364

[103] Sung M-K, Reitsma JM, Sweredoski MJ, Hess S, Deshaies RJ. Ribosomal proteins produced in excess are degraded by the ubiquitin-proteasome system. Molecular Biology of the Cell. 2016;**27**:2642-2652

[104] Rubbi CP, Milner J. Disruption of the nucleolus mediates stabilization of p53 in response to DNA damage and other stresses. The EMBO Journal. 2003;**22**:6068-6077

[105] Yang K, Wang M, Zhao Y, Sun X, Yang Y, Li X, et al. A redox mechanism underlying nucleolar stress sensing by nucleophosmin. Nature Communications. 2016;**7**:13599

[106] Xu X, Xiong X, Sun Y. The role of ribosomal proteins in the regulation of cell proliferation, tumorigenesis, and genomic integrity. Science China. Life Sciences. 2016;**59**:656-672

[107] Yang K, Yang J, Yi J. Nucleolar Stress: Hallmarks, sensing mechanism and diseases. Cell Stress. 2018;**2**:125-140

[108] Dai M-S, Lu H. Inhibition of MDM2-mediated p53 ubiquitination and degradation by ribosomal protein L5. The Journal of Biological Chemistry. 2004;**279**:44475-44482

[109] Zhang Y, Wolf GW, Bhat K, Jin A, Allio T, Burkhart WA, et al. Ribosomal protein L11 negatively regulates oncoprotein MDM2 and mediates a p53-dependent ribosomal-stress checkpoint pathway. Molecular and Cellular Biology. 2003;**23**:8902-8912

[110] Lohrum MAE, Ludwig RL, Kubbutat MHG, Hanlon M, Vousden KH. Regulation of HDM2 activity by the ribosomal protein L11. Cancer Cell. 2003;**3**:577-587

[111] Jin A, Itahana K, O'Keefe K, Zhang Y. Inhibition of HDM2 and activation of p53 by ribosomal protein L23. Molecular and Cellular Biology. 2004;**24**:7669-7680

[112] van Riggelen J, Yetil A, Felsher DW. MYC as a regulator of ribosome biogenesis and protein synthesis. Nature Reviews. Cancer. 2010;**10**:301-309

[113] Zhou X, Hao Q, Liao J-M, Liao P, Lu H. Ribosomal protein S14 negatively regulates c-Myc activity. The Journal of Biological Chemistry. 2013;**288**:21793-21801

[114] Zhang Z, Wang H, Li M, Rayburn ER, Agrawal S, Zhang R. Stabilization of E2F1 protein by MDM2 through the E2F1 ubiquitination pathway. Oncogene. 2005;**24**:7238-7247

[115] Zhang J, Bai D, Ma X, Guan J, Zheng X. hCINAP is a novel regulator of ribosomal protein-HDM2-p53 pathway by controlling NEDDylation of ribosomal protein S14. Oncogene. 2012;**33**:246-254

[116] Jones J, Wu K, Yang Y, Guerrero C, Nillegoda N, Pan Z-Q, et al. A targeted proteomic analysis of the ubiquitin-like modifier Nedd8 and associated proteins. Journal of Proteome Research. 2008;**7**:1274-1287

[117] Ohki Y, Funatsu N, Konishi N, Chiba T. The mechanism of poly-NEDD8 chain formation in vitro. Biochemical and Biophysical Research Communications. 2009;**381**:443-447

[118] Zuo W, Huang F, Chiang YJ, Li M, Du J, Ding Y, et al. c-Cbl-mediated neddylation antagonizes ubiquitination and degradation of the TGF-β type II receptor. Molecular Cell. 2013;**49**:499-510

[119] Castagnoli L, Mandaliti W, Nepravishta R, Valentini E, Mattioni A, Procopio R, et al. Selectivity of the CUBAN domain in the recognition of ubiquitin and NEDD8. The FEBS Journal. 2019;**98**:1-37

[120] Bian Y, Song C, Cheng K, Dong M, Wang F, Huang J, et al. An enzyme assisted RP-RPLC approach for in-depth analysis of human liver phosphoproteome. Journal of Proteomics. 2014;**96**:253-262

[121] Weinert BT, Schölz C, Wagner SA, Iesmantavicius V, Su D, Daniel JA, et al. Lysine succinylation is a frequently occurring modification in prokaryotes and eukaryotes and extensively overlaps with acetylation. Cell Reports. 2013;**4**:842-851

[122] Rigbolt KTG, Prokhorova TA, Akimov V, Henningsen J, Johansen PT, Kratchmarova I, et al. System-wide temporal characterization of the proteome and phosphoproteome of human embryonic stem cell differentiation. Science Signaling. 2011;**4**:rs3-rs3

[123] Hershko A, Ciechanover A. The ubiquitin system for protein degradation. Annual Review of Biochemistry. 1992;**61**:761-807

[124] Goldberg AL, Rock KL. Proteolysis, proteasomes and antigen presentation. Nature. 1992;**357**:375-379

[125] Hochstrasser M. Origin and function of ubiquitin-like proteins. Nature. 2009;**458**:422-429

[126] Hurley JH, Lee S, Prag G. Ubiquitin-binding domains. The Biochemical Journal. 2006;**399**:361-372

[127] Dikic I, Wakatsuki S, Walters KJ. Ubiquitin-binding domains—From structures to functions. Nature Reviews. Molecular Cell Biology. 2009;**10**:659-671

[128] Hicke L, Schubert HL, Hill CP. Ubiquitin-binding domains. Nature Reviews. Molecular Cell Biology. 2005;**6**:610-621

[129] Haririnia A, Verma R, Purohit N, Twarog MZ, Deshaies RJ, Bolon D, et al. Mutations in the hydrophobic core of ubiquitin differentially affect its recognition by receptor proteins. Journal of Molecular Biology. 2008;**375**:979-996

[130] Bomar MG, D'Souza S, Bienko M, Dikic I, Walker GC, Zhou P. Unconventional ubiquitin recognition by the ubiquitin-binding motif within the Y family DNA polymerases iota and Rev1. Molecular Cell. 2010;**37**:408-417

[131] Kamadurai HB, Souphron J, Scott DC, Duda DM, Miller DJ, Stringer D, et al. Insights into ubiquitin transfer cascades from a structure of a UbcH5B approximately ubiquitin-HECT(NEDD4L) complex. Molecular Cell. 2009;**36**:1095-1102

[132] Winget JM, Mayor T. The diversity of ubiquitin recognition: Hot spots and varied specificity. Molecular Cell. 2010;**38**:627-635

[133] Gill G. SUMO and ubiquitin in the nucleus: Different functions, similar mechanisms? Genes & Development. 2004;**18**:2046-2059

[134] Geiss-Friedlander R, Melchior F. Concepts in sumoylation: A decade on. Nature Reviews. Molecular Cell Biology. 2007;**8**:947-956

[135] Tanaka T, Kawashima H, Yeh ETH, Kamitani T. Regulation of the NEDD8 conjugation system by a splicing variant, NUB1L. The Journal of Biological Chemistry. 2003;**278**:32905-32913

[136] Kamitani T, Kito K, Fukuda-Kamitani T, Yeh ETH. Targeting of NEDD8 and its conjugates for proteasomal degradation by NUB1. The Journal of Biological Chemistry. 2001;**276**:46655-46660

[137] Kito K, Yeh ETH, Kamitani T. NUB1, a NEDD8-interacting protein, is induced by interferon and down-regulates the NEDD8 expression. The Journal of Biological Chemistry. 2001;**276**:20603-20609

[138] Tanji K, Tanaka T, Kamitani T. Interaction of NUB1 with the proteasome subunit S5a. Biochemical and Biophysical Research Communications. 2005;**337**:116-120

[139] Bandau S, Knebel A, Gage ZO, Wood NT, Alexandru G. UBXN7 docks on neddylated cullin complexes using its UIM motif and causes HIF1α accumulation. BMC Biology. 2012;**10**:36

[140] Besten den W, Verma R, Kleiger G, Oania RS, Deshaies RJ. NEDD8 links cullin-RING ubiquitin ligase function to the p97 pathway. Nature Structural & Molecular Biology. 2012;**19**:511-6-S1

[141] Singh RK, Zerath S, Kleifeld O, Scheffner M, Glickman MH, Fushman D. Recognition and cleavage of related to Ubiquitin 1 (Rub1) and Rub1-Ubiquitin chains by components of the ubiquitin-proteasome system. Molecular & Cellular Proteomics. 2012;**11**:1595-1611

[142] Shamay M, Greenway M, Liao G, Ambinder RF, Hayward SD. De novo DNA methyltransferase DNMT3b interacts with NEDD8-modified proteins. The Journal of Biological Chemistry. 2010;**285**:36377-36386

[143] Noguchi K, Okumura F, Takahashi N, Kataoka A, Kamiyama T, Todo S, et al. TRIM40 promotes neddylation of IKK and is downregulated in gastrointestinal cancers. Carcinogenesis. 2011;**32**:995-1004

[144] Antenos M, Casper RF, Brown TJ. Interaction with Nedd8, a ubiquitin-like protein, enhances the transcriptional activity of the aryl hydrocarbon receptor. The Journal of Biological Chemistry. 2002;**277**:44028-44034

[145] Li T, Guan J, Huang Z, Hu X, Zheng X. RNF168-mediated H2A neddylation antagonizes ubiquitylation of H2A and regulates DNA damage repair. Journal of Cell Science. 2014;**127**:2238-2248

[146] Lorenz S. Structural mechanisms of HECT-type ubiquitin ligases. Biological Chemistry. 2017;**399**:1-40

[147] Kelsall IR, Duda DM, Olszewski JL, Hofmann K, Knebel A, Langevin F, et al. TRIAD1 and HHARI bind to and are activated by distinct neddylated Cullin-RING ligase complexes. The EMBO Journal. 2013;**32**:2848-2860

[148] Scott DC, Monda JK, Grace CRR, Duda DM, Kriwacki RW, Kurz T, et al. A dual E3 mechanism for Rub1 ligation to Cdc53. Molecular Cell. 2010;**39**:784-796

[149] Ikeda F, Dikic I. Atypical ubiquitin chains: New molecular signals. EMBO Reports. 2008;**9**:493-493

[150] Mirza SP, Halligan BD, Greene AS, Olivier M. Improved method for the analysis of membrane proteins by mass spectrometry. Physiological Genomics. 2007;**30**:89-94

[151] García-Mayoral MF, Hollingworth D, Masino L, Díaz-Moreno I, Kelly G, Gherzi R, et al. The structure of the C-terminal KH domains of KSRP reveals a noncanonical motif important for mRNA degradation. Structure. 2007;**15**:485-498

[152] Baber JL, Libutti D, Levens D, Tjandra N. High precision solution structure of the C-terminal KH domain of heterogeneous nuclear ribonucleoprotein K, a c-myc transcription factor. Journal of Molecular Biology. 1999;**289**:949-962

[153] Carter S, Bischof O, Dejean A, Vousden KH. C-terminal modifications regulate MDM2 dissociation and nuclear export of p53. Nature Cell Biology. 2007;**9**:428-435

[154] Brooks CL, Gu W. p53 Ubiquitination: Mdm2 and Beyond. Molecular Cell. 2006;**21**:307-315

[155] Xirodimas DP, Sundqvist A, Nakamura A, Shen L, Botting C, Hay RT. Ribosomal proteins are targets for the NEDD8 pathway. EMBO Reports. 2008;**9**:280-286

[156] Sundqvist A, Liu G, Mirsaliotis A, Xirodimas DP. Regulation of nucleolar signalling to p53 through NEDDylation of L11. EMBO Reports. 2009;**10**:1132-1139

[157] Nagano T, Hashimoto T, Nakashima A, Kikkawa U, Kamada S. X-linked inhibitor of apoptosis protein mediates neddylation by itself but does not function as a NEDD8-E3 ligase for caspase-7. FEBS Letters. 2012;**586**:1612-1616

[158] Renaudin X, Guervilly J-H, Aoufouchi S, Rosselli F. Proteomic analysis reveals a FANCA-modulated neddylation pathway involved in CXCR5 membrane targeting and cell mobility. Journal of Cell Science. 2014;**127**:3546-3554

[159] Lee M-R, Lee D, Shin SK, Kim YH, Choi CY. Inhibition of APP intracellular domain (AICD) transcriptional activity via covalent conjugation with Nedd8. Biochemical and Biophysical Research Communications. 2008;**366**:976-981

Chapter 8

ADP-Ribosylation of the Ubiquitin C-Terminus by Dtx3L/Parp9

Teddy Kamata and Bryce Paschal

Abstract

Ubiquitylation is a post-translational modification that regulates a wide range of cellular pathways including protein degradation, autophagy, mitophagy, cell signaling, DNA damage response, and protein trafficking. This post-translational modification is characterized by covalent attachment of ubiquitin to lysine residues on target proteins by E3 ubiquitin ligases. These enzymes can catalyze both mono- and polyubiquitylation of target substrates. Because of the presence of multiple ubiquitylation acceptor sites on ubiquitin, polyubiquitin chains differing by linkage type and branching patterns can be generated. Post-translational modifications on ubiquitin including glutamine deamidation, lysine SUMOylation, lysine acetylation, and serine, threonine, and tyrosine phosphorylation add to the range of ubiquitin structures that can be synthesized in cells. Recently, ADP-ribosylation was discovered as a new post-translational modification on ubiquitin in two different biological contexts. The bacterial SidE proteins ADP-ribosylate ubiquitin to activate it for a unique mode of ubiquitylation. The human Dtx3L (E3 ubiquitin ligase)/Parp9 (ADP-ribosyltransferase) complex ADP-ribosylates ubiquitin which inhibits conjugation. In this review, we describe the discovery of ubiquitin ADP-ribosylation in the bacterial context, provide an overview of the biological roles of Dtx3L/Parp9, and discuss how NAD^+ levels and ubiquitin ADP-ribosylation could regulate the E3 output of Dtx3L/Parp9.

Keywords: ubiquitin, Dtx3L, Parp9, ADP-ribosyltransferase, ADP-ribosylation, SidE, *Legionella pneumophila*, DNA damage, cancer

1. Introduction

Ubiquitylation (also known as ubiquitination) is a major type of post-translational modification that plays diverse roles in cells and involves covalent attachment of the 76 amino acid protein ubiquitin to target substrates [1]. The process of ubiquitylation involves the sequential actions of three classes of enzymes: E1 ubiquitin-activating enzyme, E2 ubiquitin-conjugating enzyme, and E3 ubiquitin ligase [2, 3]. Activation of ubiquitin by E1 takes place in an ATP-dependent manner where an ubiquitin-adenylate intermediate is initially formed. This is followed by release of AMP and the formation of a thioester bond between the C-terminal carboxyl group of ubiquitin and the sulfhydryl group of a cysteine in the active site of E1. The activated ubiquitin is transferred from E1 to a cysteine in E2 in a *trans*-thioesterification step. Finally, E3 ubiquitin ligase catalyzes the transfer of ubiquitin from E2 onto a lysine residue in the target substrate, forming an isopeptide bond between the C-terminal carboxyl group of ubiquitin and the lysine epsilon-amino group. In mammalian cells, there are over 600 E3 ubiquitin ligases which are subdivided into three groups: really interesting new gene (RING), homologous to E6-AP carboxyl terminus (HECT), and ring between ring (RBR) [2]. RING class E3 ubiquitin ligases act as scaffolds to bring together E2 and substrate to mediate ubiquitylation. In contrast, HECT class E3 ubiquitin ligases catalyze a two-step ubiquitylation reaction where the ubiquitin is first transferred from E2 onto a cysteine on HECT E3 ubiquitin ligase before it is transferred to the target substrate. RBR has characteristics of both RING and HECT E3 ubiquitin ligases because of the presence of RING domains as well as a key cysteine that acts as an acceptor for ubiquitin from E2, analogous to HECT E3 ubiquitin ligases [3].

E3 ubiquitin ligases can catalyze the attachment of a single ubiquitin to a substrate (monoubiquitylation) where the linkage is usually formed between the carboxyl group of the C-terminal glycine 76 from ubiquitin and the epsilon-amino group of a lysine from the substrate. Monoubiquitylation can occur on multiple lysine sites on a target protein, which is referred to as multi-monoubiquitylation [1]. In addition, E3 ubiquitin ligases can catalyze attachment of ubiquitin to a monoubiquitylated substrate and generate an ubiquitin chain through successive rounds of ubiquitylation (polyubiquitylation). The basis for chain formation is that ubiquitin contains seven lysines (residues 6, 11, 27, 29, 33, 48, 63) that can serve as acceptor sites for additional ubiquitin to form a series of covalently-linked ubiquitins. In addition, the N-terminal primary amine group from methionine 1 of ubiquitin can serve as an additional site of linkage with another ubiquitin [1, 4]. Thus, there are eight possible linkages for ubiquitin that can be utilized for polyubiquitylation. Homotypic polyubiquitin chains only contain a single type of ubiquitin linkage, while heterotypic polyubiquitin chains contain more than one type of ubiquitin linkage [1, 4, 5]. The latter can be further subdivided into mixed and branched types. In mixed chains, while different linkage types exist, each ubiquitin molecule is modified by ubiquitin at a single linkage site, resulting in an unbranched polyubiquitin chain. On the other hand, ubiquitin found in branched types may be modified by more than one ubiquitin, thereby serving as a branch point for chain formation. Overall, a range of polyubiquitin structures that differ in the type of linkage present within the polyubiquitin chain as well as branching pattern can exist within cells.

An additional layer of complexity to ubiquitylation is the discovery of post-translation modifications that can occur on ubiquitin. Based on mass spectrometry data from proteomic studies, ubiquitin can be modified by SUMOylation and acetylation of lysine residues, and phosphorylation of serine, threonine, and tyrosine residues [4, 5]. Some of these post-translational modifications have been shown to impact ubiquitin conjugation properties. One of the best characterized examples of ubiquitin post-translational modification is serine 65 phosphorylation catalyzed by the protein kinase PINK1 during mitophagy [4]. PINK1 can phosphorylate both ubiquitin monomers and polyubiquitin chains *in vitro*, making the latter more resistant to degradation by deubiquitinases [6], and polyubiquitin chains with serine 65 phosphorylation have been detected in cells [7]. In all 13 E2s tested, no major differences in terms of ubiquitin loading onto E2 by E1 was observed between unmodified ubiquitin and serine 65 phosphorylated ubiquitin, suggesting that in general, this modification on ubiquitin does not affect the initial E1 and E2 processing steps [6]. However, depending on the specific E3, the subsequent assembly of polyubiquitin chains can be inhibited by serine 65 phosphorylation on ubiquitin [6]. In terms of the effects that lysine acetylation has on ubiquitin conjugation properties, the most extensive characterization has been conducted for lysine 6 and 48 [8]. Using *E. coli* with an expanded genetic code, lysine 6 and lysine

48 acetylated ubiquitin was generated and used in *in vitro* experiments to show that ubiquitin charging of E1 and E2 as well as monoubiquitylation of histone H2B was unaffected by acetylation [8]. These two acetylation modifications were detected in cells by mass spectrometry from substrate-conjugated ubiquitin as well as ubiquitin monomers, suggesting that a pool of free acetylated ubiquitin is present in cells that can be utilized for acetyl-ubiquitylation of substrates. Potential substrates for acetylated ubiquitin are histone H2A and H2B, as enrichment of lysine 6 and lysine 48 acetylation marks were observed in monoubiquitylated histone H2A and H2B fraction versus total ubiquitin conjugate fraction [8]. Acetylation on ubiquitin can influence the linkage type present within polyubiquitin chains because acetylation and ubiquitylation compete for the same lysine sites. In support of this idea, acetylation of ubiquitin lysine 48 can inhibit the formation of lysine 48-linked polyubiquitin chains [8]. In addition, lysine 48 acetylation can repress polyubiquitylation at lysine 11 and 63, and lysine 6 acetylation can repress polyubiquitylation at lysine 11, 48, and 63. The lysine acetylation data show that modification at one lysine site can influence polyubiquitylation at another lysine, and suggest potential cross-talk between post-translational modification sites on ubiquitin that affect the ubiquitin conjugation properties. Deamidation of glutamine 40 on ubiquitin has been observed as an additional type of post-translational modification [9]. This is catalyzed by the bacterial effector Cif homolog from *Burkholderia pseudomallei* and blocks polyubiquitin formation. Based on the characterization of serine 65 phosphorylation, lysine 6 acetylation, lysine 48 acetylation, and glutamine 40 deamidation, it is clear that post-translational modifications can affect ubiquitin conjugation. For the majority of ubiquitin post-translational modifications, further studies are needed to understand the full impact that ubiquitin modifications have on conjugation properties as well as the functional significance. Follow-up studies would help address the question of whether these post-translational modifications occur on free ubiquitin monomers or conjugated ubiquitin (either mono- or polyubiquitylation).

The existence of post-translational modifications on ubiquitin leaves open the possibility of expanded diversity in terms of polyubiquitin chains containing a unique pattern of post-translational modifications on the constituent ubiquitins. Each of these polyubiquitin chains could promote specific biological outcomes for the modified substrate protein. This concept of post-translational modification patterns encoding biological information is reminiscent of the role of histone post-translational modifications in chromatin regulation. Histones, which are core components of nucleosomes, can undergo a large array of post-translational modifications including phosphorylation, methylation, ADP-ribosylation, acetylation, ubiquitylation, and SUMOylation [10–12]. The term 'histone code' has been put forth for the concept that various combinations of post-translational modifications on histones encode biological information that regulate underlying chromatin processes [13, 14]. Furthermore, there are specific recognition modules (e.g., bromodomain) and enzymes (e.g., histone acetyltransferases and histone deacetylases) that function as readers, writers, or erasers of histone modifications and allow cells to interpret and change this histone code [11, 12, 15]. Thus, histone modifications constitute an important regulatory mechanism for almost all chromatin-related processes. Analogous to the histone code, the term 'ubiquitin code' has been coined to reflect the rich biological information that could be encoded in polyubiquitin chains through different combinations of post-translational modifications, polyubiquitin linkage, and branching patterns [16].

The discovery of ADP-ribosylation of ubiquitin by bacterial proteins adds to the list of post-translational modifications for ubiquitin and further expands the ubiquitin code [17]. This was soon followed by the example of a heterodimeric complex comprised of the human proteins Deltex-3-like (Dtx3L, an E3 ubiquitin ligase) and Parp9 (an ADP-ribosyltransferase (ART)) catalyzing ADP-ribosylation of ubiquitin [18], suggesting this type of ubiquitin post-translational modification could play broad roles in biology. In the following section, we discuss the initial observation of ADP-ribosylation of ubiquitin occurring in the setting of bacterial infection. We then describe the background on Dtx3L/Parp9 biological roles and the key data characterizing the ADP-ribosylation of ubiquitin by this complex. We highlight differences as well as common features between the two ways in which ADP-ribosylation of ubiquitin occurs. Finally, we pose some questions that remain to be resolved for this newly identified modification on ubiquitin.

2. ADP-ribosylated ubiquitin is generated by SidE effector proteins of *Legionella pneumophila*

ADP-ribosylation of ubiquitin was first discovered in the context of infection by the bacteria *L. pneumophila*, the causative agent for Legionnaires' disease [17]. During infection, the bacterial Dot/Icm type IV secretion system translocates into the host cell effector proteins that play a role in the formation of *Legionella*-containing vacuoles which support bacterial survival and replication [19]. Among these effectors are the SidE family of proteins consisting of four members (SdeA, SidE, SdeB and SdeC) which ubiquitylate ER-associated Rab small GTPases and Rtn4 [17, 20–25]. Investigation of Rab33b, one of the Rab small GTPases targeted by SidE proteins, showed that ubiquitylation causes modest decrease in both GTP loading and GTP hydrolysis [17]. It is not known whether these findings extend to other Rab small GTPases. Exactly how these biochemical changes in Rab33b, along with ubiquitylation of Rtn4, contribute to *Legionella*-containing vacuole formation remain to be determined. Notably, the ubiquitylation of Rab small GTPases and Rtn4 by SidE proteins does not involve E1 or E2, but requires NAD^+ in order to generate ADP-ribosylated ubiquitin as an activated form of ubiquitin. NAD^+ is a small molecule with diverse roles in biology including its role as a cofactor in redox reactions, as an ADP-ribose donor in ART-catalyzed processes, and as an enzyme substrate for the histone deacetylases, sirtuins [26]. SidE proteins use NAD^+ as an ADP-ribose donor to ADP-ribosylate and activate ubiquitin for subsequent conjugation. This stands in contrast to the canonical activation mechanism for ubiquitin that is mediated by E1 in an ATP-dependent manner. ADP-ribosylation of ubiquitin occurs on arginine 42 and is mediated by the mono-ART (mART) domain within SidE proteins [17, 20–25, 27]. Once ADP-ribosylated ubiquitin is generated by the mART domain, the modified ubiquitin is utilized by a phosphodiesterase (PDE) domain within SidE proteins to complete the ubiquitylation of target substrates on serine residues [20, 21, 23–25]. A catalytic mechanism has been proposed where AMP is initially hydrolyzed and the resulting phosphoribosylated ubiquitin is subsequently attached to serine residue of target substrates to complete the ubiquitylation process [20, 21, 23, 24]. In support of this mechanism, an enzyme reaction intermediate where the phosphoribosylated ubiquitin is covalently linked to a key catalytic histidine residue in the PDE domain of SdeA, a member of the SidE protein family, has been observed by mass spectrometry [23].

The mART and PDE domain active sites face away from each other and act independently to catalyze the two enzyme steps involved in ubiquitylation [20, 25]. Mixing together independent mART and PDE domains can recapitulate the ubiquitylation of substrate protein, although the efficiency is reduced because the separated domains cannot form proper inter-domain interactions that would be present in the intact protein [20, 25]. Supplying the PDE domain alone with

ADP-ribosylated ubiquitin recapitulates the ubiquitylation of substrate, supporting the notion that ADP-ribosylation of ubiquitin by mART and ubiquitylation of substrates by PDE using ADP-ribosylated ubiquitin are independent steps [20, 22, 23, 25]. Overall, ADP-ribosylation activates ubiquitin to allow ubiquitylation mediated by SidE proteins to take place in an E1- and E2-independent manner. Ubiquitylation of Rab small GTPases and Rtn4 by SidE proteins through a unique serine-phosphoribose linkage is important for mediating the formation of *Legionella*-containing vacuoles and supporting bacterial infection. ART mutants of SidE proteins displayed defects in terms of formation of *Legionella*-containing vacuoles within host cells, highlighting the essential role that ADP-ribosylation of ubiquitin plays in this novel mode of ubiquitylation [17, 24].

3. Dtx3L, an E3 ubiquitin ligase, forms a heterodimeric complex with Parp9, an ADP-ribosyltransferase

Deltex-3-like (Dtx3L, also known as B-lymphoma- and B aggressive lymphoma-associated protein (BBAP)) is a member of the Deltex family of E3 ubiquitin ligases [28]. Sequence analysis shows that Dtx3L shares amino acid sequence identity with other Deltex proteins in the C-terminal region containing a RING domain and the Deltex C-terminal domain, while the N-terminal region is distinct [28, 29]. Dtx3L was initially identified in a yeast two-hybrid screen as an interacting partner for Parp9 (also known as ART diphtheria toxin-like 9 (ARTD9), and B-aggressive lymphoma 1 (BAL1)), a key risk factor gene for an aggressive subset of diffuse large B cell lymphoma (DLBCL) [28]. Based on homology of the C-terminus to the Parp catalytic domain, Parp9 is classified as an ART, a class of enzymes that transfer ADP-ribose from NAD^+ onto target substrates [30–32]. In addition to its C-terminal Parp catalytic domain, Parp9 has two macrodomains in the N-terminus. Macrodomains are protein modules that bind to ADP-ribose [33, 34], and the Parp9 macrodomains have been shown to bind to poly-ADP-ribose (PAR) [18, 35, 36]. The interaction between Dtx3L and Parp9 is mediated through the N-terminus of Dtx3L and the C-terminus of Parp9 [28, 36]. Neither PAR-binding by Parp9 nor the E3 ubiquitin ligase activity of Dtx3L is required for formation of the heterodimeric complex [36].

4. Both Dtx3L and Parp9 are overexpressed in cancer

Both Dtx3L and Parp9 are overexpressed in an aggressive subset of DLBCL, and are coordinately expressed from a common, interferon (IFN) γ-inducible bidirectional promoter containing interferon regulatory factor (IRF) and signal transducer and activator of transcription (STAT) binding sites [37]. In addition, Dtx3L and Parp9 are highly overexpressed in metastatic prostate cancer cell lines that have increased IFNγ/STAT1 signaling activity [38]. Analysis of gene expression datasets from The Cancer Genome Atlas showed elevated levels of both Dtx3L and Parp9 in prostate cancer as well as breast cancer, bladder urothelial carcinoma, colorectal adenocarcinoma, head and neck squamous cell carcinoma, clear cell renal cell carcinoma, papillary renal cell carcinoma, lung adenocarcinoma, stomach adenocarcinoma, thyroid carcinoma, and uterine corpus endometrial carcinoma [18]. Dtx3L expression is increased in melanoma compared to benign melanocytic tumors [39], and in gliomas where Dtx3L expression level positively correlates with the grade of glioma [40]. In summary, these expression changes point toward Dtx3L/Parp9 having specific biological roles that support tumor behavior. Current evidence suggests that the Dtx3L/Parp9 complex plays a role in DNA damage response [18, 35, 41] and the regulation of IFNγ/STAT1 signaling pathway [38, 43]. Dtx3L also may function independently of Parp9 in promoting cell migration and metastasis [38, 39, 40].

5. Dtx3L/Parp9 mediates DNA damage response

The Dtx3L/Parp9 complex is involved in DNA damage response. Dtx3L or Parp9 knockdown leads to increased sensitivity to the DNA damaging agent doxorubicin, suggesting the Dtx3L/Parp9 complex plays a role in DNA repair [35, 41]. Using a plasmid-based GFP reporter assay that measures non-homologous end joining (NHEJ) DNA repair, Yang et al. found that knockdown of either Dtx3L or Parp9 leads to an approximate two-fold reduction in NHEJ repair [18]. Conversely, overexpression of Dtx3L/Parp9 had increased efficiency of NHEJ. Recruitment of Dtx3L and Parp9 to laser micro-irradiated sites was observed in cells, supporting a role for this complex in DNA damage response [18, 35]. PAR synthesized by Parp1 and 2 at DNA damage sites serves as a scaffold for recognition by factors involved in DNA damage repair [42]. Because Parp9 contains two macrodomains that bind to PAR, heterodimerization with Dtx3L helps target this E3 ubiquitin ligase to sites of DNA damage. Consistent with this idea, depletion of Parp9 or treatment of cells with PJ-34, a Parp inhibitor that blocks PAR synthesis, prevented Dtx3L localization to sites of laser-induced DNA damage [35].

Once recruited to DNA damage sites, Dtx3L initiates an early wave of ubiquitylation to help coordinate recruitment of DNA damage response proteins. In cells exposed to DNA damage, early recruitment of DNA damage response proteins tumor suppressor p53-binding protein 1 (53BP1) and BRCA1 were reduced when Dtx3L or Parp9 was depleted via siRNA knockdown [35, 41]. Dtx3L knockdown caused decreased chromatin association of the histone methyltransferase SET8 as well as decreased histone H4 lysine 20 mono- and dimethylation [41]. The mechanism for Dtx3L-dependent recruitment of SET8 to chromatin remains unknown. Furthermore, histone H4 monoubiquitylation was reduced by Dtx3L knockdown, and lysine 91 was identified as the site of monoubiquitylation on histone H4 by Dtx3L [41]. Thus, histone H4 lysine 91 monoubiquitylation mediated by Dtx3L is a prerequisite for methylation of histone H4 lysine 20 and subsequent recruitment of 53BP1. In terms of BRCA1, recruitment to DNA damage sites depends on the adaptor protein RAP80 which contains ubiquitin interacting motifs that could recognize ubiquitylation marks generated by Dtx3L. Because of the DNA damage response role for this complex, Dtx3L/Parp9 overexpression in various cancers could have important clinical consequences in terms of therapy resistance to DNA damaging strategies such as chemotherapy and radiation.

6. IFN signaling regulation by Dtx3L/Parp9

In addition to the DNA damage response, the Dtx3L/Parp9 complex has pro-tumorigenic functions through regulation of IFN signaling. In prostate cancer cells, both Dtx3L and Parp9 were shown to be critical factors for mediating cell proliferation and chemoresistance [38]. One mechanism for how Dtx3L/Parp9 supports tumor growth is the repression of IRF-1, an important transcription factor within the IFN signaling pathway that mediates anti-proliferative and pro-apoptotic responses. Expression of IRF-1 was negatively correlated with expression of Dtx3L and Parp9 in prostate cancer cell lines [38]. Consistent with this observation, knockdown of either protein led to increased expression of IRF-1, and overexpression of Dtx3L or Parp9 had repressive effect on expression of a luciferase reporter under the control of a IRF-1 promoter [38]. The repression was further enhanced when STAT1β, a transcriptionally repressive isoform of STAT1, was co-expressed with Dtx3L and Parp9. This closely mirrors the result from a study in DLBCL where Parp9 together with STAT1β represses IRF-1 expression, thereby supporting cancer cell survival and proliferation [43]. Thus, overexpression of Dtx3L and Parp9 represents a strategy in DLBCL and prostate cancer to repress the expression of the tumor suppressor IRF-1, thereby blocking IFNγ signaling from acting in an anti-proliferative and pro-apoptotic manner.

7. Dtx3L promotes cell migration and metastasis

In addition to acting in a complex with Parp9 to promote tumor growth, Dtx3L may have Parp9-independent roles in mediating cancer metastasis. Knockdown of Dtx3L, but not Parp9, reduced prostate cancer cell migration [38]. Both STAT1 and STAT3 are involved in Dtx3L-mediate cell migration as co-knockdown of either STATs with Dtx3L did not lead to a further decrease in prostate cancer cell migration [38]. The role of Dtx3L in regulating cell migration may be relevant in other types of cancer as well. Analysis of the mouse B16 melanoma cell and its more invasive sublines showed that Dtx3L levels positively correlate with how invasive the melanoma cells are [39]. Knockdown of Dtx3L in human melanoma cells reduced their invasive properties, and decreased lung metastasis was observed when Dtx3L-depleted mouse melanoma cells were injected into tail veins of nude mice, suggesting Dtx3L promotes melanoma metastasis [39]. Additionally, depletion of Dtx3L inhibited migration of glioma cells in a transwell migration assay [40]. The role of Parp9 in cancer cell migration and metastasis was not investigated in the setting of melanoma and glioma. Overall, the available evidence supports the role of Dtx3L in cancer metastasis, and whether Dtx3L acts independently of Parp9 in settings other than prostate cancer remains an open question.

8. Enhancement of anti-viral response by Dtx3L/Parp9

Outside of the context of cancer, the Dtx3L/Parp9 complex promotes anti-viral response through regulation of the IFN/STAT1 signaling pathway. As shown in DLBCL and prostate cancer cells, both Dtx3L and Parp9 are coordinately expressed via IFN signaling pathway in order to allow infected cells to mount an effective anti-viral response [36]. Co-expression of Dtx3L and Parp9 augmented IFN/STAT1-mediated anti-viral response, resulting in reduced viral load for encephalomyocarditis virus, Sindbis virus, and influenza A virus [36]. Furthermore, Dtx3L and Parp9 co-expression enhanced luciferase reporter expression under the control of IFN-stimulated response element and IFNγ-activated site promoter and increased nuclear localization of STAT1 [36]. Both Dtx3L and Parp9 can interact with STAT1, suggesting an interplay between all three proteins to augment anti-viral IFN response and enable cells to better control viral replication [36]. Here, enhanced binding of STAT1 to the IRF-1 promoter by Dtx3L/Parp9 co-expression is observed, which contrasts with the repressive role against IRF-1 expression that this complex plays in cancer. The enhanced STAT1-dependent IFN-stimulated gene (ISG) expression requires Dtx3L-mediated histone H2BJ monoubiquitylation at the promoter. Deposition of this histone modification leads to a concomitant increase in chromatin accessibility and histone H3 lysine 4 trimethylation, an epigenetic mark for active transcription, at the ISG IFN-induced protein with tetratricopeptide repeats

1 (IFIT1) promoter. In addition to working with Parp9 to enhance ISG expression, Dtx3L also has a Parp9-independent role in directly suppressing viral replication by ubiquitylating and targeting the encephalomyocarditis viral 3C protease for degradation [36]. Thus, for effective viral defense, Dtx3L can function with Parp9 in a complex to enhance IFN/STAT1 signaling response, and also acts independently of Parp9 to target viral protease for degradation through ubiquitylation.

9. ADP-ribosylation of the ubiquitin C-terminal glycine 76 carboxyl group by Dtx3L/Parp9

The Dtx3L/Parp9 heterodimeric complex represents the first example of mammalian proteins that catalyze ADP-ribosylation of ubiquitin [18] (**Figure 1**). Parp9 was originally thought to be inactive in terms of ART activity based on previous studies where the ability of Parp9 to automodify itself was tested [30, 44]. However, Parp9 displayed comparable NADase activity to other ART family members when Dtx3L was also added to the *in vitro* reaction, suggesting Parp9 functions as an active enzyme when complexed with Dtx3L [18]. Furthermore, when ubiquitin was added to this

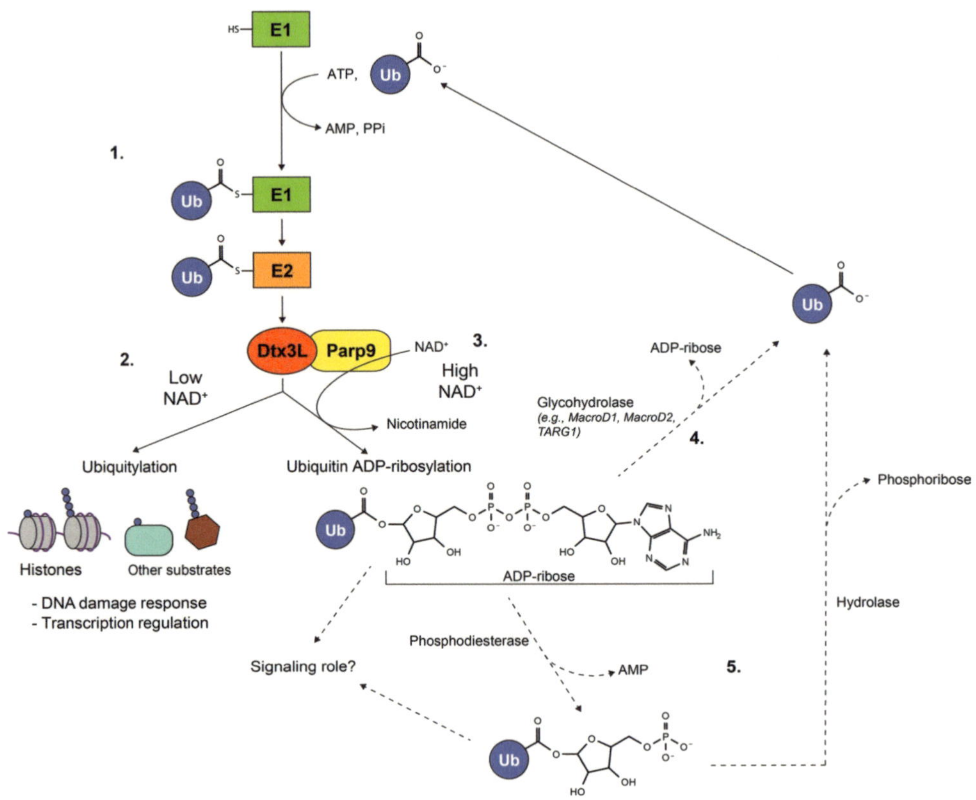

Figure 1.
ADP-ribosylation of ubiquitin by Dtx3L/Parp9 and proposed pathways for regeneration of free ubiquitin. (1) Ubiquitin (Ub, blue circle) processing by E1 and E2. (2) Under low NAD^+, ubiquitylation by Dtx3L is favored that plays a role in DNA damage response and transcription regulation. (3) Under high NAD^+, ADP-ribosylation of ubiquitin at the C-terminus by Parp9 is promoted, thereby blocking conjugation to substrate and restraining Dtx3L E3 ubiquitin ligase activity. ADP-ribosylated ubiquitin could undergo processing through two ways in order to regenerate free ubiquitin: a single step cleavage of ADP-ribose by a glycohydrolase (4), or a two-step process where a phosphodiesterase cleaves AMP to generate phosphoribosylated ubiquitin, followed by a hydrolase cleavage step to regenerate free ubiquitin (5). Both ADP-ribosylated and phosphoribosylated ubiquitin may play signaling roles. Dotted lines indicate speculative steps.

reaction, Parp9-mediated ADP-ribosylation of ubiquitin was detected. The ADP-ribosylation of ubiquitin was dependent on processing of ubiquitin by E1 and E2 as well as the RING domain in Dtx3L. Binding of an ubiquitin antibody that recognizes an epitope in the C-terminus was reduced when ubiquitin was ADP-ribosylated, indicating that the site of ADP-ribosylation was at the C-terminal portion of ubiquitin. ADP-ribosylation of proteins takes place on a variety of amino acids including glutamate, aspartate, serine, phosphoserine, threonine, lysine, arginine, asparagine, and cysteine [45, 46]. Arginine residues 72 and 74 at the C-terminus of ubiquitin were excluded as the sites of ADP-ribosylation because mutations targeting these residues did not prevent ubiquitin ADP-ribosylation by Dtx3L/Parp9 [18]. An independent approach for interrogating ADP-ribosylation sites takes advantage of the differential sensitivity to chemical-based ADP-ribose release, depending on the type of amino acid that is ADP-ribosylated [47]. Neutral hydroxylamine which removes ADP-ribose from acidic R group, efficiently removed ADP-ribose from ubiquitin, suggesting that the site of ADP-ribosylation on ubiquitin is an acidic R group [18]. Based on these observations, the site of ADP-ribosylation was narrowed down to the C-terminal carboxyl group of glycine 76, the only available acidic moiety that could serve as an acceptor for ADP-ribose in the C-terminus of ubiquitin (**Figure 1**).

10. Dtx3L E3 output is regulated by ADP-ribosylation of ubiquitin and PAR-binding to Parp9 macrodomains

Given that conjugation of ubiquitin onto target substrates occurs through its C-terminus, ADP-ribosylation of the C-terminus of ubiquitin would block ubiquitylation from taking place. Thus, modulation of Parp9-mediated ADP-ribosylation would be expected to change the ubiquitylation output by the heterodimer partner Dtx3L. This prediction was tested by adding NAD^+ to an *in vitro* ubiquitylation reaction, which would promote ADP-ribosylation of ubiquitin. As expected, increasing concentration of NAD^+ led to reduced ubiquitylation of histone H2A and histone H3 by Dtx3L [18]. This supports the idea that Parp9 ADP-ribosylation of ubiquitin effectively blocks ubiquitin from being utilized in Dtx3L-mediated ubiquitylation, and that Parp9 negative regulation of Dtx3L can be modulated through changes in free NAD^+ concentration (**Figure 1**). In addition to regulation of Dtx3L ubiquitylation by the Parp9 catalytic domain, Parp9 can also regulate Dtx3L through binding to PAR via its two macrodomains. Addition of PAR in an *in vitro* ubiquitylation reaction increased Dtx3L-generated polyubiquitylated product as well as monoubiquitylation of histone H2A [18]. This suggests PAR binding to the Parp9 macrodomains has a stimulatory role in terms of Dtx3L function and points toward a regulatory mechanism where a conformational change from PAR-binding to Parp9 macrodomains is transmitted to Dtx3L to enhance E3 ubiquitin ligase activity. A precedent for PAR activation of E3 function has been established where RNF146, an E3 ubiquitin ligase, is stimulated by PAR binding through its Trp-Glu-Glu (WWE) domain [48, 49]. On the other hand, PAR binding to Parp9 had no effect in terms of generation of ADP-ribosylated ubiquitin. As Dtx3L and Parp9 play a role in NHEJ DNA repair, the regulatory role of Parp9 ADP-ribosylating ubiquitin in this process was examined. Yang et al. found that co-expression of catalytically inactive Parp9 with Dtx3L had enhanced DNA damage repair, compared to co-expression of wild-type Parp9 and Dtx3L [18]. This is consistent with the model that Parp9 ADP-ribosylation of ubiquitin precludes ubiquitylation of substrates by Dtx3L that would be important for the recruitment of DNA damage repair proteins. Reducing the ART activity in Parp9 by mutation relieves this inhibitory effect, thereby promoting DNA damage repair dependent on Dtx3L E3 ubiquitin ligase activity.

Based on experimental evidence discussed earlier, we put forth the concept that in the context of DNA damage, NAD^+ functions as an important regulator of E3 output for Dtx3L/Parp9 via two distinct mechanisms. When cells are exposed to DNA damage, depletion of NAD^+ occurs because as part of the DNA damage response, Parp1 and 2 utilize NAD^+ as an ADP-ribose donor molecule for the synthesis of PAR at sites of DNA damage [42, 50]. As the NAD^+ level is low and PAR is present, the local environment at the DNA damage site would be optimal for Dtx3L E3 ubiquitin ligase activity that is critical for DNA damage response. Low NAD^+ means that ADP-ribosylation of ubiquitin by Parp9 is occurring minimally; hence, ubiquitin is available to participate in Dtx3L-mediated ubiquitylation (**Figure 1**). Moreover, PAR synthesized from NAD^+ helps recruit the Dtx3L/Parp9 complex to DNA damage sites and stimulates Dtx3L E3 ubiquitin ligase activity through recognition by the Parp9 macrodomains. As the repair of DNA damage proceeds, the balance shifts from synthesis of PAR toward disassembly of PAR and restoration of NAD^+ to basal level. The heterodimeric complex becomes disengaged from PAR, and hence, the stimulatory impact of PAR-binding to promote Dtx3L ubiquitylation is reduced. High NAD^+ levels would favor ADP-ribosylation of ubiquitin by Dtx3L/Parp9, and this attachment of ADP-ribose to the C-terminal glycine 76 would block utilization of ubiquitin in Dtx3L-mediated ubiquitylation (**Figure 1**). In this setting, the combination of high NAD^+ and low PAR levels would not be conducive for Dtx3L E3 ubiquitin ligase function, which is appropriate as DNA damage is resolved and Dtx3L-mediated ubiquitylation to coordinate early DNA damage response is no longer necessary. Thus, NAD^+ and its derived macromolecule PAR serve as critical regulatory mechanisms for controlling Dtx3L-mediated ubiquitylation in the context of DNA damage. As part of this regulation of Dtx3L, ADP-ribosylation of ubiquitin plays a restraining function for Dtx3L E3 ubiquitin ligase activity when DNA damage has been repaired.

11. Distinct biological properties and roles for ADP-ribosylated ubiquitin generated by Dtx3L/Parp9 and SidE proteins

While both the Dtx3L/Parp9 complex and the SidE proteins can mediate the formation of ADP-ribosylated ubiquitin, clear distinctions are evident between the two cases. The ADP-ribosylation of ubiquitin by Parp9 requires that ubiquitin undergo E1 and E2 processing as well as the presence of the E3 ubiquitin ligase Dtx3L. This requirement for strict coupling ensures that Parp9 ADP-ribosylation of ubiquitin is specifically tied to activated ubiquitin that is in the process of being handled by the heterodimer partner, Dtx3L, and ensures that ADP-ribosylation of ubiquitin does not take place promiscuously. In contrast, the mART domain of SidE proteins is sufficient to generate ADP-ribosylated ubiquitin in the absence of ubiquitin processing by E1 and E2. The roles of the generated ADP-ribosylated ubiquitin are different between Dtx3L/Parp9 and SidE proteins. For Dtx3L/Parp9, ADP-ribosylation of ubiquitin takes place to short-circuit the normal ubiquitylation process by Dtx3L, thereby acting in a negative regulatory manner. In essence, ADP-ribosylation at the C-terminus of ubiquitin prevents the modified ubiquitin from being conjugated onto target substrates, and thus appears to be a regulatory feature for Dtx3L E3 output in the context of DNA damage repair as discussed earlier. It is possible that the control of Dtx3L-mediated ubiquitylation by Parp9-mediated ADP-ribosylation of ubiquitin operates in the context of active IFN signaling and viral infection as well. Depletion of NAD^+ has been observed in cells treated with IFNγ [51], and this could further augment Dtx3L E3 output and downstream control of viral infection. Decreasing levels of NAD^+ would lead to less Parp9-mediated ADP-ribosylation of ubiquitin; hence, unmodified ubiquitin is available for Dtx3L-mediated ubiquitylation of histone

H2BJ which results in increased ISG expression. It would be interesting to test this hypothesis through co-expression of catalytically inactive Parp9 mutant and Dtx3L and examine how transcription of ISG is impacted. If our model is correct, then ISG transcription should be increased because the negative regulatory mechanism on Dtx3L E3 activity is relieved by inactivation of Parp9 catalytic activity. Similar to ubiquitin ADP-ribosylated at the C-terminal glycine 76, ubiquitin ADP-ribosylated on arginine 42 and its phosphoribosylated derivative are unable to be utilized in E1 and E2 processing, thereby disrupting host ubiquitylation processes [21]. However, the main role of ADP-ribosylated ubiquitin generated by SidE proteins is to serve as an activated intermediate for non-canonical serine ubiquitylation of host proteins that is important for supporting bacterial infection. Because Dtx3L/Parp9-mediated ADP-ribosylation of ubiquitin occurs on the C-terminus, this enzyme complex would not be able to ADP-ribosylate pre-existing polyubiquitin chains already conjugated to a target substrate, as no C-terminal carboxyl group from ubiquitin is available for the attachment of ADP-ribose. Unconjugated polyubiquitin chains are present in cells [4], so it is conceivable that the glycine 76 carboxyl group at the C-terminus of the polyubiquitin chain can be modified by Dtx3L/Parp9. Because of the tight coupling of Dtx3L/Parp9-mediated ADP-ribosylation of ubiquitin to E1 and E2 processing, this scenario would likely require that free polyubiquitin chains are first processed by E1 and E2 before ADP-ribosylation by Dtx3L/Parp9 could take place. However, to the best of our knowledge, handling of free polyubiquitin chains by E1 and E2 has not been observed, and thus, ADP-ribosylation of free polyubiquitin chains by Dtx3L/Parp9 remains highly speculative. On the other hand, SidE proteins can modify existing polyubiquitin chains. Incubation of lysine 48- or methionine 1-linked ubiquitin tetramers with SdeC led to ADP-ribosylation of the incorporated ubiquitin [24]. Further investigation showed that ADP-ribosylation and phosphoribosylation of lysine 63-, lysine 48-, lysine 11-, or methionine 1-linked diubiquitin chains could be catalyzed by SdeA [52]. No preference for modification of either ubiquitin within the diubiquitin was observed. In most cases, the ADP-ribosylated or phosphoribosylated diubiquitins were resistant to hydrolysis by deubiquitinases, suggesting SidE proteins could have broad effects in host cell by changing deubiquitinase susceptibility of polyubiquitin chains on various target substrates [52].

12. Some open questions for ADP-ribosylation of ubiquitin

While it is clear that ADP-ribosylation of ubiquitin by Dtx3L/Parp9 prevents substrate conjugation, the biological processes impacted by this form of regulation outside of Dtx3L-mediated NHEJ DNA repair remain to be identified. One can also speculate that ADP-ribosylated ubiquitin could be playing a signaling role as a second messenger molecule, similar to how free ubiquitin or a polyubiquitin chain could serve signaling roles [4] (**Figure 1**). Another major question is the fate of ADP-ribosylated ubiquitin after it is generated. It does appear that cells have enzymatic activity to reverse this post-translational modification on ubiquitin as treatment of ADP-ribosylated ubiquitin with cell lysate restores detection of the C-terminus by the C-terminal ubiquitin antibody [18]. There are at least two mechanisms by which cells can reverse the ADP-ribosylation and regenerate free ubiquitin (**Figure 1**). One is that a glycohydrolase cleaves the ADP-ribose from the C-terminus of ubiquitin in a single step reaction. MacroD1, macroD2, and terminal ADP-ribose protein glycohydrolase 1 (TARG1) are examples of glycohydrolases that remove ADP-ribose from acidic residues [45, 46], and are potential candidates for regenerating the C-terminal carboxyl group of ubiquitin. The other possibility is removal of ADP-ribose from the C-terminus of ubiquitin involves a phosphodiesterase that releases AMP to generate phosphoribosylated ubiquitin, though regeneration of ubiquitin would still require a hydrolase to remove the phosphoribosyl group. It is conceivable that both mechanisms are operating within the cell. The latter mechanism where phosphoribosylated ubiquitin is generated opens up the question of whether this modified species of ubiquitin has biological functions, rather than it occurring as a recycling intermediate. As speculated earlier for ADP-ribosylated ubiquitin, the phosphoribosylated ubiquitin could have cell signaling roles. This idea is applicable to the phosphoribosylated ubiquitin generated by SidE proteins as well where it is currently unknown whether and how phosphoribosylated ubiquitin is processed by host cells. Although the sites of modification are different, there could be overlap in terms of the host cell mechanisms that handle the reversal of phosphoribosylated ubiquitin generated by SidE proteins versus phosphoribosylated ubiquitin generated as a result of breaking down ADP-ribosylated ubiquitin generated by the Dtx3L/Parp9 complex. Thus, we currently have little understanding of what other roles ADP-ribosylated ubiquitin may play aside from being a starting point for bacterial protein-catalyzed ubiquitylation or a regulatory mechanism against the Dtx3L E3 ubiquitin ligase, as well as how cells regenerate free ubiquitin from ADP-ribosylated or phosphoribosylated ubiquitin.

13. Concluding remarks

ADP-ribosylation of ubiquitin is an exciting new addition to the possible post-translational modifications for ubiquitin. To date, SidE proteins and the Dtx3L/Parp9 complex are the only proteins known to generate ADP-ribosylated ubiquitin. These two cases of ADP-ribosylation of ubiquitin represent distinct mechanisms by which ubiquitin conjugation activity is regulated (**Figure 2**). Dtx3L/Parp9-mediated ADP-ribosylation of the C-terminal glycine 76 of ubiquitin effectively prevents the modified ubiquitin from being utilized in ubiquitylation processes until the blocking ADP-ribose is removed. SidE-mediated ADP-ribosylation of ubiquitin occurs as an activation step for subsequent E1- and E2-independent ubiquitylation process. These are still early days in understanding the important roles ADP-ribosylation of ubiquitin plays in both the Dtx3L/Parp9 and SidE context and many questions remain. How exactly does abrogating the Parp9 ART activity enhance the NHEJ DNA repair mediated by Dtx3L? What are the Dtx3L target substrates involved in this process? While histone H4 is one of the important candidate substrates based on previous data [41], there may be other DNA damage response factors that are ubiquitylated by Dtx3L during NHEJ DNA repair. For SidE proteins, it would be interesting to see whether other bacterial effector proteins as well as eukaryotic counterparts exist that can catalyze the E1- and E2-independent ubiquitylation on serine. Furthermore, by utilizing ADP-ribosylated ubiquitin, SidE-mediated ubiquitylation results in a phosphoribose linkage between serine and ubiquitin that is distinct from the canonical isopeptide linkage between lysine and ubiquitin. How this novel ubiquitin linkage changes the biological properties of Rab and Rtn4, the targets of SidE proteins, is a future area of research. An intriguing possibility is that perhaps other bacterial effector proteins specifically recognize this unique serine-phosphoribose-linked ubiquitin and co-opt the modified Rab and Rtn4 for the generation of *Legionella*-containing vacuoles. Finally, much work remains to be done on characterizing the cellular mechanisms involved in processing and handling of ADP-ribosylated and phosphoribosylated ubiquitin after they are generated. Answering these questions regarding ADP-ribosylation of ubiquitin would provide additional insight into the ubiquitin code.

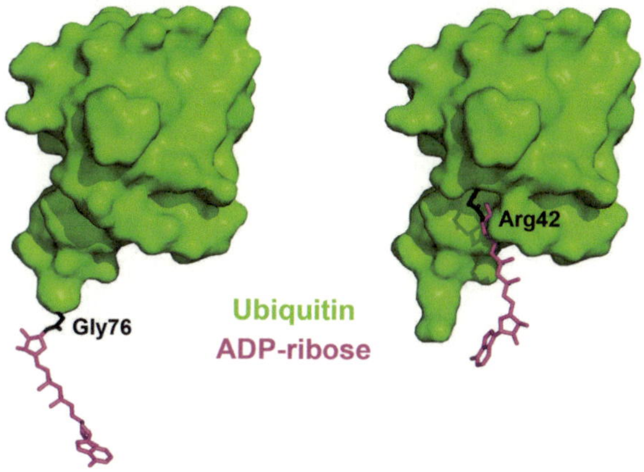

	Dtx3L (E3 ubiquitin ligase)/ Parp9 (ADP-ribosyltransferase)	SidE (ADP-ribosyltransferase)
Cofactor	NAD^+	NAD^+
ADP-ribosylation site	Glycine 76	Arginine 42
Effect on ubiquitin	Blocks conjugation	Alternative activation
Subsequent Processing	Hydrolase	SidE (phosphodiesterase)

Figure 2.
NAD^+ regulation of ubiquitin conjugation activity by two distinct mechanisms. Models of ADP-ribosylated ubiquitin were generated in Pymol by attaching ADP-ribose (magenta) to either arginine 42 (black) or C-terminal glycine 76 (black) in the solved ubiquitin crystal structure (PDB: 1UBQ, green).

Acknowledgements

This work was funded by R01CA214872.

Conflict of interest

The authors report no conflict of interest.

Author details

Teddy Kamata[1,2] and Bryce Paschal[1,2]*

1 Center for Cell Signaling, University of Virginia, Charlottesville, VA, USA

2 Department of Biochemistry and Molecular Genetics, University of Virginia, Charlottesville, VA, USA

*Address all correspondence to: paschal@virginia.edu

References

[1] Kwon YT, Ciechanover A. The ubiquitin code in the ubiquitin-proteasome system and autophagy. Trends in Biochemical Sciences. 2017;**42**:873-886. DOI: 10.1016/j.tibs.2017.09.002

[2] Kleiger G, Mayor T. Perilous journey: A tour of the ubiquitin–proteasome system. Trends in Cell Biology. 2014;**24**:352-359. DOI: 10.1016/j.tcb.2013.12.003

[3] McDowell GS, Philpott A. New insights into the role of ubiquitylation of proteins. In: Jeon KW, editor. Int. Rev. Cell Mol. Biol. Vol. 325. Cambridge (MA): Academic Press; 2016. pp. 35-88. DOI: 10.1016/bs.ircmb.2016.02.002

[4] Swatek KN, Komander D. Ubiquitin modifications. Cell Research. 2016;**26**:399-422. DOI: 10.1038/cr.2016.39

[5] Yau R, Rape M. The increasing complexity of the ubiquitin code. Nature Cell Biology. 2016;**18**:579-586. DOI: 10.1038/ncb3358

[6] Wauer T, Swatek KN, Wagstaff JL, Gladkova C, Pruneda JN, Michel MA, et al. Ubiquitin Ser65 phosphorylation affects ubiquitin structure, chain assembly and hydrolysis. The EMBO Journal. 2015;**34**:307-325. DOI: 10.15252/embj.201489847

[7] Ordureau A, Sarraf SA, Duda DM, Heo J-M, Jedrychowski MP, Sviderskiy VO, et al. Quantitative proteomics reveal a feedforward mechanism for mitochondrial PARKIN translocation and ubiquitin chain synthesis. Molecular Cell. 2014;**56**:360-375. DOI: 10.1016/j.molcel.2014.09.007

[8] Ohtake F, Saeki Y, Sakamoto K, Ohtake K, Nishikawa H, Tsuchiya H, et al. Ubiquitin acetylation inhibits polyubiquitin chain elongation. EMBO Reports. 2015;**16**:192-201. DOI: 10.15252/embr.201439152

[9] Cui J, Yao Q, Li S, Ding X, Lu Q, Mao H, et al. Glutamine deamidation and dysfunction of ubiquitin/NEDD8 induced by a bacterial effector family. Science. 2010;**329**:1215-1218. DOI: 10.1126/science.1193844

[10] Zhao Y, Garcia BA. Comprehensive catalog of currently documented histone modifications. Cold Spring Harbor Perspectives in Biology. 2015;**7**:a025064. DOI: 10.1101/cshperspect.a025064

[11] Bannister AJ, Kouzarides T. Regulation of chromatin by histone modifications. Cell Research. 2011;**21**:381-395. DOI: 10.1038/cr.2011.22

[12] Rothbart SB, Strahl BD. Interpreting the language of histone and DNA modifications. Biochimica et Biophysica Acta. 1839;**2014**:627-643. DOI: 10.1016/j.bbagrm.2014.03.001

[13] Jenuwein T, Allis CD. Translating the histone code. Science. 2001;**293**:1074-1080. DOI: 10.1126/science.1063127

[14] Strahl BD, Allis CD. The language of covalent histone modifications. Nature. 2000;**403**:41-45. DOI: 10.1038/47412

[15] Yun M, Wu J, Workman JL, Li B. Readers of histone modifications. Cell Research. 2011;**21**:564-578. DOI: 10.1038/cr.2011.42

[16] Komander D, Rape M. The ubiquitin code. Annual Review of Biochemistry. 2012;**81**:203-229. DOI: 10.1146/annurev-biochem-060310-170328

[17] Qiu J, Sheedlo MJ, Yu K, Tan Y, Nakayasu ES, Das C, et al. Ubiquitination independent of E1 and E2 enzymes by bacterial effectors.

Nature. 2016;**533**:120-124. DOI: 10.1038/nature17657

[18] Yang C-S, Jividen K, Spencer A, Dworak N, Ni L, Oostdyk LT, et al. Ubiquitin modification by the E3 ligase/ADP-ribosyltransferase Dtx3L/Parp9. Molecular Cell. 2017;**66**:503-516.e5. DOI: 10.1016/j.molcel.2017.04.028.

[19] Hubber A, Roy CR. Modulation of host cell function by *Legionella pneumophila* type IV effectors. Annual Review of Cell and Developmental Biology. 2010;**26**:261-283. DOI: 10.1146/annurev-cellbio-100109-104034

[20] Akturk A, Wasilko DJ, Wu X, Liu Y, Zhang Y, Qiu J, et al. Mechanism of phosphoribosyl-ubiquitination mediated by a single *Legionella* effector. Nature. 2018;**557**:729-733. DOI: 10.1038/s41586-018-0147-6

[21] Bhogaraju S, Kalayil S, Liu Y, Bonn F, Colby T, Matic I, et al. Phosphoribosylation of ubiquitin promotes serine ubiquitination and impairs conventional ubiquitination. Cell. 2016;**167**:1636-1649.e13. DOI: 10.1016/j.cell.2016.11.019

[22] Dong Y, Mu Y, Xie Y, Zhang Y, Han Y, Zhou Y, et al. Structural basis of ubiquitin modification by the *Legionella* effector SdeA. Nature. 2018;**557**:674-678. DOI: 10.1038/s41586-018-0146-7

[23] Kalayil S, Bhogaraju S, Bonn F, Shin D, Liu Y, Gan N, et al. Insights into catalysis and function of phosphoribosyl-linked serine ubiquitination. Nature. 2018;**557**:734-738. DOI: 10.1038/s41586-018-0145-8

[24] Kotewicz KM, Ramabhadran V, Sjoblom N, Vogel JP, Haenssler E, Zhang M, et al. A single *Legionella* effector catalyzes a multistep ubiquitination pathway to rearrange tubular endoplasmic reticulum for replication. Cell Host & Microbe. 2017;**21**:169-181. DOI: 10.1016/j.chom.2016.12.007

[25] Wang Y, Shi M, Feng H, Zhu Y, Liu S, Gao A, et al. Structural insights into non-canonical ubiquitination catalyzed by SidE. Cell. 2018;**173**:1231-1243.e16. DOI: 10.1016/j.cell.2018.04.023

[26] Poltronieri P, Čerekovic N, Poltronieri P, Čerekovic N. Roles of nicotinamide adenine dinucleotide (NAD$^+$) in biological systems. Challenges. 2018;**9**:3. DOI: 10.3390/challe9010003

[27] Kim L, Kwon DH, Kim BH, Kim J, Park MR, Park Z-Y, et al. Structural and biochemical study of the mono-ADP-ribosyltransferase domain of SdeA, a ubiquitylating/deubiquitylating enzyme from *Legionella pneumophila*. Journal of Molecular Biology. 2018;**430**:2843-2856. DOI: 10.1016/j.jmb.2018.05.043

[28] Takeyama K, Aguiar RCT, Gu L, He C, Freeman GJ, Kutok JL, et al. The BAL-binding protein BBAP and related Deltex family members exhibit ubiquitin-protein isopeptide ligase activity. The Journal of Biological Chemistry. 2003;**278**:21930-21937. DOI: 10.1074/jbc.M301157200

[29] Obiero J, Walker JR, Dhe-Paganon S. Fold of the conserved DTC domain in Deltex proteins. Proteins. 2012;**80**: 1495-1499. DOI: 10.1002/prot.24054

[30] Aguiar RCT, Takeyama K, He C, Kreinbrink K, Shipp MA. B-aggressive lymphoma family proteins have unique domains that modulate transcription and exhibit poly(ADP-ribose) polymerase activity. The Journal of Biological Chemistry. 2005;**280**:33756-33765. DOI: 10.1074/jbc.M505408200

[31] Hottiger MO, Hassa PO, Lüscher B, Schüler H, Koch-Nolte F. Toward a unified nomenclature for mammalian ADP-ribosyltransferases. Trends in

Biochemical Sciences. 2010;**35**:208-219. DOI: 10.1016/j.tibs.2009.12.003

[32] Schreiber V, Dantzer F, Ame J-C, De Murcia G. Poly(ADP-ribose): Novel functions for an old molecule. Nature Reviews. Molecular Cell Biology. 2006;**7**:517-528. DOI: 10.1038/nrm1963

[33] Karras GI, Kustatscher G, Buhecha HR, Allen MD, Pugieux C, Sait F, et al. The *macro* domain is an ADP-ribose binding module. The EMBO Journal. 2005;**24**:1911-1920. DOI: 10.1038/sj.emboj.7600664

[34] Till S, Ladurner AG. Sensing NAD metabolites through macro domains. Frontiers in Bioscience. 2009;**14**:3246-3258

[35] Yan Q, Xu R, Zhu L, Cheng X, Wang Z, Manis J, et al. BAL1 and its partner E3 Ligase, BBAP, link poly(ADP-ribose) activation, ubiquitylation, and double-strand DNA repair independent of ATM, MDC1, and RNF8. Molecular and Cellular Biology. 2013;**33**:845-857. DOI: 10.1128/MCB.00990-12

[36] Zhang Y, Mao D, Roswit WT, Jin X, Patel AC, Patel DA, et al. PARP9-DTX3L ubiquitin ligase targets host histone H2BJ and viral 3C protease to enhance interferon signaling and control viral infection. Nature Immunology. 2015;**16**:1215-1227. DOI: 10.1038/ni.3279

[37] Juszczynski P, Kutok JL, Li C, Mitra J, Aguiar RCT, Shipp MA. BAL1 and BBAP are regulated by a gamma interferon-responsive bidirectional promoter and are overexpressed in diffuse large B-cell lymphomas with a prominent inflammatory infiltrate. Molecular and Cellular Biology. 2006;**26**:5348-5359. DOI: 10.1128/MCB.02351-05

[38] Bachmann SB, Frommel SC, Camicia R, Winkler HC, Santoro R, Hassa PO. DTX3L and ARTD9 inhibit IRF1 expression and mediate in cooperation with ARTD8 survival and proliferation of metastatic prostate cancer cells. Molecular Cancer. 2014;**13**:125. DOI: 10.1186/1476-4598-13-125

[39] Thang ND, Yajima I, Kumasaka MY, Iida M, Suzuki T, Kato M. Deltex-3-like (DTX3L) stimulates metastasis of melanoma through FAK/PI3K/AKT but not MEK/ERK pathway. Oncotarget. 2015;**6**:14290-14299

[40] Xu P, Xu P, Tao X, Tao X, Zhao C, Zhao C, et al. DTX3L is upregulated in glioma and is associated with glioma progression. International Journal of Molecular Medicine. 2017;**40**:491-498. DOI: 10.3892/ijmm.2017.3023

[41] Yan Q, Dutt S, Xu R, Graves K, Juszczynski P, Manis JP, et al. BBAP monoubiquitylates histone H4 at lysine 91 and selectively modulates the DNA damage response. Molecular Cell. 2009;**36**:110-120. DOI: 10.1016/j.molcel.2009.08.019

[42] Liu C, Vyas A, Kassab MA, Singh AK, Yu X. The role of poly ADP-ribosylation in the first wave of DNA damage response. Nucleic Acids Research. 2017;**45**:8129-8141. DOI: 10.1093/nar/gkx565

[43] Camicia R, Bachmann SB, Winkler HC, Beer M, Tinguely M, Haralambieva E, et al. BAL1/ARTD9 represses the anti-proliferative and pro-apoptotic IFNγ-STAT1-IRF1-p53 axis in diffuse large B-cell lymphoma. Journal of Cell Science. 2013;**126**:1969-1980. DOI: 10.1242/jcs.118174

[44] Vyas S, Matic I, Uchima L, Rood J, Zaja R, Hay RT, et al. Family-wide analysis of poly(ADP-ribose) polymerase activity. Nature Communications. 2014;**5**:4426-4438. DOI: 10.1038/ncomms5426

[45] Cohen MS, Chang P. Insights into the biogenesis, function, and regulation

of ADP-ribosylation. Nature Chemical Biology. 2018;**14**:236-243. DOI: 10.1038/nchembio.2568

[46] Daniels CM, Ong S-E, Leung AKL. The promise of proteomics for the study of ADP-Ribosylation. Molecular Cell. 2015;**58**:911-924. DOI: 10.1016/j.molcel.2015.06.012

[47] Cervantes-Laurean D, Jacobson EL, Jacobson MK. Preparation of low molecular weight model conjugates for ADP-ribose linkages to protein. In: Methods in Enzymology. Vol. 280. Cambridge (MA): Academic Press; 1997. pp. 275-287. DOI: 10.1016/S0076-6879(97)80119-X

[48] DaRosa PA, Wang Z, Jiang X, Pruneda JN, Cong F, Klevit RE, et al. Allosteric activation of the RNF146 ubiquitin ligase by a poly(ADP-ribosyl)ation signal. Nature. 2014;**517**:223. DOI: 10.1038/nature13826

[49] Zhang Y, Liu S, Mickanin C, Feng Y, Charlat O, Michaud GA, et al. RNF146 is a poly(ADP-ribose)-directed E3 ligase that regulates axin degradation and Wnt signalling. Nature Cell Biology. 2011;**13**:623. DOI: 10.1038/ncb2222

[50] Houtkooper RH, Cantó C, Wanders RJ, Auwerx J. The secret life of NAD^+: An old metabolite controlling new metabolic signaling pathways. Endocrine Reviews. 2010;**31**:194-223. DOI: 10.1210/er.2009-0026

[51] Aune TM, Pogue SL. Inhibition of tumor cell growth by interferon-gamma is mediated by two distinct mechanisms dependent upon oxygen tension: Induction of tryptophan degradation and depletion of intracellular nicotinamide adenine dinucleotide. The Journal of Clinical Investigation. 1989;**84**:863-875. DOI: 10.1172/JCI114247

[52] Puvar K, Zhou Y, Qiu J, Luo Z-Q, Wirth MJ, Das C. Ubiquitin chains modified by the bacterial ligase SdeA are protected from deubiquitinase hydrolysis. Biochemistry. 2017;**56**:4762-4766. DOI: 10.1021/acs.biochem.7b00664

Chapter 9

Structural Insight into Regulation of the Proteasome Ub-Receptor Rpn10

Tal Keren-Kaplan, Ilan Attali, Olga Levin-Kravets, Oded Kleifeld, Shay Ben-Aroya and Gali Prag

Abstract

Ubiquitylation is a posttranslational modification that determines protein fate. The ubiquitin code is written by enzymatic cascades of E1 and E2 and E3 enzymes. Ubiquitylation can be edited or erased by deubiquitylating enzymes. Ub-receptors are proteins that read and decipher the ubiquitin codes into cellular response. They harbor a ubiquitin-binding domain and a response element. Interestingly, Ub-receptors are also regulated by ubiquitylation and deubiquitylation. However, until recently, the molecular details and the significance of this regulation remained enigmatic. Rpn10 is a Ub-receptor that shuttles ubiquitylated targets to the proteasome for degradation. Here we review recent data on Rpn10, with emphasis on its regulation by ubiquitylation.

Keywords: ubiquitin receptor, crystal structure, ubiquitylated ubiquitin receptor, regulation mechanisms, cargo shuttle

1. Introduction

Protein ubiquitylation functions as a cellular code to alter structure-function, localization, and interactions or as a destruction signal. The signal is decoded by several hundreds of ubiquitin receptors, proteins that carry a ubiquitin-binding domain(s) tethered to a response element. To precisely decode the numerous cellular ubiquitylation signals, ubiquitin receptors also carry element(s) that sense(s) the cellular context [1]. Intriguingly, ubiquitylation also regulates the function of ubiquitin receptors by their ubiquitylation.

The 26S proteasome is a multiprotein complex that degrades ubiquitylated proteins. Several proteasome Ub-receptors that mediate the recognition of ubiquitylated proteins were identified including proteasome subunits Rpn1 [2], Rpn10 [3], and Rpn13 [4] and shuttling factors Dsk2, Rad23, and Ddi1 that are not a proteasome subunit [5].

Rpn10 is one of several Ub-receptors that target ubiquitylated proteins destined for degradation by the 26S proteasome [3]. It contains a VWA (Von Willebrand factor type A domain) tethered to a ubiquitin-binding domain called UIM (ubiquitin-interacting motif). The VWA binds the proteasome, whereas the UIM binds ubiquitin non-covalently. Rpn10 is evolutionarily conserved with some species like human and plant having additional one or two UIM, respectively (**Figure 1**). Interestingly, deletion or silencing of Rpn10 in yeast and worm is dispensable for viability [3, 6]. However, in fruit fly and mice, Rpn10 deletion is lethal and has

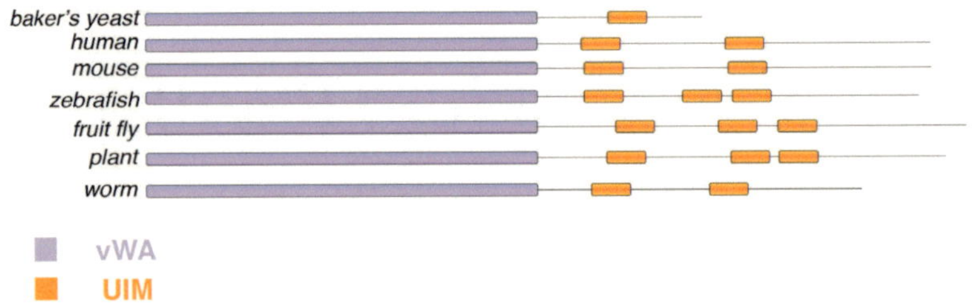

Figure 1.
Scheme of Rpn10 architecture and conservation. VWA, purple; UIM, orange.

deleterious phenotypes in plants [7–9]. Mice lacking only the UIM of Rpn10 are viable. Altogether, it is believed that in some organisms, the redundancy of proteasome Ub-receptors compensates for the lack of Rpn10.

Rpn10 can be found in a proteasome bound form and in a free cytosolic form. Its association with the proteasome is therefore dynamic [10–12]. Experiments with yeast, fly, plant, and human cells collectively suggest that free Rpn10 molecules recognize and shuttle ubiquitylated targets to degradation in the proteasome [10, 12–14]. Excess Rpn10 can bind another receptor Dsk2 and restrict its association with the proteasome. This observation raised the hypothesis that the cytosolic Rpn10 pool possesses a regulatory role on proteasome function.

It has been demonstrated that non-covalent ubiquitin binding and intramolecular monoubiquitylation are coupled [15]. Moreover, monoubiquitin binding by ubiquitin receptors is regulated to avoid occupation of the ubiquitin-binding domain (UBD) by free ubiquitin [16]. The conjugated monoubiquitin might occupy the UBD to prevent the binding to ubiquitylated partners [18]. Therefore, cleavage of conjugated ubiquitin from the receptor would expose the UBD to bind a ubiquitylated cargo in a spatially and temporally mode.

In a seminal study, Crossas and co-workers demonstrated that Rpn10 is monoubiquitylated mostly on lysine 84 (K84) in vivo in yeast by the E3 ligase Rsp5 [17]. They showed that ubiquitylation has an inhibitory effect on the ability of Rpn10 to bind ubiquitylated substrates, suggesting that in the ubiquitylated from, the UIM is blocked by the conjugated ubiquitin. Rpn10 monoubiquitylation levels were reduced under cellular stress conditions where protein degradation was enhanced, thus supporting a connection between monoubiquitylation of Rpn10 and proteasome function. It was later shown that Rpn10 monoubiquitylation leads to its dissociation from the proteasome [11]. Ub-Rpn10 molecules are much less associated with the proteasome compared to *apo*-Rpn10. Proteasomes lacking Rpn10 were still functional suggesting that Rpn10 removal does not destabilize the proteasome [11, 17]. Several groups suggested that Rpn10 monoubiquitylation serves to decrease Rpn10-associated proteasome and increase Dsk2-associated proteasome. Elevation of Ub-Rpn10 decreases Dsk2 association with the proteasome, supporting a role for Rpn10 ubiquitylation as a way to fine-tune the substrates that reach the proteasome [10, 11]. Although Rpn10 loosely associates with the proteasomes, it remains an enigma how monoubiquitylation mediates the dissociation of Ub-Rpn10 from the proteasome.

In this chapter, we will review the purification process of Ub-Rpn10 for crystallization, determine the structure by X-ray diffraction, and present the structural models of Ub-Rpn10 as *apo* and in the context of the proteasome. Moreover, we will discuss a postulated mechanism of action derived from the structures and series of in vitro and in vivo experiments that corroborate this mechanism.

2. *E. coli*-based expression and purification system for ubiquitylated proteins

To obtain large quantity of ubiquitylated proteins for downstream biochemical and biophysical studies (including X-ray crystallography), we constructed an *E. coli*-based system that synthetically expresses a functional ubiquitylation apparatus [18]. The system consists of two or three compatible plasmids that express His_6-Ub, E1, E2, E3, and MBP fusion of protein target of interest for ubiquitylation (**Figure 2A**). We constructed the system in a polycistronic manner. In the case of Ub-Rpn10 purification, we specifically expressed His_6-Ub, UBA1 (E1), and Ubc4 (E2) from one plasmid and Rsp5 (E3) and MBP-Rpn10 from other two plasmids. As the ubiquitylated protein possesses both six-histidine- and maltose-binding protein tags, we used Ni^{2+} and amylose affinity chromatography columns to purify the modified protein. We then cleaved the affinity tags by His_6-tobacco etch virus (TEV) protease and removed tags and the protease using the same affinity columns. Final purification was accomplished by an additional ion-exchange chromatography steps. The expression system and purification process were found to be very efficient as we obtained milligrams of purified Ub-Rpn10 using this purification protocol (**Figure 2B**) [18, 19]. Biochemical and mass-spectrometry analysis clearly showed that Rpn10 is authentically monoubiquitylated at K84 in *E. coli* as was previously demonstrated to be modified originally in *Saccharomyces cerevisiae* [17].

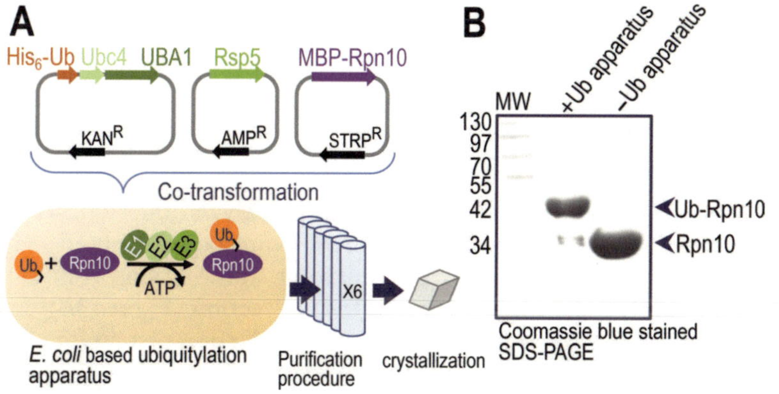

Figure 2.
Expression and purification of ubiquitylated Rpn10. (A) Scheme of the expression and purification of Ub-Rpn10 and (B) Coomassie-stained SDS-PAGE showing the purified sample of Ub-Rpn10 and apo Rpn10 for comparison.

3. Structure of Ub-Rpn10

The structure of Ub-Rpn10 highlighted two key findings [20]: (i) it revealed that in addition to the UIM, the VWA domain may also function as a UBD, and (ii) it allowed assessment of the Ub-Rpn10 structure in the context of the structure of the proteasome. We will now elaborate on the findings from the structure.

4. The VWA domain of Rpn10 is a *bona fide* UBD

The structure shows that the covalently bound Ub-moiety at K84 interacts with the neighboring VWA molecule in a non-covalent manner (**Figure 3**). The interaction interface centered at the famous ubiquitin I44 patch. This observation suggests that binding of ubiquitin by the VWA might be biologically important. In silico algorithm that screens for potential UBDs corroborated this finding [21]. Indeed,

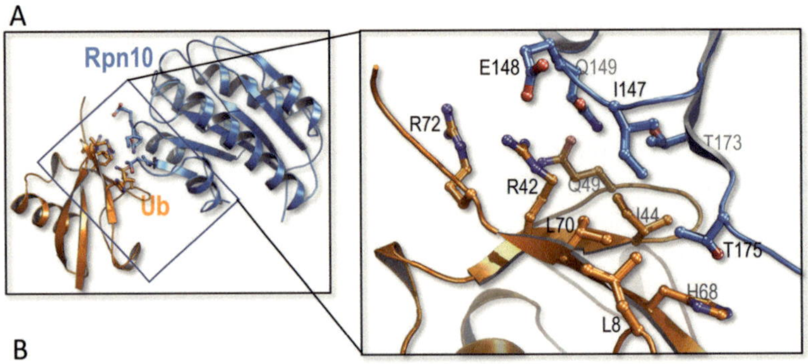

B

Protein type	K_D (SPR Analyses) [µM]	Ubiquitylation dependent bacterial growth efficiency [Normalized]
Rpn10 Full length	274±13.3	0.981
UIM	184±29.5	1.00
vWA	330±38.0	0.469
vWA-Q149A	613+100	0.273
Ub I44E	Not done	0.258
vWA Q149R	Not done	0.197
Ub H68A	Not done	0.067
vWA T173,175R	1138±233	0.024
vWA I147E	No binding	No growth
Ub L8E,I44E,V70D	No binding	No growth

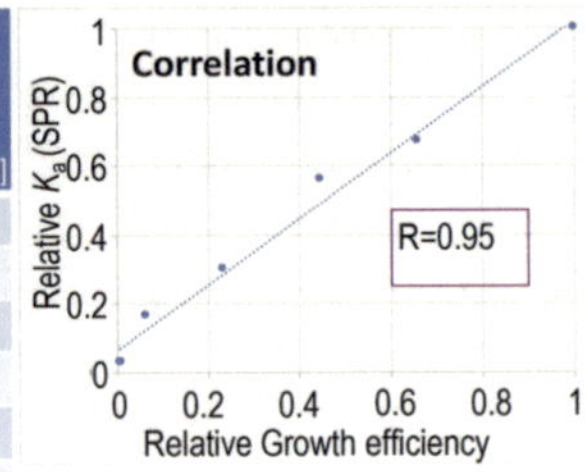

Figure 3.
A non-covalent binding interface between ubiquitin and Rpn10: (A) shows the major residues forming the interaction network and (B) shows the affinity values as measured by SPR (BIACORE) and quantitative growth efficiency as measured from bacterial spots. Correlation between the measurements is shown (right).

surface plasmon resonance experiments showed that VWA binds wild-type ubiquitin. Moreover, ubiquitin mutants at the interaction interface presented significant lower affinity. Similarly, structural-based mutations at the ubiquitin-binding patch on VWA significantly reduced or abrogated the interaction (**Figure 3B**). In an orthogonal study, we harnessed a bacterial genetic selection system for ubiquitylation to quantify the effect of these mutants on *E. coli* growth. In this system *E. coli* cells co-express split antibiotic protein tethered to ubiquitin and the VWA domain along with functional ubiquitylation cascade. The non-covalent interaction of ubiquitin with the VWA domain promotes the ubiquitylation of the latter, results in a functional assembly of the reporter, which give rise to bacterial growth under selective conditions (i.e., in the presence of antibiotic). The data obtained using this system also indicated that the VWA domain binds ubiquitin. Moreover, structural-based mutants abrogated the ubiquitylation and the growth. **Figure 3B** demonstrates the strong correlation between the two orthogonal studies. Altogether, these data indicate that the Rpn10-vWA domain is a *bona fide* ubiquitin-binding domain.

5. Ub-Rpn10 clashes with proteasome subunit Rpn9

The second finding raised from superpositioning of the Ub-Rpn10 with the proteasome cryo-EM structures [22–25]. This operation revealed that whereas the structures of the VWA domain from the two complexes are perfectly aligned, the conjugated ubiquitin collides with the proteasome adjacent subunit Rpn9. This suggests that Ub-Rpn10 cannot reside on the proteasome. The hypothesis was evaluated by three different methodologies: first, biochemical experiments with purified proteasome lacking Rpn10, supplemented with purified Rpn10 and enzymes mix of its ubiquitylation cascade. These experiments showed that while Rpn10 can associate with the proteasome, Ub-Rpn10 cannot [20]. Second, pulldown experiments show that Rpn9 binds Rpn10 but not Ub-Rpn10. Third, in orthogonal study in vivo in yeast together with the group of Ben Aroya, we demonstrated that native chromosomal expression of Rpn10 harboring the K84R mutation, which cannot undergo ubiquitylation at this site, tightly binds the proteasome subunit Rpn9. However, wild-type Rpn10 that can undergo ubiquitylation at K84 shows no interaction with Rpn9. Together, these experiments demonstrate that Rpn10 but not Ub-Rpn10 interact with Rpn9.

The structure, the biochemical, the biophysical, and the genetic experiments therefore support a model where upon ubiquitylation, Ub-Rpn10 dissociates from the proteasome, allowing a new molecule of Rpn10 to bind (**Figure 5**). Crosas and co-workers independently showed that ubiquitylation of Rpn10 leads to reduction of Rpn10-associated proteasomes, supporting our observation [11].

6. Ub-Rpn10 modifies Dsk2 interactions at the proteasome

At the same time we determined and analyzed the structure of Ub-Rpn10, Crosas and his co-workers found that monoubiquitylation of Rpn10 dissociates Ub-Rpn10 from the proteasome [11]. Moreover, they found that the proteasome shuttle Dsk2 interacts with Rpn10 but this interaction is precluded by the ubiquitylation of Rpn10. They examined the localization of Dsk2 in a model of constitutively ubiquitylated Rpn10 in which Ub is fused to Rpn10 (Ub-Rpn10). They demonstrated that Dsk2 no longer interacts with Ub-Rpn10 but instead it associates the proteasome by interaction with Rpn1. Their data indicate that under these conditions, more Dsk2 is associated with the proteasome suggesting that ubiquitylation of Rpn10 also regulates the interaction of Dsk2 with the proteasome.

7. Conclusions

It is now becoming clear that ubiquitylation signal goes beyond degradation and serves as a regulation mechanism for protein-protein interaction. Recent study in our laboratory demonstrated that a similar mechanism also regulates the activity of HECT E3 ligases [26] and other Ub-receptors (unpublished data). We, therefore, postulate that other cellular Ub-receptors and perhaps shuttling factors are regulated by coupled monoubiquitylation.

It is convincing to see that different studies conducted by several laboratories found the same data in which monoubiquitylation of Rpn10 at K84 induces dissociation of the receptor from the proteasome. While each laboratory examined different outcomes, it seems that there is no contradiction between the models, but they actually provide a more complete and comprehensive view on the regulation mechanism of Rpn10 and proteasome function. It would be interesting to explore the mechanism that determines the timing of Rpn10 ubiquitylation on the proteasome. Moreover, it is yet to be explored if and when deubiquitylation promotes the recycle of Rpn10-dependent substrate degradation.

Acknowledgements

We thank the members of the Prag lab and our collaborators for insightful comments and experiments that facilitated the described research. Data collection

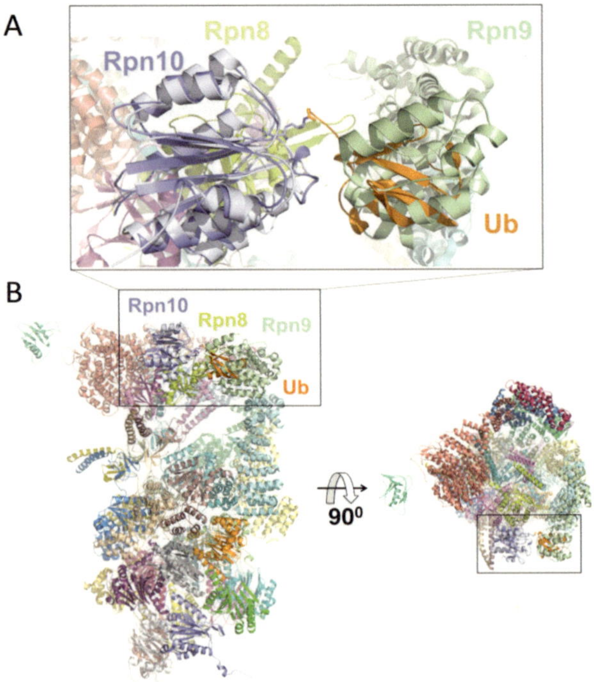

Figure 4.
Structure of Ub-Rpn10 at the proteasome context. Superimposition of the Ub-Rpn10 structure with the proteasome complex cryo-EM structure. (A) Zoom into the interaction interface between Rpn10, Rpn9, and Rpn8. The Ub-moiety [from Ub-Rpn10 molecule (colored in orange)] clearly clashes with Rpn9 subunit (light green). There is no interaction between the Ub-moiety and subunit Rpn8. (B) A view in context of the proteasome.

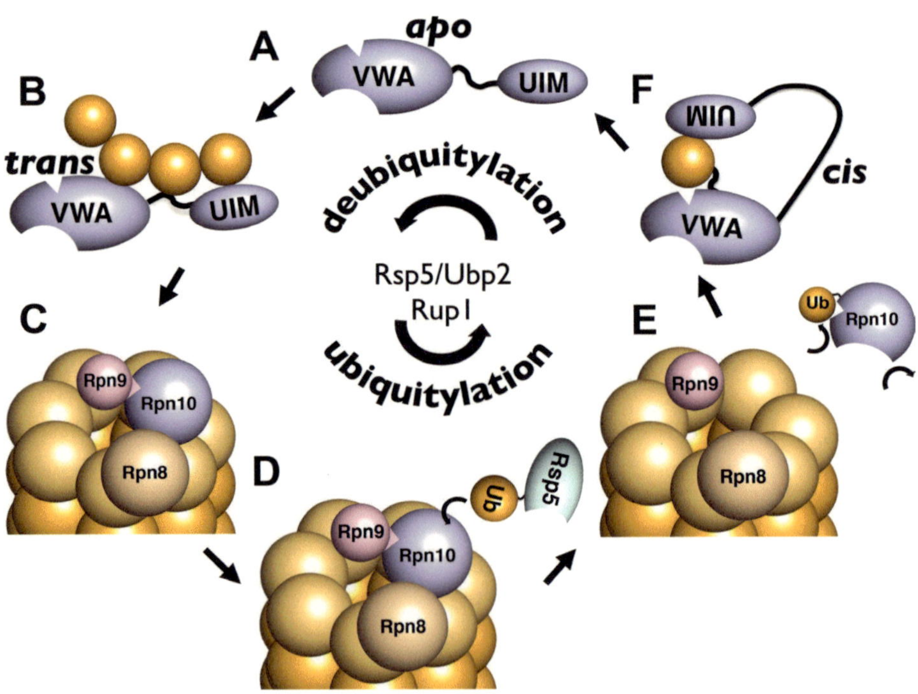

Figure 5.
A model for the regulation of Rpn10 by monoubiquitylation.

of Ub-Rpn10 was performed at the ID14-4 beamline at ESRF, Grenoble, France. The research entailed in this chapter was funded by grants from the Israel Science Foundation 1695/08 and 464/11 to G.P.; from the EC FP7 Marie Curie International Reintegration Grant (PIRG03-GA-2008-231079) to G.P.; and from the Israeli Ministry of Health (5108) to G.P. **Figures 2B**, **4** and **5** were adapted from [27].

Conflict of interest

The authors declare no conflict of interest.

Author details

Tal Keren-Kaplan[1], Ilan Attali[1], Olga Levin-Kravets[1], Oded Kleifeld[2], Shay Ben-Aroya[3] and Gali Prag[1*]

1 The George S. Wise Faculty of Life Sciences, School of Neurobiology, Biochemistry and Biophysics, Tel Aviv University, Tel Aviv, Israel

2 Faculty of Biology, Technion-Israel Institute of Technology, Haifa, Israel

3 Faculty of Life Sciences, Bar-Ilan University, Ramat-Gan, Israel

*Address all correspondence to: prag@post.tau.ac.il

References

[1] Hurley JH, Lee S, Prag G. Ubiquitin-binding domains. The Biochemical Journal. 2006;**399**(3):361-372. DOI: 10.1042/BJ20061138

[2] Shi Y et al. Rpn1 provides adjacent receptor sites for substrate binding and deubiquitination by the proteasome. Science. 2016;**351**(6275):831. DOI: 10.1126/science.aad9421

[3] Deveraux Q et al. A 26 S protease subunit that binds ubiquitin conjugates. The Journal of Biological Chemistry. 1994;**269**(10):7059-7061

[4] Husnjak K et al. Proteasome subunit Rpn13 is a novel ubiquitin receptor. Nature. 2008;**453**(7194):481-488. DOI: 10.1038/nature06926

[5] Jiang TX, Zhao M, Qiu XB. Substrate receptors of proteasomes. Biological Reviews of the Cambridge Philosophical Society. 2018;**93**(4):1765-1777. DOI: 10.1111/brv.12419

[6] Takahashi M et al. Reverse genetic analysis of the *Caenorhabditis elegans* 26S proteasome subunits by RNA interference. Biological Chemistry. 2002;**383**(7-8): 1263-1266. DOI: 10.1515/BC.2002.140

[7] Szlanka T et al. Deletion of proteasomal subunit S5a/Rpn10/p54 causes lethality, multiple mitotic defects and overexpression of proteasomal genes in *Drosophila melanogaster*. Journal of Cell Science. 2003;**116**(Pt 6):1023-1033

[8] Lin YL et al. The defective proteasome but not substrate recognition function is responsible for the null phenotypes of the Arabidopsis proteasome subunit RPN10. Plant Cell. 2011;**23**(7):2754-2773. DOI: 10.1105/tpc.111.086702

[9] Hamazaki J et al. Rpn10-mediated degradation of ubiquitinated proteins is essential for mouse development. Molecular and Cellular Biology. 2007;**27**(19):6629-6638. DOI: 10.1128/MCB.00509-07

[10] Matiuhin Y et al. Extraproteasomal Rpn10 restricts access of the polyubiquitin-binding protein Dsk2 to proteasome. Molecular Cell. 2008;**32**(3):415-425. DOI: 10.1016/j.molcel.2008.10.011

[11] Zuin A et al. Rpn10 monoubiquitination orchestrates the association of the ubiquilin-type DSK2 receptor with the proteasome. The Biochemical Journal. 2015;**472**(3): 353-365. DOI: 10.1042/BJ20150609

[12] van Nocker S et al. The multiubiquitin-chain-binding protein Mcb1 is a component of the 26S proteasome in *Saccharomyces cerevisiae* and plays a nonessential, substrate-specific role in protein turnover. Molecular and Cellular Biology. 1996;**16**(11):6020-6028

[13] Smalle J et al. The pleiotropic role of the 26S proteasome subunit RPN10 in Arabidopsis growth and development supports a substrate-specific function in abscisic acid signaling. Plant Cell. 2003;**15**(4):965-980

[14] Lipinszki Z et al. Ubiquitylation of drosophila p54/Rpn10/S5a regulates its interaction with the UBA-UBL polyubiquitin receptors. Biochemistry. 2012;**51**(12):2461-2470. DOI: 10.1021/bi3001006

[15] Shih SC et al. A ubiquitin-binding motif required for intramolecular monoubiquitylation, the CUEe domain. The EMBO Journal. 2003;**22**(6):1273-1281

[16] Prag G et al. Mechanism of ubiquitin recognition by the CUE domain of Vps9p. Cell. 2003;**113**:609-620

[17] Isasa M et al. Monoubiquitination of RPN10 regulates substrate recruitment to the proteasome. Molecular Cell.

2010;**38**(5):733-745. DOI: 10.1016/j.molcel.2010.05.001

[18] Keren-Kaplan T et al. Synthetic biology approach to reconstituting the ubiquitylation cascade in bacteria. The EMBO Journal. 2012;**31**(2):378-390. DOI: 10.1038/emboj.2011.397

[19] Keren-Kaplan T, Prag G. Purification and crystallization of mono-ubiquitylated ubiquitin receptor Rpn10. Acta Crystallographica. Section F, Structural Biology and Crystallization Communications. 2012;**68**(Pt 9):1120-1123. DOI: 10.1107/S1744309112034331

[20] Keren-Kaplan T et al. Structure of ubiquitylated-Rpn10 provides insight into its autoregulation mechanism. Nature Communications. 2016;7:12960. DOI: 10.1038/ncomms12960

[21] Keren-Kaplan T et al. Structure-based in silico identification of ubiquitin-binding domains provides insights into the ALIX-V: Ubiquitin complex and retrovirus budding. The EMBO Journal. 2013;**32**(4):538-551. DOI: 10.1038/emboj.2013.4

[22] Sakata E et al. Localization of the proteasomal ubiquitin receptors Rpn10 and Rpn13 by electron cryomicroscopy. Proceedings of the National Academy of Sciences of the United States of America. 2012;**109**(5):1479-1484. DOI: 10.1073/pnas.1119394109

[23] Lasker K et al. Molecular architecture of the 26S proteasome holocomplex determined by an integrative approach. Proceedings of the National Academy of Sciences of the United States of America. 2012;**109**(5):1380-1387. DOI: 10.1073/pnas.1120559109

[24] Beck F et al. Near-atomic resolution structural model of the yeast 26S proteasome. Proceedings of the National Academy of Sciences of the United States of America. 2012;**109**(37):14870-14875. DOI: 10.1073/pnas.1213333109

[25] Unverdorben P et al. Deep classification of a large cryo-EM dataset defines the conformational landscape of the 26S proteasome. Proceedings of the National Academy of Sciences of the United States of America. 2014;**111**(15):5544-5549. DOI: 10.1073/pnas.1403409111

[26] Attali I et al. Ubiquitylation-dependent oligomerization regulates activity of Nedd4 ligases. The EMBO Journal. 2017;**36**(4):425-440. DOI: 10.15252/embj.201694314

[27] Keren-Kaplan T. Structural insight into the regulation mechanisms of ubiquitin-receptors by ubiquitylation [PhD thesis]. Israel: Tel-Aviv University; 2014

Chapter 10

Zinc-Binding B-Box Domains with RING Folds Serve Critical Roles in the Protein Ubiquitination Pathways in Plants and Animals

Michael Anthony Massiah

Abstract

Protein ubiquitination is an essential cellular process that maintains protein homeostasis, regulates protein, and cell functions, and removes aggregated and misfolded protein. Disruption in function of any of the protein components of the ubiquitination pathway is associated with human diseases including cancers. An important member in the ubiquitination cascade is the very large E3 ligase family that directs substrate modification. The RING-type E3 ligases possess a cysteine/histidine-rich zinc-binding RING domain that confers ligase functionality. RING domains adopt a canonical ββα-fold. TRIM proteins represent a novel class of RING-type E3 ligase. TRIM proteins consist of an N-terminal RING domain followed by one or two B-box domains. The two types of B-box domains play essential roles in protein ubiquitination by contributing to substrate targeting, ligase activity enhancement, and redundancy of ligase activity. This review presents a general background of the B-box domains, a structural and functional comparison with RING domains, and a summary of recent work demonstrating their role in proteolysis. We discuss new findings that reveal B-box domains which are ubiquitous and are found in non-TRIM plant proteins without the adjacent RING domain, indicating that B-boxes are members of RING-class E3 ligases.

Keywords: zinc-finger, ubiquitination, E4 ligases, MID1, TRIM, BBX

1. Introduction

Protein ubiquitination is an essential cellular process that maintains protein homeostasis (proteostasis) and removes aggregated and misfolded protein that could recruit other proteins away from their normal cellular functions. It serves to regulate protein and cellular functions. Dysregulation of any of the protein component usually result in human diseases including cancers and birth defects. As such, the focus on protein ubiquitination has grown significantly in the past 2 decades, leading to extensive knowledge and new insights. A member of the ubiquitination cascade that has received considerable attention involves the large and growing family of E3 ligases. This family directs the last step in the reaction cascade by facilitating the ubiquitination of protein substrates. It is generally accepted that each E3 ligase has a specific or group of substrates (**Figure 1**).

There are several subgroups of E3 ligases, with the largest consisting of proteins with a RING domain that confer ligase functionality. The RING E3 ligase domains are cysteine- and histidine-rich sequences that bind two zinc ions in a unique cross-brace manner and adopt a canonical ββα-fold (**Figure 1**). *Tri*partite *mo*tif (TRIM) proteins represent a new class of RING E3 ligases. TRIM proteins are characterized by their N-terminal RING domain followed by one or two cysteine/histidine-rich

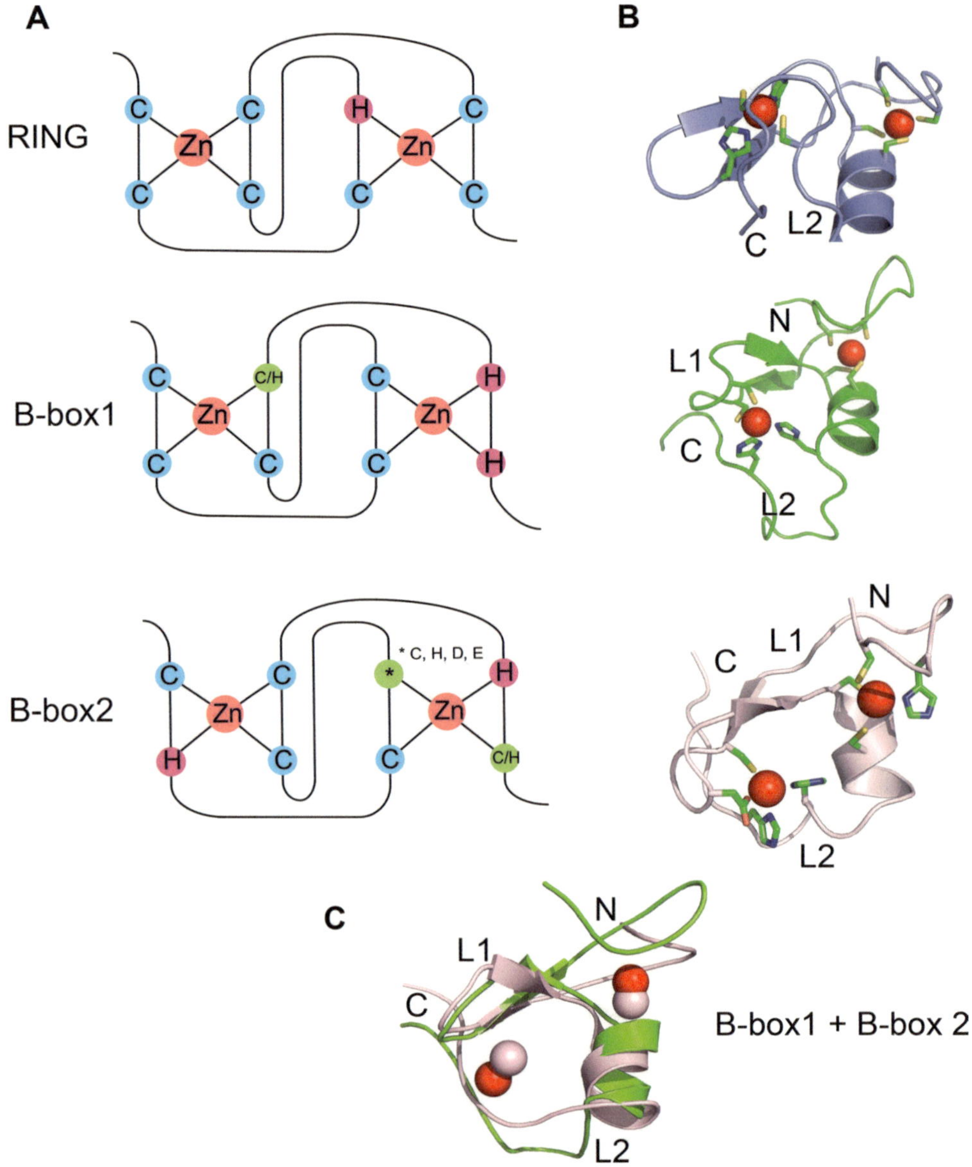

Figure 1.
Zinc-binding and structure of RING and B-box domains. A. Cross-brace zinc-binding mechanism by Cys/His-rich sequences of RING and B-box domains. Unlike other zinc-finger domains in which the zinc ion is coordinated by Cys and His ligands in a sequential manner, the cross-brace mechanism involves coordination by alternating pairs of ligands. The consensus zinc-binding sequences are different for RING and the B-box domains. B. Ribbon drawings of a representative RING (pdb 2hdp) domain and MID1 B-box1 (2ffw) and B-box2 (2dq5) domains. The ββα-canonical RING fold consists of a short α-helix and two loops, both of which contribute ligands to bind the zinc. Loop1 (L1) is usually less mobile than loop2 (L2). The relative locations of the zinc ions (red spheres) are similar among RING domains and the B-box domains. C. The superposition of structures of the B-box1 and B-box2 domains shows that the overall structures are similar. The structure is rendered to smooth out variation of the loop to simplify the image.

regions called B-box domains. The two types of B-box domains play important roles in protein ubiquitination, contributing to substrate targeting, enhancement of ligase activity, and redundancy in ligase activity. This review presents a historical background of the B-box domains commonly found in TRIM proteins, a structural and functional comparison with RING domains, and a summary of recent work demonstrating their role in protein ubiquitination. A brief discussion on the current understanding of RING E3 ligases mechanism of function is presented. We also discuss findings that reveal B-box domains are found in non-TRIM plant proteins without an accompanying RING domain. Finally, we argue that the B-box domains represent a new addition to the RING-class E3 ligases with a more versatile role than RING E3 ligase, namely they bind substrates, regulate E3 ligase activity of the adjacent RING domain, enhance E3 ligase activity of TRIM proteins, and actually function as E3 ligases.

2. Background

B-box and RING domains fall under the category of zinc-finger domains, which are present in a diverse family of proteins that includes transcription factors, ribonucleoproteins, proto-oncoproteins, and E3 ligases [1]. Zinc-finger domains or proteins are characterized as having cysteine and histidine residues arranged in one of several motifs that are relatively conserved in other proteins [2]. The thiol group (S^-) of the cysteine and a nitrogen atom of the histidine imidazole side-chain tetrahedrally bind a zinc ion [2].

Most zinc-finger proteins typically coordinate either a single zinc ion or two zinc ions, depending on the number of cysteine and histidine residues and their position within the sequence [3]. A defining property of zinc-finger domains is that zinc coordination is required to stabilize the tertiary structure. Loss of zinc coordination by a mutation of any of its cysteine or histidine residue results in complete unfolding of the protein structure. Protonation of the cysteine or histidine by decreasing the pH of the protein solution will also result in unfolding. For domains that bind two zinc ions, disruption of coordination of one zinc ion is usually accompanied by the loss of binding of the other zinc ion, causing the domain to become unfolded rather than partially folded with one zinc ion [4, 5].

Zinc-finger domains were identified in the mid-1980s within the *Xenopus* nuclear factor 7 (XNF7), first by Aaron Klug [6–8]. The subgroups of A-box and B-box cysteine/histidine-region domains were first identified a few years later with the *Xenopus* transcription protein TFIIIA ([9]). The A-box domain precedes the B-box region by 20–45 amino acids. Subsequently, other proteins were observed to have a similar A-box motif without an accompanying B-box region. The A-box regions constitute 60–80 amino acids with a zinc-binding consensus sequence of $C\text{-}X_2\text{-}C\text{-}X_{[9–39]}\text{-}C\text{-}X_{[1–3]}\text{-}H\text{-}X_{[2–3]}\text{-}C\text{-}X_2\text{-}C\text{-}X_{[4–48]}\text{-}C\text{-}X_2\text{-}C$. This consensus sequence reveals eight zinc-binding ligands (cysteines) that can coordinate two zinc ions. There are also variations in the zinc ligands as A-box domains are observed to have more than one histidine. By the early 1990s, the A-box domain was renamed *r*eally *i*nteresting *n*ew *g*ene (RING). The uncreative moniker remains for the B-box domain [10].

For the next 2 decades, the number of proteins observed with the RING and B-box domain pairs would have increased, with most belonging to TRIM proteins that are defined by their N-terminal RING, B-box, and coiled-coil (RBCC) domains [11, 12]. This RBCC domain arrangement is conserved and found in all multicellular organisms [12]. In humans, the RBCC domain is observed in a family of over 50 proteins; although few have been characterized in detail, their importance is underscored by the fact that some are oncoproteins (e.g., PML, RFP, and TIF1a), while others, when mutated, give rise to various congenital abnormalities [13, 14]. Members of this large protein family are found to play regulatory roles in a variety of cellular processes, including sperm vesicle exocytosis and intracellular release of HIV [15, 16]. The RBCC domain arrangement indicates at the very least that TRIM proteins have an overall common function. The RBCC proteins can have quite diverse C-terminal domain arrangements [17].

Interestingly, many TRIM proteins possess three consecutive cysteine/histidine-rich regions, the first being the RING domain. The other two domains are referred to as B-box1 and B-box2 domains ([18]). While the nomenclatures suggest that the two types of B-box domains are homologous, they do not share any discernable sequence similarity with each other or with RING domains. A single B-box domain in TRIM proteins is always of the type 2 form (B-box2), while TRIMs with two have the B-box1 domain preceding the B-box2 domain. The name may have persisted to prevent confusion in distinguishing the presence of the two types of B-box domains in TRIM proteins. The B-box1 domain is slightly larger (50–60 aa) with a zinc-binding consensus sequence of C-X_2-C-$X_{7–12}$C-X_2-C-X_4-C-X_2-[C/H]-$X_{3–4}$-H-$X_{4–9}$-H **[C5(C/H)H2]**. The B-box2 domains are 35–45 and have a consensus sequence of C-X_2-H-$X_{7–9}$-C-X_2-[C/D/E]-X_4-C-X_2-C-$X_{3–6}$-H-$X_{2–4}$-[C/H] **[CHC(C/D/E) C2H(C/H)]** [4]. Comparison of the consensus sequences of RING and the B-box domains reveals two regions in which the number of amino acids between zinc-binding pairs is different. RING domains have the longest length (X $_{[9-39]}$, X $_{[4-48]}$) following by the B-box1 domain ($X_{7–12}$,$X_{4–9}$) and B-box2 domain ($X_{7–9}$, $X_{3–6}$).

3. Description of B-box domain structures

Despite their prevalence and location downstream of RING domains in TRIM proteins, very little was initially done to characterize the structures and functions of B-box domains. We postulate that this might have been so because of difficulties in obtaining sufficient quantities of the B-box domains for structural and functional studies. Indeed, each type of the B-box domain has proven to be quite challenging to express and purify using *E. coli*. The same can be said for RING domains because there are only a few dozen structures solved, and given their uniqueness in defining substrate specificity, structural comparisons would be important to identify differences in mechanism of function and activity. Our experience revealed that the B-box and RING domains tend to form inclusion bodies [19]. In identifying conditions to obtain large quantities of these domains, we have established a purification protocol that not only works for B-box domains but also for other proteins that are prone to forming inclusion bodies [19, 20]. The protocol can extract natively folded proteins from inclusion bodies without refolding. Despite some of these initial challenges, there are now several structures of both types of B-box domains in the protein database.

The first comprehensive structural studies of B-box domains were based on the TRIM18/MID1 protein [13, 21]. Human MID1 is required for proper midline development during embryogenesis ([22–26]). Mutations of MID1, some of which are found within the B-box domains, are associated with X-Linked Opitz G/BBB syndrome (XLOS), a congenital disorder characterized by clefts of the lip and palate, cardiac structural defects, and genital anomalies [14, 27].

The structure of the B-box1 domain (residues Gln87-Pro165) was solved in 2006 by analyzing multidimensional data acquired by nuclear magnetic resonance (NMR) spectroscopy. The B-box1 domain was observed to coordinate two zinc ions in a cross-brace manner with six cysteine and two histidine residues (**Figure 1**) [13]. Residues

Ala115 to Pro165 form the core of the structure, while the preceding 30 amino acids are unstructured and initially included to aid in solubility. The structure consists of a two-turn α-helix that is preceded by a long structured loop consisting of two short β-strands separated by a type-2 β-turn. Two cysteine residues within the first part of the structured lasso-like loop1 coordinate one zinc ion with two other cysteine residues located within the first helical turn of the helix. Two cysteine residues that are part of the β-turn and two histidine residues, one located at the end of the α-helix and the other on the loop2 that follows the helix, coordinate the second zinc ion. The overall structure is very similar to the ββα-canonical RING fold (**Figure 1**).

The structure of MID1 B-box2 domain was solved a year later, using NMR data. In contrast to the MID1 B-box1 domain, the MID1 B-box2 consists of seven classical cysteine and histidine zinc-binding residues, suggesting that only one zinc might be coordinated by four of these residues (**Figure 1**). Sequence alignment of TRIM B-box2 domains reveals that approximately half of B-box2 domains consist of aspartate residues and the other half a cysteine residues in the same location. This observation suggests that Asp must be a highly conserved change [21] that should be performing the same role as the cysteine residue. Indeed, the MID1 B-box2 domain coordinates two zinc ions in a similar cross-brace manner as the B-box1 and RING domains. Two histidine, cysteine, and aspartate residues coordinate one zinc ion. The carboxylate oxygen of this conserved aspartate side chain participates in zinc coordination. The aspartate residue forms the necessary zinc-knuckle conformation with a cysteine residue two positions away (CxxD) to tetrahedrally coordinate the zinc ion [2, 21]. Although carboxylate groups are involved in binding catalytic zinc ions, for example, carbonic anhydrase [28–30], or other non-structural metals, this was the first demonstration in a zinc-finger protein. The B-box2 domain adopts a two-turn α-helix, two short β-strands separated by a type-2 β-turn, and two structured loops adjacent to the helix. Despite a lack of sequence similarity, the structures of the two types of B-box domains are remarkably similar (**Figure 1C**). The positions of the two zinc ions are in similar locations, namely near the N-terminus of the helix and to the bottom left of the helix (given the specific orientation shown in **Figure 1B**). Importantly, the mechanism of zinc coordination (cross-brace) and the ββα-fold are comparable to those of RING domains.

Structures of the B-box1 domain from TRIM19 and the B-box2 domain from TRIM1/MID2, TRIM5α, TRIM21, TRIM29, TRIM39, TRIM41, TRIM54, and TRIM63/MuRF1 have been solved. All the B-box domain structures are similar. Consequently, we conclude that MID1 consists of three consecutive domains with RING folds. Thus, the TRIM protein family must represent a new class of E3 RING-type ligase, consisting of two or three consecutive RING folds.

To identify a possible role of the two adjacent B-box domains, the structure of both was determined in their native tandem form (res A110-E214). The two B-box domains maintained their original structures and pack against each other with the interface formed by residues located on the structured loop-1 near the two antiparallel β-strands. The surface area of the interface is 188 $Å^2$ (17% of the total surface). Interestingly, the tandem globular structure is very reminiscent of the intermolecular association observed for heterodimeric RING structures, such as the BARD1 and BRCA1 domains (12) and the polycomb group protein (Bmi-1) and Ring1B polycomb group (14), and the homodimeric RINGs, such as HDM2 [31], RNF4 and 8 [32, 33], and cIAP2 [34]. The TRIM19 B-box1 and TRIM54 B-box2 domains were solved as symmetric dimers by X-ray crystallography. The structures of RING dimers reveal the domains interacting via residues located on and near loop-1. The BRCA1-BARD and RNF8 dimers also include adjacent structures, such as helical dimers. The area of the interface of the hetero- and homo-RING dimers is approximately 150–200 $Å^2$, similar to that observed for the MID1 B-box1,2 heterodimer. In spite of their interactions, it appears that in the case of MID1, unfolding of the B-box1 structure, via a mutation of one its zinc-binding residues, had little effects on the structure and stability of the B-box2 domain [4]. This observation suggests the possibility that each B-box domain could function independently or have redundant E3 ligase function.

The structures of two B-box domains were solved in complex with another TRIM domain. The TRIM5α B-box2 was crystallized with its coiled-coil domain, which contributes to oligomerization. Binding studies using NMR and dynamic light scattering using a TRIM5α proteins with a native and mutant B-box2 domain reveal that the B-box2 domain contribute to higher order self-association [35]. Given that the B-box2 domain is required for substrate ubiquitination, self-association may contribute to enhanced E3 ligase activity and substrate targeting [36, 37] of the native TRIM5α. The B-box domains of TRIM27 are also determined to be crucial for multimerization by possibly helping to orient the coiled-coil domain in a way that maintained the multimer interaction [38]. In contrast, the B-box2 domain of TRIM21 was crystallized with the N-terminal RING domain, and the structure reveals that the B-box2 domain interacts with the RING domain on a surface that is important for RING-E2 interaction. In this case, the structure suggests that TRIM21 B-box2 may have an autoinhibitory effect, although further studies are required. It is possible that the structures of these complexes may be affected by protein packing within the crystal lattice. More work needs to be done to understand the mechanism of function of these B-box domains, which we postulate are now key players in the ubiquitination field.

4. A brief description of ubiquitination and the role of E3 ligases

In order to appreciate the function of RING and B-box domains, a brief summary of protein ubiquitination is provided. All living organisms employ a fairly common mechanism to recycle proteins so as to regulate protein function, proteostasis, and cell cycle. Eukaryotic cells employ protein ubiquitination, a post-translational modification using the highly stable 76-amino acid ubiquitin protein (Ub) [39–41]. Bacteria use prokaryotic ubiquitin-like protein (Pup) in an analogous manner [42]. Polyubiquitinated proteins, usually with a chain of at least four-linked Ub, are targeted to the proteasome where they are proteolytically cleaved into peptides [39, 40, 43]. Ubiquitin chains can form via any of its seven lysine residues or combinations of the seven; homogenous chain links, example K48, promote protein degradation [44–46], but some have signaling functions [47, 48]. The Ubs are cleaved by deubiquitinating enzymes (DUBS) and recycled [49]. Although mono- and diubiquitinated proteins are directed to the proteasome [50], there is evidence that this level of modification serves a signaling role, in which modified proteins can have their functions and cellular location altered. The monoubiquitination of cytosolic proteins results in translocation to the nucleus to participate in DNA repair, transcription regulation, and inflammatory response [39, 51–54].

Ubiquitination involves three classes of enzymes. First, the E1-activating enzyme (E1) catalyzes the adenylation of the C-terminal glycine of Ub (Ub~AMP). This phosphoester bond undergoes a nucleophilic attack by the sulfhydryl group of the active site cysteine residue on the E1. In the next step, the Ub is transferred to an active site cysteine residue on a family of E2 conjugating enzymes (E2) to form an activated thioester-linked E2~Ub complex [39, 43]. Typically most types of E2 enzymes require the concerted action of an E3 ligase (E3) to target and facilitate substrate ubiquitination [40, 43]. There are several classes of E3 ligases: the homologous to the E6-AP Carboxyl Terminus (HECT), RING-InBetweenRing-RING families, and RING class [39–41, 50, 55]. The HECT and RING-IBR-RING families accept the Ub via a trans-thiolation reaction from the E2 before transferring it to the target protein. The RING E3 ligase, which includes the Skp-Cullin-F-box (SCF) complex, U-box, and now the B-box, represents the overwhelming majority of E3 ligases. While the mechanism is unclear, the RING-type ligase binds and places in close proximity to the target protein and the E2 enzyme. With the SCF complex, the RING domain (aka RBX) recruits the E2~Ub, while another SCF subunit (usually, the F-box) binds the substrate. In the last reaction, the E2~Ub thioester bond undergoes a nucleophilic attack (thiolysis) by a lysine residue of the target protein whereby the side-chain amino group forms an isopeptide bond with the C-terminal carboxylate group of Ub. Subsequent Ubs can be attached to other lysine residues of the substrate, but more commonly observed to form a polyubiquitin chain with linkages to one of seven lysine residues of the Ub [39, 40, 43]. Chains can be formed with Lys 6, 11, 27, 29, 33 and the N-terminus (M1) amino group, but the two more common reported linkages involve Lys48 and Lys63 [44]. We argue that the level or amount of polyubiquitination or Ub processivity can be an assessment of the level of E3 ligase activity by a RING protein.

Typically, confirmation that a RING domain/protein possesses E3 ligase activity is accomplished by performing *in vitro* autoubiquitination assays consisting of all the protein components except the substrate. In most cases, the protein substrate is unknown. In these reactions, the proximity of the RING domain with the activated

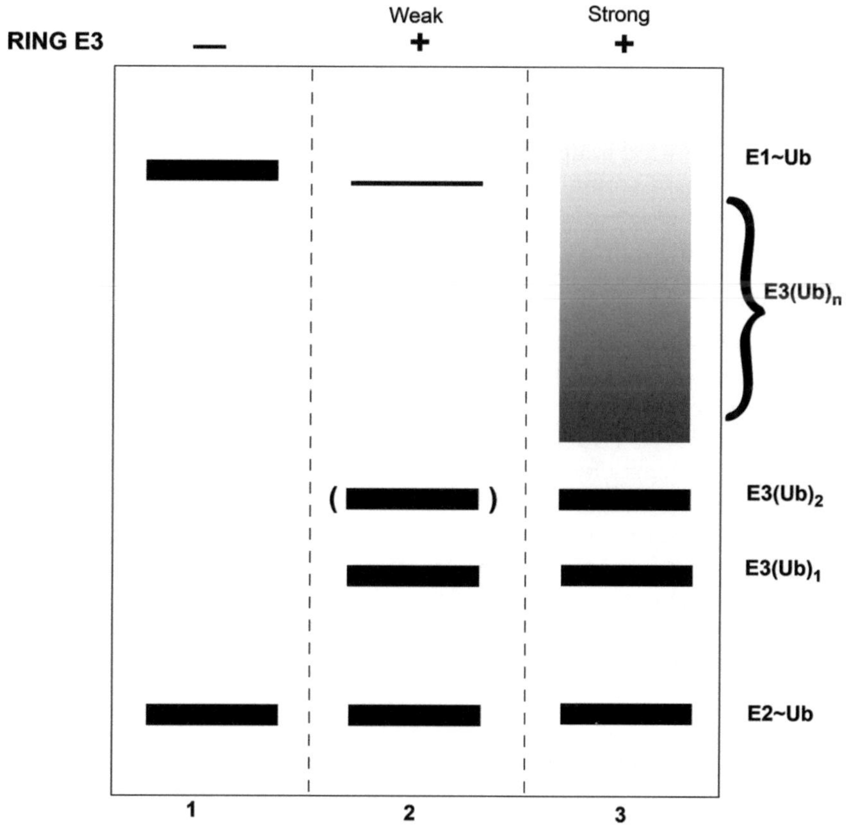

Figure 2.
Expected RING E3 ligase results. Cartoon representation of a Western blot image showing autoubiquitination results in no E3 ligase (ln1), a RING-type E3 with "weak" E3 ligase activity forming mono- and sometimes (di-ubiquitinated) products, and a RING E3 with stronger activities (ln3). The smearing represents polyubiquitinated products with various amounts of Ub on the protein. Substrate ubiquitination will usually mirror the results shown.

E2~Ub complex promotes auto- or self-ubiquitination, suggesting that a RING protein is a substrate of its own E3 ligase activity. How autoubiquitination affects substrate ubiquitination is a subject of intense studies. A protein mixture with E1, E2, E3, Ub, and ATP is incubated, and ubiquitinated E3 is probed by Western blot with an antibody against the Ub or the E3 ligase (**Figure 2**). Protein bands corresponding to a RING domain with covalently attached Ubs typically indicate E3 ligase activity [39, 40]. The level of E3 ligase activity can be estimated by the rate and amount of autoubiquitinated products observed as a function of assay time [56, 57]. Western blot images of various levels of polyubiquitinated products are often indicated with a smearing of high molecular weight products, while a less active enzyme is indicated by less smearing or the presence of mono- or di-ubiquitinated products (**Figure 2**). However, it should be noted that even though an E3 ligase catalyzes mono- and di-ubiquitination does not mean it is a "weak" ligase; it is wholly possible that it has evolved to function at this level, and as noted, mono- and di-ubiquitination may serve as signaling events [54, 58]. There are lots of questions about how and why RING domains exhibit different levels of substrate ubiquitination.

5. Possible mechanism of action of RING E3 ligases

As noted, the mechanism of function of RING E3 ligases is unclear, but considerable progress has been made to provide insights. The structures of several E2-RING complexes reveal that the RING domain is positioned ~15 Å from the active site and the thioester linkage between the E2 and Ub [59, 60]. Based on these structures, it is unclear how the RING domain affects reactivity or electrophilicity on the E2~Ub linkage. To gain insights on the role of RING E3 ligases, Klevit and co-workers [61, 62] used molecular dynamic and NMR studies to show that the bound RING E3 ligase promoted a "closed" E2~Ub conformation, whereby the Ub populates one interaction mode with slightly greater frequency. In the absence of a bound RING domain, the covalently attached Ub is highly mobile and does not favor any specific surface of E2 to interact [62]. There is no fixed or stable structure between the Ub and E2 proteins. Promoting the positioning of the Ub to the "back" surface of the E2 reduces steric hindrance for nucleophilic attack by the incoming lysine residue. In addition, key amino acid interactions at the E2-RING binding interface appear to contribute to the activation of the thioester bond. For example, residue Gln92 of UbcH5 (Ube2D1), which is located on a helical turn adjacent to active site Cys-85, forms a hydrogen bond with an arginine or lysine residue on loop-2 of the RING domain. Disruption of this interaction through mutation of Gln92 or the arginine severely disrupts the rate of Ub transfer [62, 63]. In fact, the rate of thiolysis with free lysine, as the substrate, for either mutant is comparable to that of E2~Ub without a bound RING domain. Thus, it is generally accepted, given what is known from the various E2-RING interaction studies that the RING domain contributes allosterically to electrophilicity of the thioester bond.

6. The B-box domains are new members of the RING E3 ligase

Given that the B-box domains have similar RING folds, it was postulated to function similarly. First confirmation that the B-box domains possess E3 ligase activity was demonstrated with MID1 [64]. The *in vitro* assays revealed that each type of B-box domain exhibited weak E3 ligase activity. Mono-ubiquitinated products were observed. Varying degrees of RING E3 ligases activities have been reported in the literature, but no explanation has been provided; for some instances, RING hetero- or homodimerization is necessary. Despite the similar overall fold of RING and B-box domains, we postulate that subtle amino acid differences and structural variations between RING folds may contribute to differences in the levels of activities. Furthermore, it is possible that the cohort E2 enzyme used may provide a specific level of activity [59]. Most RING E3 ligases are confirmed using the common $UbcH5_{a-c}$ (Ube2D1–3) E2 family. For MID1 B-box domains, mono-ubiquitination activities were observed with at least 12 different E2 enzymes [64]. As noted above, it is possible that the level of activities of the B-box domains may be physiological and evolutionarily determined. Below, the structures of RING and B-box domains are compared to provide rationales for the differences in activities.

Intriguingly, the tandem B-box domains also exhibit weak E3 ligase activity, with no greater level of autoubiquitination activity than that observed with the B-box1 domain [64]. This is in contrast to hetero- and homodimeric RING dimers, which exhibited greater activity than the mono form [32, 33, 65]. The BRCA1-BARD1 complex, in which BRCA1 (breast cancer 1) heterodimerizes with BARD1 (BRCA1-associated RING domain), exhibited enhanced activities compared to BRCA1 alone; BARD1 does not exhibit ligase activity [65–68]. Enhancements of activities were also observed for MDM2/HDX [69], RNF4 [33], inhibitor of apoptosis (IAP) proteins [70], BMI1-RING1 [71], and membrane-associated RING-CH family of E3 ubiquitin ligases (MARCH1) RING dimers [72], to name a few. The mechanism of E3 ligase enhancement by RING dimers or the lack of enhancement by the MID1 B-box domains is unclear. However, there are several publications that proposed rationales of the role of RING dimers [32, 73], but they will not be discussed here.

Despite the MID1 B-box domains not showing strong ligase activity, studies with TRIM16 revealed that its B-box domains exhibited greater level of activities. There were substantial amount of polyubiquitinated products, as demonstrated by the intensity of the smearing observed by Western blot analysis [74]. Both *in vivo* and *in vitro* ubiquitination assays with constructs containing B-box1 and B-box2 domain deletions resulted in the loss of the polyubiquitin smearing. TRIM16 is a pseudo-TRIM that lacks a RING domain; it is possible that TRIM16 B-box1 and 2 domains have evolved to possess increased activity in light of the missing RING domain. However, results from TRIM16 indicate that TRIM16 can dimerize with other TRIM proteins, and it may be that these interactions contribute to *in vivo* increased activity. This observation suggests a more intricate mechanism of action for TRIM proteins, namely that they can homo- and heterodimerize, and this can affect the levels of E3 ligase activity. Furthermore, recent *in vitro* and *in vivo* studies with TRIM27/rtf, a protein with RING–B-box2-CC domain, showed that the B-box2 domain, and not the RING domain, is responsible for substrate binding and ubiquitination [75].

7. Tandem RING and B-box domains are more active: could it be E4 ligases?

To understand the role of the B-box domains in the context of being adjacent to the RING domains, as they are commonly found in TRIM proteins, autoubiquitination assays were performed with the MID1 RING domain in tandem with B-box1 (RING-B-box1 (RB1)) and both B-box domains (RING–B-box1–B-box2 (RB1B2)). The goal was to determine whether each domain functions independently or if they have synergistic contribution to justify a possible evolutional reason for their presence in tandem. In the case of RB1, greater amount of polyubiquitinated products were observed compared with the results of the ubiquitination assay with the RING domain alone [14, 64, 76]. The rate of product formation was qualitatively faster. Whereas polyubiquitinated products were observed with the MID1 RING domain after 120 minutes, polyubiquitinated products were observed within the first 5–10 minutes of the assay with RB1. Similarly, the MID1 RB1B2 protein construct exhibited comparably rapid ligase activity as the RB1 domain construct. Within experimental error, it was difficult to determine whether there was greater or lesser amount of polyubiquitinated products. Therefore, it was difficult to identify the contribution of the B-box2 domain within the RB1B2 construct.

To probe whether the B-box2 domain contributes to ligase activity as part of the RB1B2 construct, a C142S mutation was introduced within the B-box1 domain (RB1*B2). Cysteine-142 coordinates one of the two zinc ions, and its mutations to serine resulted in the loss of coordination of both zinc ions and unfolding of the B-box1 domain [4]. By Western blot, the E3 ligase activity of the RB1*B2 protein construct was indistinguishable from the RB1B2 construct, indicating that the B-box2 domain can compensate for the loss of function of the B-box1 domain. To confirm that the B-box2 domain has the same enhancing role as the B-box1 domain, an RB1* protein construct was designed, and the activity was observed to be similar to that of just the RING domain [64].

These observations indicate that the B-box domains, by some unknown mechanism, appear to enhance the E3 ligase functionality of the adjacent RING domain. It is wholly possible that the enhancement observed could be that both RING and the B-box domains have gained E3 ligase activities, there is some synergy in activities, or that the B-box domains may function as E4 ligases [77–80]. E4 ligases are domains with a RING fold that enhance the ligase activity of RING E3 ligases. Possible examples of E4-enhancing ligases are the BARD1 and HDMX RING domains. In the mid-2000s, the U-box domain, which adopts a similar ββα-RING fold but without the coordination of zinc ions ([81]), was initially shown to play an E4-enhancing role for RING E3 domains [82]. Subsequently, it was concluded that U-box domains can function as E3 ligases and now represent a new member of the RING-type E3 ligases with a similar mechanism of action as RING domains [61, 83, 84].

The function of the B-box domains of TRIM5α, TRIM25, and TRIM32 is also studied [85, 86]. TRIM5α possesses anti-viral/anti-HIV activities [56, 87]. TRIM25 plays a crucial anti-viral role by ubiquitinating the N-terminal caspase activation and recruitment domains (CARDs) of the recognition receptor RIG-I [86, 88]. Mutations of TRIM32 are associated with limb-girdle muscular dystrophy type 2H. TRIM25 consists of RING, B-box1, and B-box2, while TRIM5α and TRIM32 have the RING and B-box2 domains. TRIM25 possesses both Ub and interferon-stimulated gene 15 (ISG15)-E3 ligase activities [86, 88]. ISGylation serves more of a signaling role, as ISG15-modified proteins have altered functions [89]. For these proteins, the B-box domains are required for enhanced activities. Using thiolysis assays (nucleophilic attack on the thioester of charged E2~Ub by lysine), the role of the B-box domains was assessed for TRIM25 and 32 [86]. The constructs with the RING domain alone marginally activated the reaction, but those including the B-box domains significantly accelerated thiolysis.

In the case of TRIM5α, autoubiquitination assays with monomeric RING alone did not produce polyubiquitinated products/chains. In contrast, the RB2 and RB2CC protein constructs showed considerable increases in activities, which were attributed to the presence of the B-box domain. As control, the RB2*CC protein construct with a destabilizing B-box2 mutant resulted similar levels of ubiquitination products as just with the RING domain [36].

In summary, the results from the RB1, RB1B2, RB1*B2, and RB1* autoubiquitination assays of MID1/TRIM18, TRIM25, and TRIM32 suggest that TRIM proteins with RING and two B-box domains have some redundancies in the enhancement role of the B-box domains. Furthermore, given that the RING-less TRIM16 tandem B-box possesses strong ligase activity adds to the support that the B-box domains can possibly have dual roles, functioning as E3 and E4 ligases.

8. The B-box domains are required for substrate polyubiquitination

While the *in vitro* assays confirmed E3 ligase activities of B-box domains and their possible role in E3 ligase enhancement, it is important to resolve whether observations of autoubiquitination activities translate to substrate ubiquitination. In the case of MID1, there are three known substrates: the catalytic subunit of protein phosphatase 2A (PP2Ac) [14, 90], alpha4 [76], and the fused kinase (FK) [91]. PP2Ac is part of a heterotrimeric PP2A complex consisting of the scaffolding subunit PP2Aa (PR65) and one of several regulatory subunits (PP2Ab) that defines cellular location and substrate specificity. PP2A functions as a master switch to control metabolism (review [92]), cell cycle progression (via cdc2 kinase activation), DNA replication, transcription and translation, cell proliferation, cytoskeleton dynamics and cell mobility, and apoptosis [93–100]. PP2A is considered a tumor suppressor, deactivating oncogenic MEK1 and ERK within the RAS-RAF-MEK-ERK/MAP kinase cascade [92, 96, 101]. The alpha4 protein [102–106] regulates PP2A within the target of rapamycin (TOR) signaling pathway that controls transcription, protein synthesis, and cell cycle progression in response to nutrients and extracellular stimuli [104, 105, 107–112]. Alpha4 binds the PP2Ac and induces a conformational change that keeps PP2Ac in an inactive conformation until needed [14, 113, 114]. The fused kinase is a key regulator within the Sonic Hedgehog pathway important for cell polarization and body symmetry [115]. Specifically, FK is shown to activate Hh- and Ci-dependent transcriptional activation in *Drosophila* Schneider 2 cells.

For all three substrates, full-length MID1 was shown to catalyze their polyubiquitination [14, 76, 91]. The role of the B-box domains for substrate ubiquitination was demonstrated for PP2Ac and alpha4 [14, 76]. With just the RING domain, a weak band was observed on the Western blot, indicating low amount of mono-ubiquitinated products [14, 76]. In contrast, polyubiquitinated products were observed with the RB1 and RB1B2 protein constructs. The results of the assays with the RB1* (C141S) protein construct yielded monoubiquitinated PP2Ac and alpha4 products, confirming the B-box1 domain is important for substrate targeting and polyubiquitination. The RB1*B2 protein construct catalyzed the polyubiquitination of PP2Ac but not alpha4. These results indicate a few things: the B-box binding of protein substrates is a critical role for polyubiquitination, the B-box2 domain can compensate for the unfolded B-box1 domain, and the B-box2 domain can contribute to some B-box1 redundancies in MID1's overall E3 ligase activity. The levels of ubiquitination of PP2Ac and alpha4 parallel the results observed with the autoubiquitination assays, confirming the various roles of the B-box domains. Similar results were observed with TRIM5α TRIM25, TRIM19/PML, and TRIM63.

TRIM5α targets the HIV capsid protein. Deletion of the B-box2 domain affected oligomerization state, E3 ligase activity, and substrate ubiquitination [35, 37, 56].

TRIM25 targets RIG-I. *In vivo* studies with HEK293T cells were performed with wild-type TRIM25 and with the B-box domains deleted individually and as a pair. Protein constructs consisting of either B-box domain yielded polyubiquitinated RIG-I products, supporting key observations with the MID1 protein [88]. TRIM19/PML is a tumor suppressor protein that is associated with a wide variety of cancers. TRIM19 is shown to function as a small ubiquitin-like modifier (SUMO) protein E3 ligase targeting MDM2, which functions as a regulator of protein p53. Sumoylation affects protein stability and cellular localization and sometimes serves as precursor to prime the protein substrate for ubiquitination. The results of *in vivo* and *in vitro* sumoylation assays revealed that the B-box domains are required for MDM2 sumoylation. Similarly, mutations destabilizing of the B-box2 domain of TRIM28/KAP-1 completely eliminated KRAB domain binding, demonstrating that the B-box2 domain is important for binding and targeting of the KRAB [116]. TRIM63/MuRF1 is found in striated muscle and observed to be upregulated with muscle atrophy. The muscle-type creatine kinase (M-CK) is one substrate of TRIM63 and co-precipitated with TRIM63 only with an intact B-box2 domain [117]. Ubiquitination of M-CK required the B-box2 domain. Furthermore, the B-box2 domain also contributed to the overall oligomeric TRIM63 structure [118]. It is unclear whether B-box2 domain dimerization or oligomerization is a universal observation or protein specific. In the case of MID1, there is no evidence that the B-box2 domain is dimer (unpublished data).

Finally and importantly, studies with TRIM27/rfp revealed a central role of the B-box domain in substrate binding and ubiquitination and subsequent degradation [75]. This report was the first demonstration that E3 ligase activity of TRIM27, a TRIM protein with a RING domain, is conferred to the B-box domain instead [75]. This unique finding has not, to our knowledge, been observed with any other TRIM proteins containing both RING and B-box domains.

9. MID1 B-box1 domain E3 ligase activity can be enhanced

As noted, we postulate that RING E3 ligases that exhibit different levels of *in vitro* activities might be due to evolutionary determinants that include subtle differences in amino acids and structures between the RING E3 ligases (**Figures 3** and **4**). This idea is supported with the MID1 B-box1 domain in which an XLOS-specific mutation, P151L, yielded a B-box1 domain with greater E3 ligase activity [119]. Furthermore, the RB1$^+$(P151L) protein construct possesses greater activity than the wild-type RB1 protein construct. NMR data revealed that the B-box1 P151L mutant is folded. This proline is conserved among many TRIM B-box1 domains, and it may be a key determinant for the weaker E3 ligase activity observed. Residue P151 is located at the end of the helix and beginning of loop2. Residues of loop2 participate in E2-RING interactions, specifically allowing an arginine or lysine residue to hydrogen bond with E2-Gln92, necessary for activation of the thioester bond in the case of Ube2D1. For the MID1 B-box1 domain, the rigidity of the backbone property of proline-151 positions loop2 differently, which prevents key loop2-E2 interactions (**Figure 4A**) [119]. The P151L mutation is most likely repositioned loop2 for more favorable interactions with the E2 (Ube2D1) enzyme used in the assay. In the case of TRIM16, there is no proline in the corresponding location, and this maybe the reason why it exhibits greater E3 ligase activity than the MID1 B-box1 domain. The lower activity of the B-box2 domain cannot be explained because there is no proline residue to reorient loop2. However, compared to the RING consensus sequence,

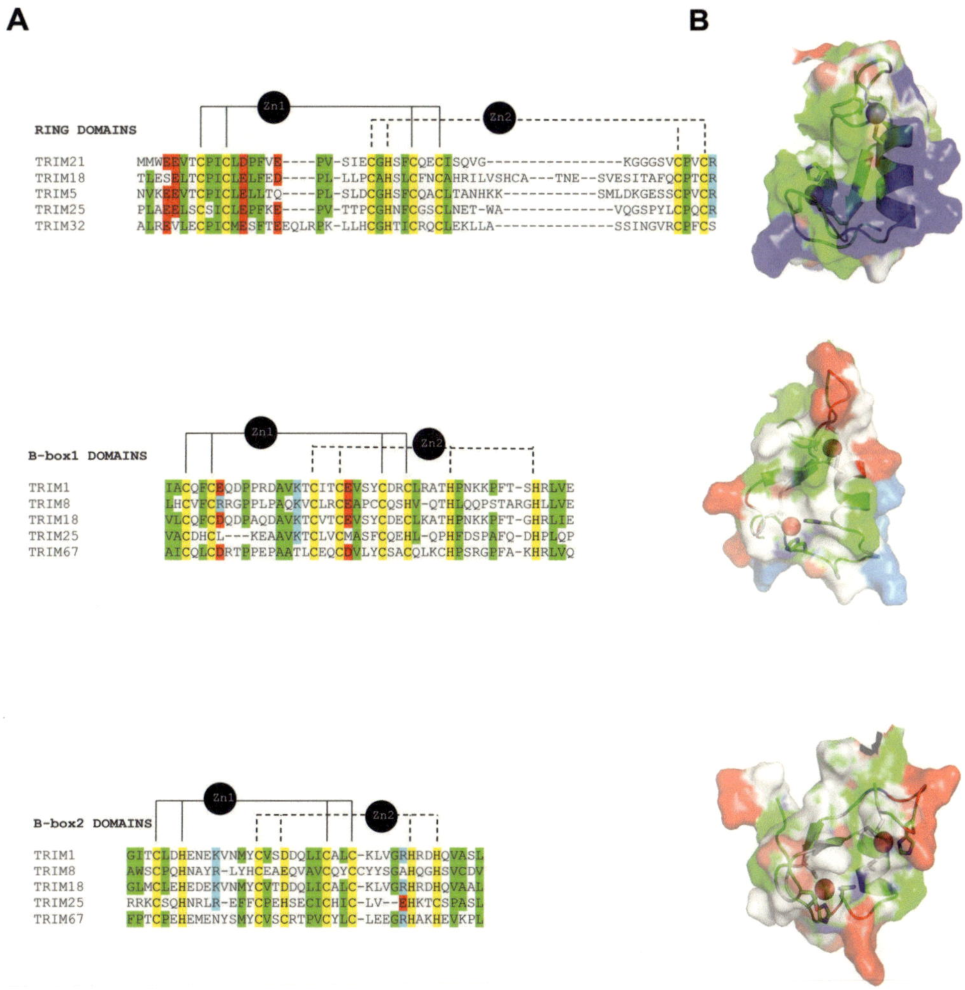

Figure 3.
RING and B-box domain comparison. A. Sequence alignments of five representative RING domains and five B-box1 and B-box2 domains from TRIM proteins. The amino acids involved the two zinc ions are identified in yellow and lines above connecting the corresponding residues consistent with the cross-brace mechanism. Some more conserved hydrophobic residues are colored green, while the acidic and basic residues are colored red and cyan. B. Surface representation of HDM2 RING and the MID1 B-box1 and B-box2 domains displaying the surface in same relative orientation for which the E2 enzyme interacts. Hydrophobic (green), basic (K, R, blue), acid (E, D, red), and uncharged polar (white) regions are shown. The pattern of amino acid type distribution is similar for many of the RING and solved B-box1 and B-box2 structures.

loop2 of MID1 B-box2 and several B-box2 domains are significantly shorter, and this could limit optimal interactions with the E2 enzyme.

Intriguingly, even though the P151L mutant B-box1 possesses greater activity, substrate ubiquitination assays revealed that the mutation disrupts binding and targeting of the alpha4 protein. This observation strongly supports our hypothesis that RING E3 ligase activities may be a compromise between level of activity and substrate binding, as defined evolutionarily. In unpublished work, we have identified several specific amino acids in RING domains that are important for RING–E2 interaction but that are not present in B-box1 domains. Introduction of these amino acids into the MID1 B-box1 domain resulted in significant increases in auto-ubiquitination activity, including polyubiquitination (unpublished).

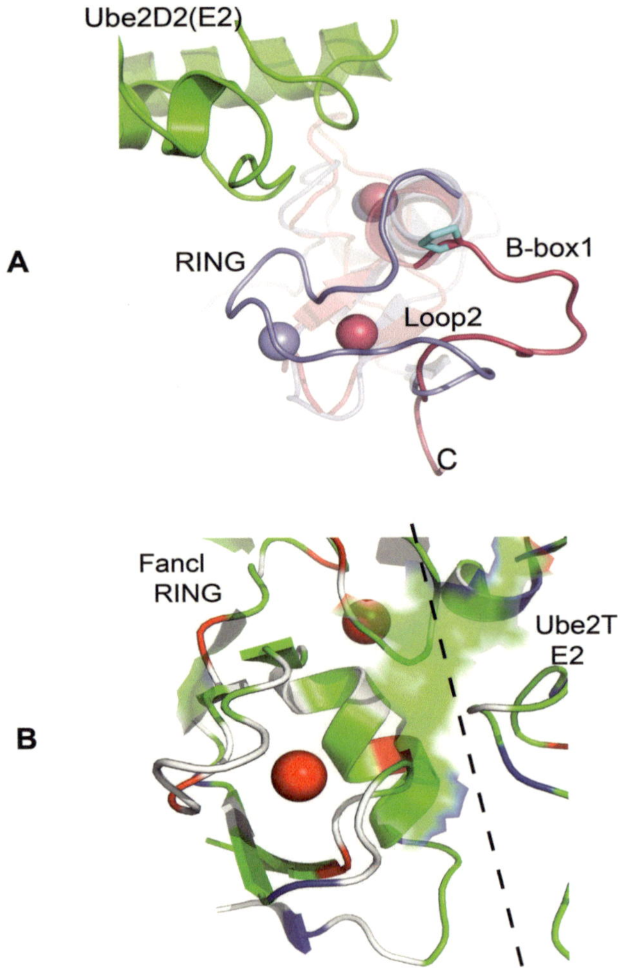

Figure 4.
E2-E3 interactions. A. Ribbon representation of the UbcH5/Ube2D1(green)-cCbl RING (purple) complex with the structure of the B-box1 domain (red) superimposed onto the RING domain. In this orientation, loop2 is positioned away from the E2-RING binding interface. Residues of RING loop2 make important E2-binding interaction; similar interactions are not observed with the B-box1 domain. B. Close-up to the E2-binding surface of the fancl RING domain; the dashed line separates the surface of the RING and E2 Ube2T (most of which is not shown for clarity). Consistent with several E2-RING complexes, the E2 enzymes bind on a fairly large hydrophobic surface on one side of the RING domain (**Figure 1**) *with specific electrostatic and hydrogen bonding interactions. Green = hydrophobic residues, blue = basic, red = electrostatic, white = uncharged polar, and red-sphere = zinc ions.*

10. Structural comparison of RING and B-box domains

In light of our findings that the E3 ligase activity of the MID1 B-box1 domain can be enhanced, we examine the structures of RING and B-box domains to understand if there may be additional features that can rationalize the difference in activities of MID1 RING and B-box domains. While it is not feasible to provide detailed analyses of all the differences between the various RING and B-box domain structures, we make general qualitative comparisons. As noted, the overall structures of RING and B-box domains are similar (**Figure 1**). In the case of the MID1 B-box domain, the position and size of loop2 may contribute to their decreased E3 ligase activities (**Figure 4A**). The sequences of a few RING and B-box domain are aligned, and distributions of amino acid types on the E2-binding surface on the RING and B-box domains are depicted for comparison (**Figures 3** and **4**). There are some key differences in amino acids between the RING and B-box domains that may also contribute to differences in the level of activity. However, those will not be discussed in detail here. The exact mechanism of E2-Bbox binding has not been characterized, and therefore, for the following discussion, we make the assumption that the B-box domains interact with the E2 enzyme in a similar manner as RING domains. The different types of residues (hydrophobic [green], acidic [red], basic [blue], and uncharged polar [gray]) are displayed. We used the HDM2 and human Fanconi anemia (fancl) RING domains [31, 120]. On the E2-binding surfaces, both RING domains (**Figures 3** and **4**) show predominantly hydrophobic residues. With the structure of the HDM2 RING domain (PDB 2hdp), there is also a large adjacent basic patch on the outer surface of the helix, but its role in E2 binding is not clear. There is also a basic residue on loop2 and a small acidic patch toward the top of the structure, and these participate in Ube2D2 E2 binding. For the fancl RING domain, the large hydrophobic patch is located in same region, but there are no large charged surfaces. Evaluation of fancl-RING-Ube2T complex (PDB 4ccg) reveals that the E2 enzyme interacts predominantly with the hydrophobic region of the RING domain (**Figure 4B**) [120]. Structures of other RING-E2 complexes reveal similar types of interactions.

In contrast, the hydrophobic patches on the MID1 B-box1 and B-box2 structures are smaller and not as contiguous as those observed with the HDM2 and Fancl RING structures (**Figure 3**). There are more charged residues on the surface. The PML/TRIM19 B-box1 domain (PDB 2mvw) [121] has more polar residues distributed instead of hydrophobic residues. Two smaller hydrophobic patches are observed on opposite sides. The distribution of residues for the TRIM5α B-box2 domain (PDB 2ecv) is very similar to that of the MID1 B-box1 structure. These comparisons reveal that there are differences in amino acid types at the canonical E2-binding site that might influence the mechanism of interactions between RING and B-box domains with their cognate E2 enzymes and hence the level of activity.

Interestingly, the structure of fancl RING domain with Ube2T E2 reveals that Ube2T does not have a corresponding Gln92 residue to form a hydrogen bond with a basic residue on loop2 of the RING domain, which is present is several RING and U-box domains. Instead, the complementary positions consist of hydrophobic residues, suggesting that allosteric effects of RING binding might be transmitted via hydrophobic interactions. In contrast, the HDM2 RING domain has an arginine that can form a hydrogen bond with Gln92 of the Ube2D2/UbcH5 E2 enzyme. This interaction is important for allosteric effects to influence cleavage of the thioester bond. It is possible that these subtle differences in binding mechanisms might provide a rationale for differences in the level of E3 ligase activities observed for RING-type E3 ligases. Differences in activity may also be due to mismatch in cognate E2-RING partners with *in vitro* ubiquitination assays. For example, some RING and B-box E3 ligases may possess sumoylation and ISGylation activities and, therefore, prefer different E2 conjugating enzymes.

11. B-box2 domain may additionally possess a regulatory role

In addition to the above noted roles of the B-box domains, it has been suggested that the presence of MID1 B-box2 domain impacts the binding efficiency of the B-box1 domain to alpha4. Binding studies with the MID1 and alpha4 proteins revealed tightest binding with a RB1 construct and reduction in binding with RB1B2 and larger MID1 constructs [90]. The apparent reduction in RB1-alpha4 binding may due to the B-box2 domain binding in an overlapping site with alpha4 [122]. Interestingly, we saw a reduction of the band intensities of the polyubiquitinated products of both auto- and alpha4-ubiquitinations with RB1B2 [76], indicating that B-box2 domain is regulating the alpha4 interaction. On the other hand, it is possible that the decrease is due to the two B-box domains affecting the RB1B2 interaction with the E2 enzyme.

Another TRIM protein for which the B-box2 domain has been ascribed a possible regulatory function is TRIM21, which is involved in immune signaling and is found in almost all cell types and tissues in mammals. In contrast to MID1, the TRIM21 RING domain exhibited greater ligase activity than the RB2 protein construct [123]. The result is confirmed by the E2~Ub thiolysis assays: the rate of Ub discharge was greater with the RING than with the RB2 domain construct. NMR experiments confirmed that the RING and B-box2 domains interact via the surface important for self-oligomerization [124].

12. The role of B-box domains in RINGless-TRIM proteins

There are currently six characterized human TRIM proteins that lack the N-terminal RING domain: TRIM14, TRIM16 (EBBP), TRIM20 (PYRIN/MEFV), TRIM29 (ATDC), TRIM44 (DIPB), and TRIM66. RINGless TRIM proteins are found in *Drosophila melanogaster (*Brat, Wech/Dappled*)* and *C. elegans* (LIN-41), but the functions of these proteins are not characterized. A detailed list of all current TRIM, RINGless TRIM, and BBX proteins are shown in **Figure 5**.

TRIM16 is a transcriptional regulator involved in regulating neuroblastoma cell growth, migration, and tumorigenicity [74]. Apparently, it can function as E3 ligase via both homodimerization and heterodimerization with TRIM18, TRIM19, and TRIM24 [125]. The tandem B-box domains are capable of very weak homodimerizing interactions in the absence of the coiled-coil domain. *In vitro* ubiquitination assay with TRIM16 consisting of domain deletions confirmed that the B-box domains confer E3 ligase activity [74]. TRIM20 is involved in innate immune response and is associated with the autoinflammatory disorder familial Mediterranean fever (FMF), characterized by pyogenic arthritis and pyoderma gangrenosum [126]. For TRIM20,

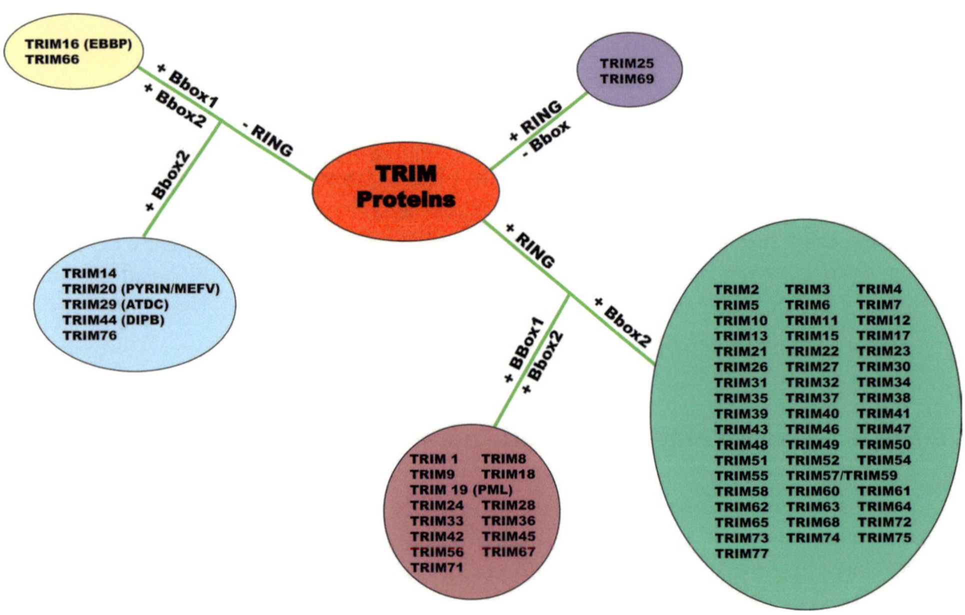

Figure 5.
B-box containing TRIM proteins. Protein family distribution of all currently known TRIM and RINGless TRIM proteins grouped by the presence of the B-box1 and B-box2 domains. Pseudo-TRIM16 and 66 lack a RING domain, and pseudo-TRIM 25 and 69 lack the B-box domains.

a pyrin domain (PYD) that belongs to the death domain family associated with apoptosis and inflammatory responses is in place of the RING domain. TRIM20 targets the proline/serine/threonine phosphatase-interacting protein 1 (PSTPIP1) by interactions via the B-box domain. The binding causes the B-box domain to unmask the PYD domain, allowing it to interact with downstream binding partners that are important for regulating inflammation. Deletion of the B-box domain resulted in constitutively active PYD and TRIM20 [127].

Very little is known about the other RINGless-TRIM proteins (**Figures 5** and **6**). TRIM29 is an oncogene that regulates p53 and is overexpressed in many different cancers including breast, lung, bladder, and pancreatic [128, 129]. It can form both homodimers and heterodimers with TRIM1/MID2, TRIM11, TRIM23, and TRIM27 [125]. Despite the lack of the RING domain, *in vivo* studies showed that TRIM27 maintained E3 ligase activities in response to viral and bacterial infections [130]. E3 ligase activity of TRIM29 was confirmed through its single B-box2 domain [131]. TRIM44 is involved in antiviral and immune response regulation, with some functions linking it to cancer. It can form a heterodimer with TRIM8, TRIM17, TRIM27, and TRIM69. Instead of a RING domain, TRIM44 contains a zinc-finger ubiquitin protease domain (UBP) that functions as a deubiquitinase [125]. Interestingly, TRIM44 is the only known deubiquitinase among the TRIM protein family [132, 133]. TRIM14 is involved in antiviral innate immune response [134, 135]. TRIM66 is correlated with the proliferation, invasion, and migration of non-small cell lung cancer [136]. TRIM66 possesses a PHD zinc-finger domain that preceded the B-box domains. The PHD domain is shown to coordinate two zinc ions and adopts RING fold, but its function as an E4 ligase is not clear.

While the mechanism of action is not clear, it is possible that RINGless-TRIM proteins function through homo- or heterodimerization via at least one of its B-box domains. Self-association/oligomerization through the B-box domains is shown to contribute to E3 ligase activity possible via an apparent localized concentration effect.

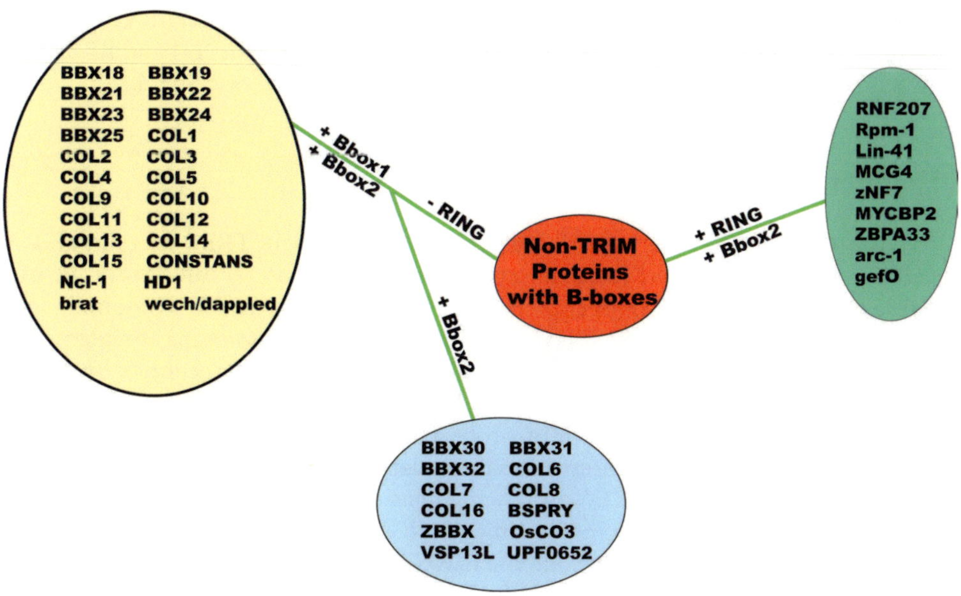

Figure 6.
B-box containing non-TRIM proteins. Protein family distribution of all currently known non-TRIM proteins arranged by the presence of the B-box1 and B-box2 domains. These proteins are generally termed BBX (B-box) and COL (Constans-like) and are found in plants. There are nine non-BBX non-TRIM proteins that contain RING and B-box domains but lack the coiled-coil domain required for TRIM definition; these are not found in plants.

13. B-box domains are found in plants

While the majority of B-box domains are found in mammals, recent publications have identified 32 B-box proteins (known as BBX proteins) in *Arabidopsis*, rice, and more than a dozen other plants species [137]. These non-TRIM proteins do not possess a RING domain, and the B-box domains are usually found at the N-terminus (**Figure 6**). The conservation of BBX proteins in multiple plant species suggests that these proteins play important roles in plant physiology similar to TRIM proteins in mammals [138]. The parallels of TRIM and BBX proteins regarding the B-box domains conferring E3 ligase activity suggest conservation of function across multiple kingdoms. Like TRIM proteins, there is a mixture of BBX proteins that consists of either a single or tandem B-box domains. Of the 32 BBX proteins, 21 of them have tandem B-box domains, in the same order observed in RING-less TRIM proteins. In rice (*Oryza sativa*), more than half of its 30 known BBX proteins contain two B-box domains [139]. The BBX proteins do not contain a coiled-coil domain but shown to interact with proteins that contain a coiled-coil domain to create the TRIM equivalent of a RBCC motif (see review [140]). Studies of BBX proteins containing tandem B-box domains revealed that only one of the B-box domains is essential for maintaining biological activity in plants, supporting redundancy observed with several TRIM proteins [141].

Defining specific functions to the B-box domains of BBX proteins is lacking, probably because of their recent realization in plants. Without specifics, a large number of these BBX proteins are postulated to be involved in the ubiquitination pathway. Several BBX proteins are shown to interact with an *Arabidopsis* RING E3 ligase called *co*nstitutive *p*hotomorphogenic-1 (COP1) through interactions with the B-box domain, during the dark cycle of plants (see review [140]), but it is unclear whether this association is as an E4 ligase to COP1, as a substrate or something else. Similar to TRIM B-box domains, the BBX B-box domains play important roles in protein-protein interactions, whether directly or indirectly [140]. They can form heterodimers and facilitate transcriptional regulation. For example, BBX21, BBX22, BBX24, and BBX25 interact with a COP1-associated protein called HY5. The BBX21 and BBX22 activated HY5 [142], while BBX24 and BBX25 repressed its activity [143, 144].

14. Summary

While it is clear that there is still much to be learned about TRIM proteins and their E3 ligase activity, there has been a great push toward better understanding of the role of the B-box domains over the past decade. Although the RING domain has received much of the spotlight, it is now clear that the B-box domains are integral for substrate binding/targeting, protein ubiquitination, and enhancement or activation of the ligase activity of TRIM and BBX protein families. B-box domains have diverse roles that include protein-protein interactions, substrate ubiquitination and sumoylation, and transcriptional regulation. What is not clear is whether B-box domain really served as E4-enhancing ligase to enhance the ligase activity of the RING domains, as observed in dimer RING E3 ligases, or whether they synergistically gain activity alongside the RING domains. Their role in contributing to oligomerization for some TRIM and BBX proteins to account for enhanced E3 ligase activity may be due to an apparent increase in the localized concentration of the TRIM or BBX protein. However, more studies are needed. Progress will require multifaceted approaches involving structure determination, protein-protein binding studies, and functional assays. Nonetheless, we hope that sufficient evidence have been provided, including those of TRIM18/MID1 and TRIM27, demonstrating that B-box domains are and should be considered E3 ligases and not as a supporting player to RING domains.

Acknowledgements

The author thanks Dr. Katharine Wright for her work in the lab and for contributing ideas for this review, and Ms. Jessica Webb for her help and contributions with the figures. This work was supported in part by a National Science Foundation grant (MAM, 1808391).

Author details

Michael Anthony Massiah
Department of Chemistry, George Washington University, Washington, DC, USA

*Address all correspondence to: massiah@gwu.edu

References

[1] Klug A. The discovery of zinc fingers and their development for practical applications in gene regulation and genome manipulation. Quarterly Reviews of Biophysics. 2010;**43**:1-21

[2] Massiah MA, Blake PR, Summers MF. Nucleic Acid Interactive Protein Domains that Require Zinc. New York, NY: Oxford University Press; 1998

[3] Berg JM. Zinc fingers and other metal-binding domains. Elements for interactions between macromolecules. The Journal of Biological Chemistry. 1990;**265**:6513-6516

[4] Wright KM, Wu K, Babatunde O, Du H, Massiah MA. XLOS-observed mutations of MID1 Bbox1 domain cause domain unfolding. PLoS One. 2014;**9**:e107537

[5] Zhao Y, Zeng C, Massiah MA. Molecular dynamics simulation reveals insights into the mechanism of unfolding by the A130T/V mutations within the MID1 zinc-binding Bbox1 domain. PLoS One. 2015;**10**:e0124377

[6] Miller J, McLachlan AD, Klug A. Repetitive zinc-binding domains in the protein transcription factor IIIA from Xenopus oocytes. The EMBO Journal. 1985;**4**:1609-1614

[7] Neuhaus D, Nakaseko Y, Nagai K, Klug A. Sequence-specific [1H] NMR resonance assignments and secondary structure identification for 1- and 2-zinc finger constructs from SW15. A hydrophobic core involving four invariant residues. FEBS Letters. 1990;**262**:179-184

[8] Churchill ME, Tullius TD, Klug A. Mode of interaction of the zinc finger protein TFIIIA with a 5S RNA gene of Xenopus. Proceedings of the National Academy of Sciences of the United States of America. 1990;**87**:5528-5532

[9] Reddy BA, Etkin LD. A unique bipartite cysteine-histidine motif defines a subfamily of potential zinc-finger proteins. Nucleic Acids Research. 1991;**19**:6330

[10] Lovering R, Hanson IM, Borden KL, Martin S, O'Reilly NJ, Evan GI, et al. Identification and preliminary characterization of a protein motif related to the zinc finger. Proceedings of the National Academy of Sciences of the United States of America. 1993;**90**:2112-2116

[11] Ogawa S, Goto W, Orimo A, Hosoi T, Ouchi Y, Muramatsu M, et al. Molecular cloning of a novel RING finger-B box-coiled coil (RBCC) protein, terf, expressed in the testis. Biochemical and Biophysical Research Communications. 1998;**251**:515-519

[12] Short KM, Cox TC. Sub-classification of the rbcc/trim superfamily reveals a novel motif necessary for microtubule binding. The Journal of Biological Chemistry. 2006;**281**(13):8970-8980

[13] Massiah MA, Simmons BN, Short KM, Cox TC. Solution structure of the RBCC/TRIM B-box1 domain of human MID1: B-box with a RING. Journal of Molecular Biology. 2006;**358**:532-545

[14] Du H, Wu K, Didoronkute A, Levy MV, Todi N, Shchelokova A, et al. MID1 catalyzes the ubiquitination of protein phosphatase 2A and mutations within its Bbox1 domain disrupt polyubiquitination of alpha4 but not of PP2Ac. PLoS One. 2014;**9**:e107428

[15] Uchil PD, Quinlan BD, Chan WT, Luna JM, Mothes W. TRIM E3 ligases interfere with early and late stages of the retroviral life cycle. PLoS Pathogens. 2008;**4**:e16

[16] Kitamura K, Tanaka H, Nishimune Y. Haprin, a novel haploid germ

cell-specific RING finger protein involved in the acrosome reaction. The Journal of Biological Chemistry. 2003;**278**:44417-44423

[17] Short KM, Cox TC. Subclassification of the RBCC/TRIM superfamily reveals a novel motif necessary for microtubule binding. The Journal of Biological Chemistry. 2006;**281**:8970-8980

[18] Klug A, Schwabe JW. Protein motifs 5. Zinc fingers. FASEB Journal: Official Publication of the Federation of American Societies for Experimental Biology. 1995;**9**:597-604

[19] Massiah MA, Wright KM, Du H. Obtaining soluble folded proteins from inclusion bodies using Sarkosyl, triton X-100, and CHAPS: Application to LB and M9 minimal media. Current Protocols in Protein Science. 2016;**84**(6):13.1-6-13.1-24

[20] Tao H, Liu W, Simmons BN, Harris HK, Cox TC, Massiah MA. Purifying natively folded proteins from inclusion bodies using sarkosyl, Triton X-100, and CHAPS. Biotechnology Techniques. 2010;**48**:61-64

[21] Massiah MA, Matts JA, Short KM, Simmons BN, Singireddy S, Yi Z, et al. Solution structure of the MID1 B-box2 CHC(D/C)C(2)H(2) zinc-binding domain: Insights into an evolutionarily conserved RING fold. Journal of Molecular Biology. 2007;**369**:1-10

[22] Cox T. Taking it to the max: The genetic and developmental mechanisms coordinating midfacial morphogenesis and dysmorphology. Clinical Genetics. 2004;**65**:163-176

[23] Cox TC, Allen LR, Cox LL, Hopwood B, Goodwin B, Haan E, et al. New mutations in MID1 provide support for loss of function as the cause of X-linked Opitz syndrome. Human Molecular Genetics. 2000;**9**:2553-2562

[24] Granata A, Quaderi NA. The Opitz syndrome gene MID1 is essential for establishing asymmetric gene expression in Hensen's node. Developmental Biology. 2003;**258**:397-405

[25] Quaderi NA, Schweiger S, Gaudenz K, Franco B, Rugarli EI, Berger W, et al. Opitz G/BBB syndrome, a defect of midline development, is due to mutations in a new RING finger gene on Xp22. Nature Genetics. 1997;**17**:285-291

[26] Schweiger S, Schneider R. The MID1/PP2A complex: A key to the pathogenesis of Opitz BBB/G syndrome. BioEssays : News and Reviews in Molecular, Cellular and Developmental Biology. 2003;**25**:356-366

[27] Schweiger S, Foerster J, Lehmann T, Suckow V, Muller YA, Walter G, et al. The Opitz syndrome gene product, MID1, associates with microtubules. Proceedings of the National Academy of Sciences of the United States of America. 1999;**96**:2794-2799

[28] Lindskog S. Structure and mechanism of carbonic anhydrase. Pharmacology & Therapeutics. 1997;**74**:1-20

[29] Bertini I, Luchinat C. The structure of cobalt(II)-substituted carbonic anhydrase and its implications for the catalytic mechanism of the enzyme. Annals of the New York Academy of Sciences. 1984;**429**:89-98

[30] Tripp BC, Smith K, Ferry JG. Carbonic anhydrase: New insights for an ancient enzyme. The Journal of Biological Chemistry. 2001;**276**:48615-48618

[31] Kostic M, Matt T, Martinez-Yamout MA, Dyson HJ, Wright PE. Solution structure of the Hdm2 $C_2H_2C_4$ RING, a domain critical for ubiquitination of p53. Journal of Molecular Biology. 2006;**363**:433-450

[32] Plechanovova A, Jaffray EG, McMahon SA, Johnson KA, Navratilova I, Naismith JH, et al. Mechanism of ubiquitylation by dimeric RING ligase RNF4. Nature Structural & Molecular Biology. 2011;**18**:1052-1059

[33] Liew CW, Sun H, Hunter T, Day CL. RING domain dimerization is essential for RNF4 function. The Biochemical Journal. 2010;**431**:23-29

[34] Mace PD, Linke K, Feltham R, Schumacher FR, Smith CA, Vaux DL, et al. Structures of the cIAP2 RING domain reveal conformational changes associated with ubiquitin-conjugating enzyme (E2) recruitment. The Journal of Biological Chemistry. 2008;**283**:31633-31640

[35] Wagner JM, Roganowicz MD, Skorupka K, Alam SL, Christensen D, Doss G, et al. Mechanism of B-box 2 domain-mediated higher-order assembly of the retroviral restriction factor TRIM5alpha. eLife. 2016;**5**:1-26. e16309

[36] Diaz-Griffero F, Li X, Javanbakht H, Song B, Welikala S, Stremlau M, et al. Rapid turnover and polyubiquitylation of the retroviral restriction factor TRIM5. Virology. 2006;**349**:300-315

[37] Li X, Sodroski J. The TRIM5alpha B-box 2 domain promotes cooperative binding to the retroviral capsid by mediating higher-order self-association. Journal of Virology. 2008;**82**:11495-11502

[38] Cao T, Borden KL, Freemont PS, Etkin LD. Involvement of the rfp tripartite motif in protein-protein interactions and subcellular distribution. Journal of Cell Science. 1997;**110**(Pt 14):1563-1571

[39] Pickart CM. Mechanisms underlying ubiquitination. Annual Review of Biochemistry. 2001;**70**:503-533

[40] Pickart CM. Ubiquitin biology: An old dog learns an old trick. Nature Cell Biology. 2000;**2**:E139-E141

[41] Callis J. The ubiquitination machinery of the ubiquitin system. Arabidopsis Book. 2014;**12**:e0174

[42] Darwin KH. Prokaryotic ubiquitin-like protein (pup), proteasomes and pathogenesis. Nature Reviews Microbiology. 2009;**7**:485-491

[43] Pickart CM. Ubiquitin in chains. Trends in Biochemical Sciences. 2000;**25**:544-548

[44] Akutsu M, Dikic I, Bremm A. Ubiquitin chain diversity at a glance. Journal of Cell Science. 2016;**129**:875-880

[45] Nishikawa H, Ooka S, Sato K, Arima K, Okamoto J, Klevit RE, et al. Mass spectrometric and mutational analyses reveal Lys-6-linked polyubiquitin chains catalyzed by BRCA1-BARD1 ubiquitin ligase. The Journal of Biological Chemistry. 2004;**279**:3916-3924

[46] Kravtsova-Ivantsiv Y, Sommer T, Ciechanover A. The lysine48-based polyubiquitin chain proteasomal signal: Not a single child anymore. Angewandte Chemie. 2013;**52**:192-198

[47] Yau R, Rape M. The increasing complexity of the ubiquitin code. Nature Cell Biology. 2016;**18**:579-586

[48] Herhaus L, Dikic I. Expanding the ubiquitin code through post-translational modification. EMBO Reports. 2015;**16**:1071-1083

[49] Komander D, Clague MJ, Urbe S. Breaking the chains: Structure and function of the deubiquitinases. Nature Reviews Molecular Cell Biology. 2009;**10**:550-563

[50] Glickman MH, Ciechanover A. The ubiquitin-proteasome proteolytic pathway: Destruction for the sake of construction. Physiological Reviews. 2002;**82**:373-428

[51] Deng L, Wang C, Spencer E, Yang L, Braun A, You J, et al. Activation of the IkappaB kinase complex by TRAF6 requires a dimeric ubiquitin-conjugating enzyme complex and a unique polyubiquitin chain. Cell. 2000;**103**:351-361

[52] Hofmann RM, Pickart CM. Noncanonical MMS2-encoded ubiquitin-conjugating enzyme functions in assembly of novel polyubiquitin chains for DNA repair. Cell. 1999;**96**:645-653

[53] Kaiser P, Flick K, Wittenberg C, Reed SI. Regulation of transcription by ubiquitination without proteolysis: Cdc34/SCF(Met30)-mediated inactivation of the transcription factor Met4. Cell. 2000;**102**:303-314

[54] Hicke L. Protein regulation by monoubiquitin. Nature Reviews Molecular Cell Biology. 2001;**2**:195-201

[55] Metzger MB, Pruneda JN, Klevit RE, Weissman AM. RING-type E3 ligases: Master manipulators of E2 ubiquitin-conjugating enzymes and ubiquitination. Biochimica et Biophysica Acta. 2014;**1843**:47-60

[56] Kim J, Tipper C, Sodroski J. Role of TRIM5alpha RING domain E3 ubiquitin ligase activity in capsid disassembly, reverse transcription blockade, and restriction of simian immunodeficiency virus. Journal of Virology. 2011;**85**:8116-8132

[57] Spratt DE, Wu K, Kovacev J, Pan ZQ, Shaw GS. Selective recruitment of an E2~ubiquitin complex by an E3 ubiquitin ligase. The Journal of Biological Chemistry. 2012;**287**:17374-17385

[58] Kwon YT, Ciechanover A. The ubiquitin code in the ubiquitin-proteasome system and autophagy. Trends in Biochemical Sciences. 2017;**42**:873-886

[59] Christensen DE, Brzovic PS, Klevit RE. E2-BRCA1 RING interactions dictate synthesis of mono- or specific polyubiquitin chain linkages. Nature Structural & Molecular Biology. 2007;**14**:941-948

[60] Plechanovova A, Jaffray EG, Tatham MH, Naismith JH, Hay RT. Structure of a RING E3 ligase and ubiquitin-loaded E2 primed for catalysis. Nature. 2012;**489**:115-120

[61] Soss SE, Klevit RE, Chazin WJ. Activation of UbcH5c~Ub is the result of a shift in interdomain motions of the conjugate bound to U-box E3 ligase E4B. Biochemistry. 2013;**52**:2991-2999

[62] Pruneda JN, Littlefield PJ, Soss SE, Nordquist KA, Chazin WJ, Brzovic PS, et al. Structure of an E3:E2~Ub complex reveals an allosteric mechanism shared among RING/U-box ligases. Molecular Cell. 2012;**47**:933-942

[63] Ulrich HD. Protein-protein interactions within an E2-RING finger complex. Implications for ubiquitin-dependent DNA damage repair. The Journal of Biological Chemistry. 2003;**278**:7051-7058

[64] Han X, Du H, Massiah MA. Detection and characterization of the in vitro e3 ligase activity of the human MID1 protein. Journal of Molecular Biology. 2011;**407**:505-520

[65] Xia Y, Pao GM, Chen HW, Verma IM, Hunter T. Enhancement of BRCA1 E3 ubiquitin ligase activity through direct interaction with the BARD1 protein. The Journal of Biological Chemistry. 2003;**278**:5255-5263

[66] Wu-Baer F, Lagrazon K, Yuan W, Baer R. The BRCA1/BARD1 heterodimer assembles polyubiquitin chains through an unconventional linkage involving lysine residue K6 of ubiquitin. The Journal of Biological Chemistry. 2003;**278**:34743-34746

[67] Brzovic PS, Rajagopal P, Hoyt DW, King MC, Klevit RE. Structure of a BRCA1-BARD1 heterodimeric RING-RING complex. Nature Structural Biology. 2001;**8**:833-837

[68] Mallery DL, Vandenberg CJ, Hiom K. Activation of the E3 ligase function of the BRCA1/BARD1 complex by polyubiquitin chains. The EMBO Journal. 2002;**21**:6755-6762

[69] Linares LK, Hengstermann A, Ciechanover A, Muller S, Scheffner M. HdmX stimulates Hdm2-mediated ubiquitination and degradation of p53. Proceedings of the National Academy of Sciences of the United States of America. 2003;**100**:12009-12014

[70] Feltham R, Bettjeman B, Budhidarmo R, Mace PD, Shirley S, Condon SM, et al. Smac mimetics activate the E3 ligase activity of cIAP1 protein by promoting RING domain dimerization. The Journal of Biological Chemistry. 2011;**286**:17015-17028

[71] Buchwald G, van der Stoop P, Weichenrieder O, Perrakis A, van Lohuizen M, Sixma TK. Structure and E3-ligase activity of the Ring-Ring complex of polycomb proteins Bmi1 and Ring1b. The EMBO Journal. 2006;**25**:2465-2474

[72] Bourgeois-Daigneault MC, Thibodeau J. Autoregulation of MARCH1 expression by dimerization and autoubiquitination. Journal of Immunology. 2012;**188**:4959-4970

[73] Fu TM, Shen C, Li Q, Zhang P, Wu H. Mechanism of ubiquitin transfer promoted by TRAF6. Proceedings of the National Academy of Sciences of the United States of America. 2018;**115**:1783-1788

[74] Bell JL, Malyukova A, Holien JK, Koach J, Parker MW, Kavallaris M, et al. TRIM16 acts as an E3 ubiquitin ligase and can heterodimerize with other TRIM family members. PLoS One. 2012;**7**:e37470

[75] Zheng Q, Hou J, Zhou Y, Yang Y, Xie B, Cao X. Siglec1 suppresses antiviral innate immune response by inducing TBK1 degradation via the ubiquitin ligase TRIM27. Cell Research. 2015;**25**:1121-1136

[76] Du H, Huang Y, Zaghlula M, Walters E, Cox TC, Massiah MA. The MID1 E3 ligase catalyzes the polyubiquitination of Alpha4 (alpha4), a regulatory subunit of protein phosphatase 2A (PP2A): Novel insights into MID1-mediated regulation of PP2A. The Journal of Biological Chemistry. 2013;**288**:21341-21350

[77] Koegl M, Hoppe T, Schlenker S, Ulrich HD, Mayer TU, Jentsch S. A novel ubiquitination factor, E4, is involved in multiubiquitin chain assembly. Cell. 1999;**96**:635-644

[78] Baranes-Bachar K, Levy-Barda A, Oehler J, Reid DA, Soria-Bretones I, Voss TC, et al. The ubiquitin E3/E4 ligase UBE4A adjusts protein Ubiquitylation and accumulation at sites of DNA damage, facilitating double-Strand break repair. Molecular Cell. 2018;**69**:866-878 e7

[79] Ferreira RT, Menezes RA, Rodrigues-Pousada C. E4-ubiquitin ligase Ufd2 stabilizes Yap8 and modulates arsenic stress responses independent of the U-box motif. Biology Open. 2015;**4**:1122-1131

[80] Chatterjee A, Upadhyay S, Chang X, Nagpal JK, Trink B, Sidransky D. U-box-type ubiquitin E4 ligase, UFD2a attenuates cisplatin mediated

degradation of DeltaNp63alpha. Cell Cycle. 2008;7:1231-1237

[81] Aravind L, Koonin EV. The U box is a modified RING finger—A common domain in ubiquitination. Current Biology. 2000;**10**:R132-R134

[82] Ohi MD, Vander Kooi CW, Rosenberg JA, Chazin WJ, Gould KL. Structural insights into the U-box, a domain associated with multi-ubiquitination. Nature Structural Biology. 2003;**10**:250-255

[83] Nordquist KA, Dimitrova YN, Brzovic PS, Ridenour WB, Munro KA, Soss SE, et al. Structural and functional characterization of the monomeric U-box domain from E4B. Biochemistry. 2010;**49**:347-355

[84] Klevit RE, Herriott JR, Horvath SJ. Solution structure of a zinc finger domain of yeast ADR1. Proteins. 1990;7:215-226

[85] Kudryashova E, Kudryashov D, Kramerova I, Spencer MJ. Trim32 is a ubiquitin ligase mutated in limb girdle muscular dystrophy type 2H that binds to skeletal muscle myosin and ubiquitinates actin. Journal of Molecular Biology. 2005;**354**:413-424

[86] Koliopoulos MG, Esposito D, Christodoulou E, Taylor IA, Rittinger K. Functional role of TRIM E3 ligase oligomerization and regulation of catalytic activity. The EMBO Journal. 2016;**35**:1204-1218

[87] Javanbakht H, Diaz-Griffero F, Stremlau M, Si Z, Sodroski J. The contribution of RING and B-box 2 domains to retroviral restriction mediated by monkey TRIM5alpha. The Journal of Biological Chemistry. 2005;**280**:26933-26940

[88] Gack MU, Shin YC, Joo CH, Urano T, Liang C, Sun L, et al. TRIM25 RING-finger E3 ubiquitin ligase is essential for RIG-I-mediated antiviral activity. Nature. 2007;**446**:916-920

[89] Villarroya-Beltri C, Guerra S, Sanchez-Madrid F. ISGylation - a key to lock the cell gates for preventing the spread of threats. Journal of Cell Science. 2017;**130**:2961-2969

[90] Trockenbacher A, Suckow V, Foerster J, Winter J, Krauss S, Ropers HH, et al. MID1, mutated in Opitz syndrome, encodes an ubiquitin ligase that targets phosphatase 2A for degradation. Nature Genetics. 2001;**29**:287-294

[91] Schweiger S, Dorn S, Fuchs M, Kohler A, Matthes F, Muller EC, et al. The E3 ubiquitin ligase MID1 catalyzes ubiquitination and cleavage of Fu. The Journal of Biological Chemistry. 2014;**289**:31805-31817

[92] Seshacharyulu P, Pandey P, Datta K, Batra SK. Phosphatase: PP2A structural importance, regulation and its aberrant expression in cancer. Cancer Letters. 2013;**335**:9-18

[93] Garcia A, Cayla X, Guergnon J, Dessauge F, Hospital V, Rebollo MP, et al. Serine/threonine protein phosphatases PP1 and PP2A are key players in apoptosis. Biochimie. 2003;**85**:721-726

[94] Janssens V, Goris J. Protein phosphatase 2A: A highly regulated family of serine/threonine phosphatases implicated in cell growth and signalling. The Biochemical Journal. 2001;**353**:417-439

[95] Sontag E. Protein phosphatase 2A: The Trojan horse of cellular signaling. Cellular Signalling. 2001;**13**:7-11

[96] Lechward K, Awotunde OS, Swiatek W, Muszynska G. Protein phosphatase 2A: Variety of forms and diversity of functions. Acta Biochimica Polonica. 2001;**48**:921-933

[97] Hunter T. Protein kinases and phosphatases: The yin and yang of protein phosphorylation and signaling. Cell. 1995;**80**:225-236

[98] Millward TA, Zolnierowicz S, Hemmings BA. Regulation of protein kinase cascades by protein phosphatase 2A. Trends in Biochemical Sciences. 1999;**24**:186-191

[99] Shenolikar S. Protein serine/threonine phosphatases--new avenues for cell regulation. Annual Review of Cell Biology. 1994;**10**:55-86

[100] Wera S, Hemmings BA. Serine/threonine protein phosphatases. The Biochemical Journal. 1995;**311**(Pt 1):17-29

[101] Cowley S, Paterson H, Kemp P, Marshall CJ. Activation of MAP kinase kinase is necessary and sufficient for PC12 differentiation and for transformation of NIH 3T3 cells. Cell. 1994;**77**:841-852

[102] Chuang E, Fisher TS, Morgan RW, Robbins MD, Duerr JM, Vander Heiden MG, et al. The CD28 and CTLA-4 receptors associate with the serine/threonine phosphatase PP2A. Immunity. 2000;**13**:313-322

[103] Inui S, Kuwahara K, Mizutani J, Maeda K, Kawai T, Nakayasu H, et al. Molecular cloning of a cDNA clone encoding a phosphoprotein component related to the Ig receptor-mediated signal transduction. Journal of Immunology. 1995;**154**:2714-2723

[104] Murata K, Wu J, Brautigan DL. B cell receptor-associated protein alpha-4 displays rapamycin-sensitive binding directly to the catalytic subunit of protein phosphatase 2A. Proceedings of the National Academy of Sciences of the United States of America. 1997;**94**:10624-10629

[105] Inui S, Sanjo H, Maeda K, Yamamoto H, Miyamoto E, Sakaguchi N. Ig receptor binding protein 1 (alpha4) is associated with a rapamycin-sensitive signal transduction in lymphocytes through direct binding to the catalytic subunit of protein phosphatase 2A. Blood. 1998;**92**:539-546

[106] Onda M, Inui S, Maeda K, Suzuki M, Takahashi E, Sakaguchi N. Expression and chromosomal localization of the human alpha 4/IGBP1 gene, the structure of which is closely related to the yeast TAP42 protein of the rapamycin-sensitive signal transduction pathway. Genomics. 1997;**46**:373-378

[107] Cygnar KD, Gao X, Pan D, Neufeld TP. The phosphatase subunit tap42 functions independently of target of rapamycin to regulate cell division and survival in Drosophila. Genetics. 2005;**170**:733-740

[108] Jiang Y, Broach JR. Tor proteins and protein phosphatase 2A reciprocally regulate Tap42 in controlling cell growth in yeast. The EMBO Journal. 1999;**18**:2782-2792

[109] Dutcher JP. Mammalian target of rapamycin (mTOR) inhibitors. Current Oncology Reports. 2004;**6**:111-115

[110] Mita MM, Mita A, Rowinsky EK. Mammalian target of rapamycin: A new molecular target for breast cancer. Clinical Breast Cancer. 2003;**4**:126-137

[111] Rao RD, Buckner JC, Sarkaria JN. Mammalian target of rapamycin (mTOR) inhibitors as anti-cancer agents. Current Cancer Drug Targets. 2004;**4**:621-635

[112] Smetana JH, Oliveira CL, Jablonka W, Aguiar Pertinhez T, Carneiro FR, Montero-Lomeli M, et al. Low resolution structure of the human alpha4 protein (IgBP1) and studies on the stability of alpha4 and of its

yeast ortholog Tap42. Biochimica et Biophysica Acta. 2006;**1764**:724-734

[113] McConnell JL, Watkins GR, Soss SE, Franz HS, McCorvey LR, Spiller BW, et al. Alpha4 is a ubiquitin-binding protein that regulates protein serine/threonine phosphatase 2A ubiquitination. Biochemistry. 2010;**49**:1713-1718

[114] Watkins GR, Wang N, Mazalouskas MD, Gomez RJ, Guthrie CR, Kraemer BC, et al. Monoubiquitination promotes calpain cleavage of the protein phosphatase 2A (PP2A) regulatory subunit alpha4, altering PP2A stability and microtubule-associated protein phosphorylation. The Journal of Biological Chemistry. 2012;**287**:24207-24215

[115] Fukumoto T, Watanabe-Fukunaga R, Fujisawa K, Nagata S, Fukunaga R. The fused protein kinase regulates hedgehog-stimulated transcriptional activation in Drosophila Schneider 2 cells. The Journal of Biological Chemistry. 2001;**276**:38441-38448

[116] Peng H, Begg GE, Schultz DC, Friedman JR, Jensen DE, Speicher DW, et al. Reconstitution of the KRAB-KAP-1 repressor complex: A model system for defining the molecular anatomy of RING-B box-coiled-coil domain-mediated protein-protein interactions. Journal of Molecular Biology. 2000;**295**:1139-1162

[117] Koyama S, Hata S, Witt CC, Ono Y, Lerche S, Ojima K, et al. Muscle RING-finger protein-1 (MuRF1) as a connector of muscle energy metabolism and protein synthesis. Journal of Molecular Biology. 2008;**376**:1224-1236

[118] Mrosek M, Meier S, Ucurum-Fotiadis Z, von Castelmur E, Hedbom E, Lustig A, et al. Structural analysis of B-box 2 from MuRF1: Identification of a novel self-association pattern in a RING-like fold. Biochemistry. 2008;**47**:10722-10730

[119] Wright KM, Du H, Massiah MA. Structural and functional observations of the P151L MID1 mutation reveal alpha4 plays a significant role in X-linked Opitz syndrome. The FEBS Journal. 2017;**284**:2183-2193

[120] Hodson C, Purkiss A, Miles JA, Walden H. Structure of the human FANCL RING-Ube2T complex reveals determinants of cognate E3-E2 selection. Structure. 2014;**22**:337-344

[121] Huang SY, Naik MT, Chang CF, Fang PJ, Wang YH, Shih HM, et al. The B-box 1 dimer of human promyelocytic leukemia protein. Journal of Biomolecular NMR. 2014;**60**:275-281

[122] Du H, Massiah MA. NMR studies of the C-terminus of alpha4 reveal possible mechanism of its interaction with MID1 and protein phosphatase 2A. PLoS One. 2011;**6**:e28877

[123] Dickson C, Fletcher AJ, Vaysburd M, Yang JC, Mallery DL, Zeng J, et al. Intracellular antibody signalling is regulated by phosphorylation of the fc receptor TRIM21. eLife. 2018;**7**:1-22. e32660

[124] Wallenhammar A, Anandapadamanaban M, Lemak A, Mirabello C, Lundstrom P, Wallner B, et al. Solution NMR structure of the TRIM21 B-box2 and identification of residues involved in its interaction with the RING domain. PLoS One. 2017;**12**:e0181551

[125] Hatakeyama S. TRIM family proteins: Roles in autophagy, immunity, and carcinogenesis. Trends in Biochemical Sciences. 2017;**42**:297-311

[126] Gumucio DL, Diaz A, Schaner P, Richards N, Babcock C, Schaller M, et al. Fire and ICE: The role of pyrin domain-containing proteins in inflammation and apoptosis. Clinical and Experimental Rheumatology. 2002;**20**:S45-S53

[127] Yu JW, Fernandes-Alnemri T, Datta P, Wu J, Juliana C, Solorzano L, et al. Pyrin activates the ASC pyroptosome in response to engagement by autoinflammatory PSTPIP1 mutants. Molecular Cell. 2007;**28**:214-227

[128] Hatakeyama S. Early evidence for the role of TRIM29 in multiple cancer models. Expert Opinion on Therapeutic Targets. 2016;**20**:767-770

[129] Henderson BR. Regulation of BRCA1, BRCA2 and BARD1 intracellular trafficking. BioEssays : News and Reviews in Molecular, Cellular and Developmental Biology. 2005;**27**:884-893

[130] Sho T, Tsukiyama T, Sato T, Kondo T, Cheng J, Saku T, et al. TRIM29 negatively regulates p53 via inhibition of Tip60. Biochimica et Biophysica Acta. 2011;**1813**:1245-1253

[131] Xing J, Weng L, Yuan B, Wang Z, Jia L, Jin R, et al. Identification of a role for TRIM29 in the control of innate immunity in the respiratory tract. Nature Immunology. 2016;**17**:1373-1380

[132] Allen MD, Bycroft M. The solution structure of the ZnF UBP domain of USP33/VDU1. Protein Science: A Publication of the Protein Society. 2007;**16**:2072-2075

[133] Chen Z, Lin TC, Bi X, Lu G, Dawson BC, Miranda R, et al. TRIM44 promotes quiescent multiple myeloma cell occupancy and survival in the osteoblastic niche via HIF-1alpha stabilization. Leukemia. 2018

[134] Su X, Wang J, Chen W, Li Z, Fu X, Yang A. Overexpression of TRIM14 promotes tongue squamous cell carcinoma aggressiveness by activating the NF-kappaB signaling pathway. Oncotarget. 2016;7:9939-9950

[135] Zhou Z, Jia X, Xue Q, Dou Z, Ma Y, Zhao Z, et al. TRIM14 is a mitochondrial adaptor that facilitates retinoic acid-inducible gene-I-like receptor-mediated innate immune response. Proceedings of the National Academy of Sciences of the United States of America. 2014;**111**:E245-E254

[136] Dai HY, Ma Y, Da Z, Hou XM. Knockdown of TRIM66 inhibits malignant behavior and epithelial-mesenchymal transition in non-small cell lung cancer. Pathology, Research and Practice. 2018;**214**:1130-1135

[137] Khanna R, Kronmiller B, Maszle DR, Coupland G, Holm M, Mizuno T, et al. The Arabidopsis B-box zinc finger family. The Plant Cell. 2009;**21**:3416-3420

[138] Zou ZY, Wang RH, Wang R, Yang SL, Yang YJ. Genome-wide identification, phylogenetic analysis, and expression profiling of the BBX family genes in pear. The Journal of Horticultural Science and Biotechnology. 2018;**93**:37-50

[139] Huang J, Zhao X, Weng X, Wang L, Xie W. The rice B-box zinc finger gene family: Genomic identification, characterization, expression profiling and diurnal analysis. PLoS One. 2012;**7**:e48242

[140] Gangappa SN, Botto JF. The BBX family of plant transcription factors. Trends in Plant Science. 2014;**19**:460-470

[141] Xu D, Jiang Y, Li J, Holm M, Deng XW. The B-box domain protein BBX21 promotes Photomorphogenesis. Plant Physiology. 2018;**176**:2365-2375

[142] Xu D, Jiang Y, Li J, Lin F, Holm M, Deng XW. BBX21, an Arabidopsis B-box protein, directly activates HY5 and is targeted by COP1 for 26S proteasome-mediated degradation. Proceedings of the National Academy of Sciences of the United States of America. 2016;**113**:7655-7660

[143] Gangappa SN, Crocco CD, Johansson H, Datta S, Hettiarachchi C, Holm M, et al. The Arabidopsis B-BOX protein BBX25 interacts with HY5, negatively regulating BBX22 expression to suppress seedling photomorphogenesis. The Plant Cell. 2013;**25**:1243-1257

[144] Gangappa SN, Holm M, Botto JF. Molecular interactions of BBX24 and BBX25 with HYH, HY5 HOMOLOG, to modulate Arabidopsis seedling development. Plant Signaling & Behavior. 2013;**8**(8):1-4. e25208